Advances in Materials Science for Environmental and Energy Technologies V

Advances in Materials Science for Environmental and Energy Technologies V

Ceramic Transactions, Volume 260

Edited by
Tatsuki Ohji
Raghunath Kanakala
Josef Matyáš
Navin Jose Manjooran
Gary Pickrell
Winnie Wong-Ng

The American Ceramic Society

WILEY

Published by John Wiley & Sons, Inc., Hoboken, New Jersey.
Published simultaneously in Canada.

Library of Congress Cataloging-in-Publication Data is available.

ISBN: 978-1-119-32361-7
ISSN: 1042-1122

10 9 8 7 6 5 4 3 2 1

Contents

Preface

The Materials Science and Technology 2015 Conference and Exhibition (MS&T'15) was held October 4–8, 2015 at the Columbus Convention Center, Columbus, Ohio. One of the major themes of the conference was Environmental and Energy Issues. Twenty papers from six symposia are included in this volume. These symposia include:

- 7th International Symposium on Green and Sustainable Technologies for Materials Manufacturing and Processing
- Materials Issues in Nuclear Waste Management in the 21st Century
- Nanotechnology for Energy, Healthcare and Industry
- Materials and Processes for CO_2 Capture, Conversion, and Sequestration
- Materials Development for Nuclear Applications and Extreme Environments
- Hybrid Organic-Inorganic Materials for Alternative Energy

The success of these symposia and the publication of the proceedings could not have been possible without the support of The American Ceramic Society and other organizers of the program. The program organizers for the above symposia are appreciated. Their assistance, along with that of the session chairs, was invaluable in ensuring the creation of this volume.

TATSUKI OHJI, AIST, JAPAN
RAGHUNATH KANAKALA, University of Idaho, USA
JOSEF MATYÁŠ, Pacific Northwest National Laboratory, USA
NAVIN JOSE MANJOORAN, Siemens AG, USA
GARY PICKRELL, Virginia Polytechnic Institute and State University, USA
WINNIE WONG-NG, NIST, USA

Green and Sustainable Technologies for Materials Manufacturing and Processing

“COMMONIZATION” OF MATERIALS: GUILTY BY ASSOCIATION

Marsha S. Bischel, Ph. D.
Amy A. Costello, PE, LEED-AP
Tawnya R. Hultgren
Armstrong World Industries, Inc.
Lancaster, PA 17603

ABSTRACT

As transparency efforts driven by green building requirements move the evaluation of “green” materials and chemicals into the offices of architects and designers and away from scientists, it is important to understand that common, generic terms may not necessarily represent the hazards and risks associated with specific chemicals or materials. Such terms, and the perceptions they create, may negatively exaggerate the actual health-based risks associated with the exposure to a specific product, often via a “guilty by association” mindset. This paper will review requirements that encourage material selection based on chemical content, and will provide examples of common building materials and how generic chemical terms influence market-place decisions.

INTRODUCTION

In 2014, the construction industry accounted for 4% of US GDP;[1] according to a recent report by McGraw Hill Construction, nearly 50% of all buildings in the US are being built to be “green.”[2] Therefore, trends in this industry segment will have a large impact on the materials science community, which ultimately provides the materials and finishes used in buildings.

In the green building community, there is a growing belief that sustainable buildings should enhance the health and well-being of occupants, resulting in a desire to construct them using “non-toxic” materials.[2] These factors have driven a movement to have manufacturers publically share the content of construction materials and finishes to levels as low as 1000 or even 100 ppm, with the presumption that such transparency will eliminate all harmful chemicals from the building. For example, Version 4 of the United States Green Building Council’s (USGBC) Leadership in Energy and Environmental Design (LEED) Building Rating System, which is arguably the most influential green building rating system in the world, offers credit for selecting products that identify material ingredients to 1000 ppm.[3] Common sense and science dictate that the mere presence of a chemical in a product does not necessarily make that product a health risk. Even the most regulated drinking water will contain trace amounts of lead and other heavy metals.

Credits such as the LEED v4 Material Ingredient Credit can inadvertently create fear in the marketplace. The intent of this LEED credit is “to encourage the use of products and materials for which life-cycle information is available and that have environmentally, economically, and socially preferable life-cycle impacts.”[3] Yet, most people in the building industry are not scientists and likely do not understand that the mere presence of a chemical in a product does equate to a product health risk. Before jumping on this bandwagon, it is important to remember that chemical risk is a function toxicity, exposure, hazard, dose and time, and is specific to each compound. Toxicity indicates what health problems are associated with various doses or concentrations, and is estimated using two sources of information: 1) any available data on effects on humans, and 2) bioassay experiments. Exposure is an estimate of how much of the chemical a person is likely to eat, drink, or absorb from water, air, or other sources.

Many decisions made to comply with the green ratings systems and requirements are based solely on the potential hazard a chemical poses rather than the actual risk a chemical poses or the risk that is created if the chemical is not utilized. OSHA has stated that hazard determination does not involve an estimation of risk and that the difference between the terms hazard and risk is often poorly understood. OSHA goes on to say that considerable differences may exist in the risk posed by a substance depending on conditions that result in or limit exposure.[4]

An easily grasped example of hazard versus risk involves the hand-held hair dryer. If "hazard" is broadly assumed to be anything that can cause harm, then these devises clearly have inherent hazards associated with their use. If "risk" is considered to be the chance that someone will actually suffer harm, then an example of risk associated with a hand-held dryer would be the risk of being electrocuted if one uses it in standing water, such as a bath tub.

In an effort to simplify the identification of potentially "harmful" chemicals for non-technical specifiers, so-called listings of "chemicals of concerns" (COC's) have been developed by a wide variety of groups. Such lists are commonly being used by the promoters of healthy buildings. Unfortunately, many of these groups have overly simplified risk and hazard and rely instead on simple lists of broad groups of chemicals that should be avoided. Generally, the worst-case hazard, or "hazard endpoint," such as "reproductive toxin" or "potential carcinogen" is given, without including information about exposure, dose or toxicity levels. In addition, these COC groupings often link materials together based on a "worst player," and ultimately target not only problematic chemicals, but ones that are perfectly safe. This "commonization" of materials through the use of generic terms and broad classes can create negative perceptions of whole groups of materials that are not based on sound science, and can ultimately create fear and frustration on the part of end-users and manufacturers.

Generic terms for chemical compounds can be extremely misleading. It is not uncommon for non-scientists to refer to an entire class of related, or even non-related, compounds using a common name. Examples of these include "chlorine," "antimicrobials," and "phthalates." To a scientist, such terms are vague, and often refer to a part of a compound or even a material's function. However, by using such common, non-exact terms, whole classes of materials are being lumped together and are presumed to be equally dangerous to human health; this phenomenon unfairly targets many safe materials. For instance sodium chloride (table salt) and chlorine gas are two extremely different compounds, but both could be listed as chlorine-containing substances; only the latter is generally considered to be dangerous.

In 2008, the United States passed legislation prohibiting the manufacture, sale, distribution, or import of any children's toy or child care article that contains concentrations of more than 0.1 percent of three (3) specific phthalate-based plasticizers due to concerns about their toxicity and ability to bioaccumulate. Although parts of the ban apply only to items that can be placed in the mouth and be sucked or chewed[5,6] the details are often overlooked. This lack of addressing the specific risk associated with exposure via oral ingestion, combined with the use of the generic term "phthalate" has penalized other products which contain other phthalate-based chemicals, since opponents assume that all phthalates are harmful via any route of exposure or contact. Although the use of these chemicals in building materials does not pose a health risk since there is no route for ingestion-based exposure, by using the common term "phthalate," all chemicals in the family become suspect.

By focusing too narrowly on a presumed health risk and a specific product function, it is possible that other benefits to building occupants or consumers are not being considered. For example, some green building schemes, including LEED V4, frown upon the addition of all anti-microbial compounds out of fears that we are unnecessarily using chemicals that will result in microbes that are resistant to modern medicines. While specific antimicrobial agents have been linked to such resistance, the use of a common term based on a material's function has created the perception that all antimicrobial substances are problematic and should be avoided. However, when these compounds are eliminated, the risks of mold and bacteria growing on materials greatly increases, the material life span can be reduced, and germs and harmful bacteria may be more easily spread. In general, the effort to eliminate a few "dangerous" compounds is not accounting for such holistic trade-offs.

Finally, titanium dioxide is an example of a commonly used material that is simultaneously considered "dangerous" by some and beneficial by others. The fine dust forms of titanium dioxide have been listed as possible carcinogens by IARC, yet other forms are approved for use in cosmetics, food and other products due to its low risk to human health. It is a well-known UV-blocking substance widely used in sunscreens, and has documented antimicrobial properties, both of which enhance human health. It has other beneficial uses in sustainable buildings as well, but none of these can be realized if the material is avoided due to a specific, narrowly defined health risk associated with only one form.

Although these trends towards transparency and simplifying chemical risks are global, the examples discussed here will pertain to the US market. The impacts of commonization by chemical family, function and form will be addressed.

GUILTY BY CHEMICAL FAMILY: "PHTHALATES"

One example where an entire family of chemicals is now perceived by some to be unsafe based on the issues of a few specific chemicals is the case of phthalate esters or "phthalates." "Phthalates" are a class of synthetic chemicals with a broad spectrum of uses including softeners in plastics, solvents in perfumes, and additives to nail polish, as well as in lubricants and insect repellents. Polyethylene terephthalate (PET) is one of the most commonly used polymers in the world, used to make polyester fibers and clear plastic bottles, among other things. However, when people use the generic term "phthalate," they not generally referring to terephthalates but to dialkyl ortho-phthalates, which are a class of about 30 commercial chemicals; these ortho-phthalates are used primarily as plasticizers for polyvinyl chloride (PVC) and as solvents.

In February 2009, three types of ortho-phthalates were permanently banned from use in children's toys and child care articles in the US: di (2-ethylhexyl) phthalate (DEHP); dibutyl phthalate (DBP); and butyl benzyl phthalate (BBP). The US Consumer Product Safety Improvement Act ("CPSIA") of 2008 prohibits the manufacture, sale, distributions, or import into the United States any children's toy or child care article used for feeding or teething that contains concentrations of more than 0.1 percent of DEHP, DBP, or BBP. Three additional phthalates were "interim banned:" diisononyl phthalate (DINP); diisodecyl phthalate (DIDP); and di-n-octylphthalate (DnOP).[5,6] The permanent CPSIA ban applies to children's toys and to child care items that can be brought to the mouth by a child so that it can be sucked and chewed. The interim ban applies only if the toy or toy part can be placed in the mouth; if it can only be licked, then it is not considered as being able to be "placed in mouth."[6] Yet, the later part of this requirement – must be able to be sucked and chewed – is usually not mentioned.

Furthermore, to assume that because these chemicals are harmful in children's toys they must be harmful in all products is not accurate. For example, for a child to be exposed to DEHP in a toy they must chew or suck on it for lengthy periods of time. In fact, the CSPC specifically states that the ban does not apply to components of children's toys that are inaccessible, nor to children's socks, shoes, and packaging for toys or feeding items.[6] These same chemicals used in building materials, such as carpets or floors, do not pose a risk, because there is no route for oral exposure. However, many manufacturers have moved away from the use of ortho-phthalates due to market demand. Other chemically different members of the phthalate family, such as PET, have been widely tested and are generally considered to be safe for human use.

But, for many people, the fact that a few specific phthalates have been banned from being used in a limited number of specific types of products has translated into a belief that all phthalates should be universally banned. As a result, products which contain other chemicals in this family are penalized because of the negative association with the term "phthalate," even though they are not part of the ban. The CPSC even states that children's items may contain other non-banned phthalates, which should suggest that other phthalates are safe to use.[6] Thus, by a using generic chemical term, an entire family of chemicals has been given a "bad reputation," even though the hazards are associated with a limited number of specific compounds.

GUILTY BY FUNCTION: "ANTIMICROBIAL" COMPOUNDS

So-called "antimicrobial" compounds are another example whereby a large number of materials have been lumped together as being dangerous. In this case, the determination has been made based on the function of the material, i.e., the ability to inhibit the growth of microbes, rather than by chemical classification. Indeed, antimicrobial products make use of a wide variety of organic and inorganic compounds, some of which are man-made, and some of which are naturally occurring. Discouraging, or even banning, the use of all "antimicrobials" without consideration of their actual chemistry in an effort to reduce the potential for drug-resistant microbes is reactionary and is frequently not based on good science.

Many of the concerns and issues surrounding antimicrobials may again be attributed to not understanding the complete risk assessment of specific chemicals. Antimicrobial use has increased dramatically, with at least 275 chemicals now registered with the US EPA.[7] As an example, Triclosan is a synthetic broad-spectrum antimicrobial that was first registered with the EPA as a pesticide in 1969.[8] In recent years, evidence of Triclosan (and other antimicrobials) has been found throughout our environment including in water, soil, breast milk, and urine. Due to the amount of Triclosan in the environment and its ability to bioaccumulate, there have been many questions raised regarding the impacts Triclosan and other antimicrobials may have on the environment and human health. These questions have resulted in the discouragement or even banning of numerous antimicrobials in green buildings without an understanding the risk or benefit of the specific chemicals, presuming that all are "guilty" simply based on their ability to kill microbes.

In many cases there is no proven benefit to adding antimicrobials to consumer goods or building products. However, for many years this was deemed to be a best practice for controlling microbial growth, and was an attribute requested by customers; as a result, in many cases antimicrobials were overused. Now the pendulum is beginning to swing the other way, and there is a movement to ban all use of antimicrobials, even though there is currently no true scientific understanding of the actual environmental and health impacts of many members of this broad group of unrelated materials.

However, there are cases in which there truly is a need for antimicrobials, such as in healthcare facilities treating individuals with compromised immune systems. And although many consumer products lack clear evidence that antimicrobial addition is beneficial, in 1997 the FDA studied Triclosan in toothpaste and found that it was effective in preventing gingivitis.[9] This is an example of an instance where we may cause more harm by NOT using antimicrobials.

Perhaps the most glaring example of harm that could occur by not using antimicrobials is in hospital settings. A report based on hospital infections in Pennsylvania estimates that in the US in 2006 there were approximately 720,000 hospital acquired infections, resulting in 74,000 deaths, and $125 billion in additional health care costs.[10,11] And yet, motivated by legitimate fears of creating anti-bacterial resistant microbes, some hospitals and the USGBC are now encouraging the elimination of antimicrobials from buildings. However, the use of antimicrobial compounds in materials that are frequently touched and that are difficult to clear is known to be an effective means of reducing the transmission of infectious disease.

In addition to organic substances, including Triclosan, inorganic materials have also been shown to have antimicrobial properties. These include copper and silver, both of which have been used for over a century to control microbial growth. Copper alloys, such as brass and bronze, have also been shown to prevent the spread of microbes in healthcare facilities.

Initial research indicates that clean, uncoated copper surfaces are perpetually able to kill certain microbes that come in contact with the metal surface; thus, copper appears to be a "self-sanitizing" compound. For example, the use of copper in hospitals on specific surfaces such as toilet seats, faucet handles and door plates has been shown to reduce the amount of three specific types of microbes by 90 -100% as compared to the control fixtures, made of plastic, chrome-plated metals, or aluminum.[10] In another study, a consulting room was retrofitted with copper in areas of high patient contact. Over six months, 71% of the bacterial load was reduced. A study from 1983 showed that the use of copper and bronze doorknobs prevented the spread of microbes in hospitals.[12] However, the use of copper fixtures is banned from LEED V4.[3] This is despite any firm evidence showing that copper leads to the development of drug-resistant microbes.

Silver has also been shown to be highly effective at killing microbes. There is a long history of applying silver compounds directly to wounds; this practice is being resurrected to avoid the use of antibiotics, which have clearly been linked to the rise of so-called "super bugs."[13] Other recent research suggests that the addition of small amounts of silver to other drugs can greatly enhance the efficacy of those drugs in killing microbes, increasing the number of bacteria killed by 10 to 1000 times.[14] Silver does have some documented toxicity issues: there have been documented cases of silver-resistant bacteria,[13] and silver coatings were shown to be toxic to heart tissue when used in heart valves. It can also cause the skin to permanently turn blue-gray.[14] However, outright bans of antimicrobials do not address the potential benefits in reducing the number of fatal infections in healthcare facilities.

Beyond the control of infectious disease, there are reasons to use materials containing antimicrobials in healthcare settings, specifically in porous substances used for sound control. Numerous studies have shown that modern hospitals are extremely noisy, and that this noise has a negative impact on patients, as well as staff. For example, studies have shown that the addition of sound absorbing materials to healthcare spaces can dramatically improve the acoustical quality of the space, directly resulting in reduced patient pain and stress, improved sleep, and reduced medical errors. Ultimately, all of these contribute to improved patient recovery rates.[15,16]

However, many of the best materials for correcting ambient noise issues are open, porous materials.[16] Because mold and bacteria can grow in these materials, due to their inability to be easily cleaned, they cannot be considered for use in healthcare settings unless they are treated. But guidelines put out by several healthcare organizations limit the use of antimicrobials in building products. By limiting the ability of manufacturers to add antimicrobial agents to porous products, it becomes even more difficult to justify the use of these acoustical absorbers in healthcare settings, potentially causing documented, negative impacts on patients due to increased noise.

Moving beyond applications in healthcare, there are valid uses for antimicrobial substances in other arenas that may not be immediately obvious to non-technical end users. For example, manufacturers may add these substances to enhance processability, for example extending the time during which a paint or binder is able to be used before it is attacked by microbes (i.e., spoils) and needs to be disposed of. Thus the use of small amounts of specific antimicrobials can avoid the generation of unnecessary waste, and the associated environmental impacts.

Manufacturers may also add specific antimicrobials to give final products particular functions. Again, the addition of small amounts to paints helps to ensure the long-term stability and durability of the paint and the surfaces below.[17] However, adding these substances to items that are cleaned regularly, such as clothing or flooring, is not functionally necessary, and could cause more harm than good. In such cases, responsible manufacturers should be assessing all the potential risks associated with the use of antimicrobial agents and using them only when the benefits outweigh the risks.

Research is also being conducted into ways of incorporating antimicrobials into coatings in such a way that they continue to inhibit the growth of microbes, but remain safe for humans. Examples include the addition of silver to zeolites, glass, epoxies and other materials. [17]

Understanding all of the risks should naturally lead to a discussion of the potential trade-offs involved with using or avoiding specific antimicrobials. However, banning entire classes of materials solely due to function is not a holistic perspective, and can result in harm being done, or benefits not being realized. It also avoids addressing the risks associated with specific chemicals, instead broadly implying that all members of the category are problematic simply based on one of the functions.

GUILTY BY FORM: TITANIUM DIOXIDE

A specific example of the importance of understanding the entire risk equation is titanium dioxide. Titanium dioxide is used to whiten many products, from food to building materials. The US Centers for Disease Control lists respirable titanium dioxide particles, i.e. dust, as an occupational carcinogen and IARC lists the powdered form as possibly carcinogenic to humans. Yet at the same time, the US FDA allows the use of titanium dioxide in cosmetics and in food products, in quantities up to 1% by weight of the food.[18] The EPA Safer Choices (formerly Design for the Environment) program considers titanium dioxide to be a "material of verified low concern."[19] So without an understanding of the entire risk equation, including the chemical form, it is easy to understand the confusion that arises when one is told in one instance that the substance is a carcinogen and in another, it is safe.

To clarify the carcinogen classification, it is important to understand that the IARC listing is related specifically to findings that fine and ultrafine powders were found to cause respiratory cancers in rats. This led to concern that workers exposed to high levels of dust could be at risk, and the subsequent listing as a possible carcinogen.[20] When used in foods, cosmetics and other products, in the appropriate amounts, the dose and routes of exposure are not ones by which humans can be harmed. Thus the form of this chemical is key to its safety or risks.

In addition, any assumptions that all forms of titanium dioxide are dangerous fail to acknowledge many known benefits for this common material. Beyond its ability to whiten cosmetics, titanium dioxide is a well-known, effective UV-blocking additive used in many sunscreens[21] and other sun-blocking cosmetics. Thus, while the loose particulate form of TiO_2 is considered a carcinogen, it is also a key ingredient in a product known to reduce the instances of skin cancers.

Within the building products industry, titanium dioxide has traditionally been used as a filler in many architectural materials due to its very high light refractive index (2.7 for the rutile form); the addition of titanium dioxide can consequently increase the light reflectance of the finish. This is turn has been shown to increase the effectiveness of natural light within a space, lowering the need for lighting, thereby reducing energy usage.[22,23,24] Many of the sustainable building schemes recognize the use of highly reflective surfaces as a strategy for enhancing daylighting and lowering energy costs and the associated environmental impacts. Increased regulation of green house gasses emitted from power plants will likely increase the desire for products that contribute to reductions in energy usage, making the use of highly reflective surfaces that incorporate TiO_2 even more desirable.

Newer uses for titanium dioxide in construction products include self-cleaning surfaces, which take advantage of the super-hydrophilic properties of the material;[25] the use of these in architectural finishes would lower or negate the use of cleaners, potentially reducing exposure of occupants to these chemicals. In exterior applications, it could be used to help keep daylight harvesting surfaces, such as solar cells and windows, clean.

In addition, titanium dioxide has beneficial photo-catalytic properties that are being exploited in a number of ways to improve the built environment.[25] ALCOA and others have developed architectural coatings which oxidize harmful nitrous oxide (NOx) fumes and turn them into nitrides;[26,27] others have used the same principles to create "smog-eating" roof tiles and other surfaces.[28,29] These products have the potential to use existing surfaces to reduce the generation of harmful smog in the environment.

Finally, the anatase form of titanium dioxide has been shown to be an anti-bacterial substance that is activated by UV light, creating the potential for surfaces that can sterilize themselves in the presence of light.[25,30] The use of this property in medical facilities and other areas is being investigated.

Thus, while titanium dioxide can be a potential carcinogen in certain situations related primarily to the generation of dust during manufacturing, it can also be used with minimal risk to humans, as evidenced by its approval for use in food and cosmetics. Furthermore, it can actually be used to improve the indoor environmental quality of buildings and the health of humans. It can also reduce the environmental impacts of buildings through various mechanisms, including energy reduction. Thus, although different forms of this material have different hazards, there are some who are assuming that all forms are problematic based on the listing of the dust. Indeed in many of the new transparency requirements and lists of chemicals of concern, only the hazard end-point of "potential carcinogen" would be listed, despite the fact that this classification, in this case, is for a narrowly defined form of titanium dioxide. Again such broad groupings, in this case ignoring form, can ignore the potential benefits associated with other, non-hazardous forms.

TRADE-OFFS

Trade-offs such as those associated with antimicrobials and titanium dioxide are frequently not being addressed by those advocating the elimination of chemicals to improve the health of building occupants. Instead, it is likely that a non-scientist would see that LEED credits that encourage the avoidance of antimicrobials are "proof" that any material that can kill microbes should be avoided. Similarly, they might see that TiO_2 is a possible carcinogen and leap to the conclusion that it should be avoided as a "toxic" chemical. Indeed, the internet is full of articles and blogs asking such questions as whether one's lipstick, hand wash, toothpaste or sunscreen is causing health issues due to the presence of titanium dioxide or antimicrobial chemicals.

In addition to balancing the trade-offs between a material's beneficial properties and potential hazards and risks, manufacturers face other trade-offs. These include costs, both in using alternate materials, including things such as capital investment and availability of materials, and productivity losses, due for example to spoiled paint. There can also be a lack of non-toxic alternatives; even the US EPA's Safer Choice program recognizes that in some cases there may be hazards associated with some of the chemicals that are currently "best options."

However, by referring to generic lists of potentially problematic materials, commonization is resulting in a situation whereby these trade-offs are generally not considered. As a consequence, some beneficial and safe products can be overlooked or omitted from use in buildings and other products.

MARKETPLACE FRUSTRATIONS

Today's climate for increasing transparency, while well intended, is creating frustration for manufactures and consumers alike. For manufacturers, much of this frustration can be linked to the difficulty in obtaining and then conveying highly technical information around chemical content and risk to their customers and stakeholders. For consumers, much of the frustration can be linked to a lack of easy-to-understand information, resulting in difficulties in assessing risks and comparing products.

Efforts to simplify such information for non-technical audiences have been made. In some cases, the use of generic terms, such as "antimicrobial" or "phthalate" have helped lead to these cases of "commonization" and have allowed perfectly safe chemicals to be assumed to be health hazards. Indeed, several of the earliest influential "red lists" of so-called toxic chemicals were primarily lists of generic families or functions, helping to propagate the issues associated with commonization. Updated programs tend to have summary lists which contain generic terms, but these are generally now expandable into more detailed lists that include specific CAS numbers, etc. However, if someone merely scans the summary lists, the commonization effect is still applicable. Additionally, the ability to thoroughly comprehend the chemical and toxicity information associated with each chemical requires a reasonably scientific background, and most consumers do not have this skill set.

While manufacturers and consumers would both like easy-to-understand ways to assess the true risks associated with products, there is currently no easy solution. And the current trend of requiring chemical transparency to lower and lower trace levels is likely to only add to the confusion and resulting frustration.

CONCLUSIONS:

As transparency requirements in building products becomes more widespread and stringent, manufacturers and consumers alike will continue to face issues related to the evaluation and understanding of a material's true risk. Efforts to simplify the transparency process have resulted in the misuse of generic terms for chemicals and materials. This commonization often oversimplifies or takes the place of a complete and through chemical evaluation and may misrepresent a chemical.

True risk assessments for chemicals should be based on the end-use and include consideration of exposure route, duration of exposure, and dose. Exposure is rarely being considered by the current schemes for evaluating chemicals, but it plays a vital role in understanding risk. These assessments should also compare costs, benefits and hazards of using the material. For building materials, such cost-value assessments should include the impact on the total indoor environmental quality and health of the occupants, the impact on the building's environmental profile, and other pertinent factors, such as manufacturability and availability of alternatives.

This commonization of materials based on chemical family, function or form has resulted in many safe materials being perceived by the public as potentially toxic. In some instances, this misconception has resulted in the green building schemes avoiding or banning whole groups of materials. These decisions tend not to have addressed a holistic, scientific assessment of the benefits and risks associated with specific materials and finished goods.

Attempts to simplify transparency have led to many misinformed decisions regarding the appropriate use of a specific chemical or products containing these chemicals. Just as the industry has adopted comprehensive life cycle analyses, ultimately allowing for the creation of environmental product declarations that attempt to address all of the environmental impacts and phases of a product, there needs to be a clear way to perform a similar assessment of a product's true impact on human health and well-being. Such systems need to be chemical and product specific, and consider the holistic impacts rather than use common terms for chemical family, function or form. They need to be based on internationally recognized, consensus-based standards that give guidance on how to evaluate the hazards, risks and benefits of specific materials, similar to the ISO suite of standards that were developed to for creating life cycle assessments.

In short, scientists need to actively steer the industry away from a system that utilizes over-simplified lists to one that assesses the true, holistic risk associated with the use of a specific chemical in a specific product. Only then will we move away from a market in which products can be guilty merely by association to one in which well-informed, science-based decisions can be made.

REFERENCES:

[1]U.S. Bureau of Economic Analysis, *Gross Output by Industry, 2006 – 2014.*

[2]Bernstein, H. M., ed., *World Green Building Trends*, McGraw Hill Construction, 2013.

[3]USGBC, LEED V4 for Building Design and Construction, 1 July 2015.

[4]OSHA U.S. Department of Labor Occupational Safety and Health Administration, "Guidance for Hazard Determination for Compliance with the OSHA Hazard Communication Standard (29 CFR 1910.1200)."

[5]U.S. Consumer Product Safety Improvement Act ("CPSIA") OFF 2008, Pub. L. No. 110-314, § 108, 122 Stat. 3036 (2008). Section 108(A) of the CPSIA.

[6]US Consumer Products Safety Commission, "Phthalates," 7 July 2015, http://www.cpsc.gov/en/Business--Manufacturing/Business-Education/Business-Guidance/Phthalates-Information/ .

[7]US Environmental Protection Agency, http://iaspub.epa.gov/apex/pesticides/f?p=CHEMICALSEARCH:46:0::NO:::, 3 Sept 2015.

[8]US Environmental Protection Agency, "Frequent Questions Associated with the RED", December 20, 2013: http://www.epa.gov/oppsrrd1/reregistration/triclosan/triclosan-questions.htm

[9]US Food and Drug Administration, http://www.fda.gov/downloads/ForConsumers/ConsumerUpdates/UCM206222.pdf

[10]Grass, G., et al., "Minireviews: Metallic Copper as an Antimicrobial Surface," *Applied and Environmental Microbiology*, 77 (2011) 1571.

[11]Pennsylvania Health Care Cost Containment Council, *Hospital-acquired Infections in Pennsylvania: Calendar Year 2006*, April 2006.

[12]Kuhn, P. J., "Doorknobs: a source of nosocomial infection?" Copper Development Association, New York, NY, 1983.

[13]Chopra, I., "The increasing use of silver-based products as antimicrobial agents: a useful development or a cause for concern?" *Journal of Antimicrobial Chemotherapy*, **59** (2007) 587.
[14]Owens, B., "Silver makes antibiotics thousands of times more effective," *Nature*, 19 June 2013.
[15]Ulrich, R., et al., "Literature Review on Evidence-Based Healthcare Design," *Health Environments Research and Design Journal*, **1** (2008) 61.
[16]Caulfield, J., "Shh! 6 ways to keep the noise down in new and existing hospitals," *Building Design + Construction*, http://www.bdcnetwork.com/shhh-6-ways-keep-noise-down-new-and-existing-hospitals
[17]Zielnik, A., "The green revolution enters the antimicrobial fray," *Journal of Architectural Coatings*, Aug/Sept 2009, 31.
[18]US Food and Drug Administration Department of Health and Human Services, "Title 21--Food and Drugs, 21CFR Part 73."
[19]US Environmental Protection Agency Safer Choice Program, http://www2.epa.gov/saferchoice
[20]Canadian Centre for Occupational Health and Safety, "Titanium Dioxide Classified as Possibly Carcinogenic to Humans," Aug. 2006, http://www.ccohs.ca/headlines/text186.html .
[21]Portier, K., "Can your sunscreen pose a health risk?" American Cancer Society, July 22, 2013, http://www.cancer.org/cancer/news/expertvoices/post/2013/07/22/can-your-sunscreen-pose-a-health-risk.aspx .
[22]Bischel, M. S. and W. E. Beakes, "Using high light reflectance acoustical ceilings to increase the energy efficiency of buildings," *Euronoise 2008 Conference Proceeding*, July 2008.
[23]Mistrick, R. and L. Chen, "Energy Saving Potential of High Reflective Flooring Material for Sustainable Interiors," 2014.
[24]Kennedy, M., "Daylighting and Flooring: Don't Overlook the Issue of Reflectivity," Armstrong World Industries, Inc., January 2015, http://www.armstrong.com/commflooringna/whitepaper.jsp?paper=daylighting .
[25]Fujishima, A., et. al, "Titanium dioxide photocatalysis," *Journal of Photochemistry and Photobiology C: Photochemistry Reviews*, **1** (2000) 1–21.
[26]ALCOA Architectural Products, "Eco-Clean Newsletter," 2011.
[27]Gobin, C. et al., *Innovative Façade Coatings with De-Soiling and De-polluting Properties*, PICADA, Feb. 2006.
[28]Wray, P., *American Ceramic Society Bulletin*, 30 July 2010, http://ceramics.org/ceramic-tech-today/titanium-dioxide-coated-anti-nox-roof-tiles-now-being-marketed-in-us
[29]Wray, P., *American Ceramic Society Bulletin*, 24 July 2010, https://ceramics.org/ceramic-tech-today/titania-concrete-combination-in-roadways-reduces-nox-levels-25-45
[30]Kliemann, O., et al., "Cold Spraying for Titanium Dioxide Coatings with High Photocatalytic Bactericidal Activity," *International Thermal Spray Conference and Exposition 2010 Conference Proceeding*, May 2010, p 341.

EXPERIMENTAL RESEARCH AND APPLICATION OF COPPER OXIDE FLOTATION USING THE COMBINED COLLECTORS OF BENZOHYDROXAMIC ACID AND BUTYL XANTHATE

Daixiong Chen[1,2], Jun Xiao [1,2],Chunming He[1,2] , Xiaodong Li [1,2]

(1. Hunan Research Institute for Non-ferrous Metals; 2. Hunan Provincial Key Laboratory for Complex Copper Lead Zinc Associated Metal Resources Comprehensive Utilization.)

ABSTRACT

Benzohydroxamic acid with double dentate can combine with copper and form cyclic hydroxyl oxime acid copper chelate, so it has good recovering capability and selectivity for copper oxide minerals, especially for malachite. The flotation results of the pure mineral malachite show that the strongest synergistic collecting effect between hydroximic acid and xanthate can be generated with the highest malachite recovery of 95.5% when certain sodium sulphide is used as vulcanizing agent in the optimal pH range and the best ratio of hydroximic acid to xanthate is 3.5:1. In the refractory real ore flotation experiments, due to the synergistic collecting effect, the flotation recovery of malachite can reach to over 92% under the condition of optimum proportion, increasing 12% compared with the conventional collectors. In industrial production, the flotation indexes of the copper grade are high, and the copper oxide resource is utilized with high efficiency.

INTRODUCTION

Beneficiation of copper oxide ore is one of the recognized puzzles in mineral processing industry, and it is also the important research field for dressing workers. Copper oxide ore is the vital component of copper resource in the world, which has abundant reserves. Majority of sulphide deposits have oxidized zone on the top, and some oxidation deposits are oxidized to become the large and medium-sized oxidation deposits [1]. However, most copper oxide ores have the problems such as high slime, poor floatability, complex ore property and low flotation recovery, so it is of great significance in beneficiation of copper oxide ore to develop an efficient collector for copper oxide ore especially for malachite[2~4]. At present, sulfuration-xanthate

Corresponding author: Dai-Xiong Chen, Professor, chief engineer and director of dressing Institute.Hunan Research Institute of Non-ferrous Metals- Hunan Provincial Key Laboratory of Complexed Copper-Lead-Zinc Associated Metal Resource Comprehensive Utilization, Changsha, China. 99 Yada Road Changsha 410100, P.R. China Phone: 0731-85239033. Email: chendaixiong888@163.com flotation is the most important method for treating copper oxide ore. In recent years, due to the decrease of mining grade, the utilization of refractory copper oxide ore draws more and more attention [5], except for sulfuration-xanthate method, chelating flotation method [6], chemical beneficiation method [7], microwave irradiation flotation method, hydrothermal sulfuration-floatation method [8] and ammonia leaching of sulfide precipitation-floatation method [9] have been occurred.

1 MATERIAL PREPARATION OF SINGLE MINERAL FLOTATION EXPERIMENTS, REAGENTS AND RESEARCH METHODS

The pure malachite samples were taken from a mine in Congo, and the purity was 98%. It was peacock green with well crystallization, and the copper content was 54.8%. The samples were crushed, handpicked, ground in porcelain mill and then sieved with a 0.074mm sieve, the through product was used in experiments.

Reagents used in this reseach are as follows: sodium sulfide ($Na_2S \cdot 9H_2O$), analytically pure, colorless cubic crystal, soluble in water; butyl xanthate ($C_4H_9OCSSNa$), industrial quality, pungent pale yellow powder; benzohydroxamic acid ($C_7H_7NO_2$), industrial quality with 70% purity, pink powder.

Single mineral flotation tests were conducted on a XFG-76 flotation machine with 50 ml volume, and the impeller speed was 1500 rpm. For each test, 2g of samples and moderate distilled water were added in the cell, and stirred 2 min to adjust the pulp, in order to examine the flotation behavior of malachite under different conditions. The flotation time was 3 min, after flotation, the froth product and the unfloated product were dried under natural conditions and weighed, and the percentage of the froth product weight to the total weight was calculated, that is the recovery, the experiment process is shown in Fig.1.

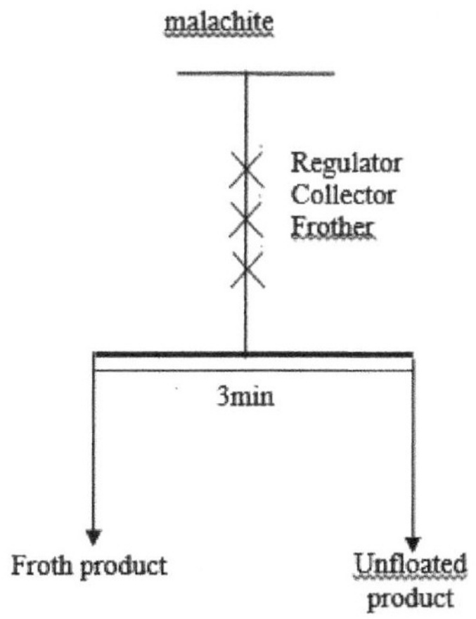

Fig.1 The process of the pure malachite mineral flotation experiments

2 RESULTS AND DISCUSSIONS

2.1 CONDITION EXPERIMENTS OF SODIUM SULFIDE CONCENTRATION

The concentration of butyl xanthate and pine oil was fixed at 220mg/L and 72mg/L, respectively. The relationship between sodium sulfide concentration and malachite recovery is shown in Fig.2.

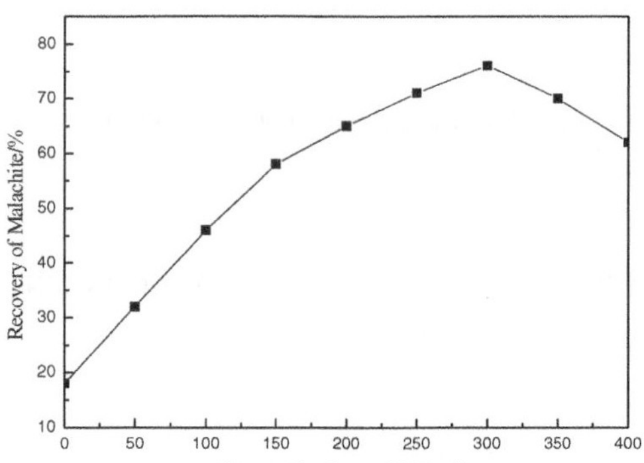

Fig.2 The relationship between sodium sulfide concentration and malachite recovery

The results showed that the recovery of malachite was extramely low when butyl xanthate was used as the collector without sodium sulfide. When sodium sulfide was used as vulcanizing agent and its concentration was in the range of 0~300mg/L, malachite recovery increased with the increase of sodium sulfide concentration, while the recovery had a decrease as the concentration was more than 300 mg/L.

Sodium sulfide, as a vulcanizing agent, can vulcanize malachite surface, mainly because sodium sulphide solution shows strong alkaline, adding sodium sulphide solution means adding OH^-, HS^- and S^{2-} in the pulp, these three ions can react with malachite according to the following reactions:

$$xCuCO_3{\cdot}yCu(OH)_2+2OH^-=(x-1)CuCO_3{\cdot}yCu(OH)_2{\cdot}Cu(OH)_2+CO_3^{2-}$$

$$(x-1)CuCO_3{\cdot}yCu(OH)_2{\cdot}Cu(OH)_2+HS^-=(x-1)CuCO_3{\cdot}yCu(OH)_2{\cdot}CuS+OH^-+H_2O$$

In the case of the presence of OH^- ions, part of mineral surface can achieve hydroxylation, the HS^- ions change the hydroxylation area into sulfide area, at the same time part of the carbonate dissolves, leading to the formation of new hydroxylation area[4].The reaction will not stay on the mineral surface, but go deep into the mineral internal, and form the CuS thin films with different thickness, finally strengthened the hydrophobicity of mineral surface, and improved the adsorption rate of collector. At last, depending on the different initial concentrations of the sodium sulfide, CuS thin films of different thickness were formed[6].

The uneven surface of malachite made its surface exist different areas of complete sulfidation, incomplete sulfidation and non sulfuration, when butyl xanthate was used as the collector, it can adsorb on the complete sulfidation area of malachite surface, so some malachite can be recovered.

2.2 CONDITION EXPERIMENTS OF BUTYL XANTHATE CONCENTRATION

The concentration of sodium sulfide and pine oil was fixed at 300mg/L and 72mg/L, respectively. Butyl xanthate was used as the collector, and the relationship between its concentration and malachite recovery is shown in Fig.3.

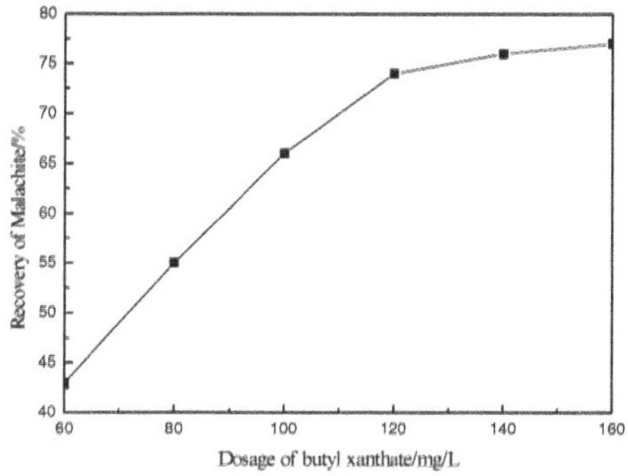

Fig.3 The relationship between butyl xanthate concentration and malachite recovery

After the sulfuration of malachite, the area of complete sulfuration can be considered as copper sufide ore, in the redox system formed by the CuS thin film, dissolved oxygen and butyl xanthate, the dissolved oxygen took place reduction reaction in the cathode, and the oxidation reaction between CuS and butyl xanthate took place in the anode. Due to the two independent and interdependent electrode reaction processes of CuS thin film on malachite surface, xanthate can adsorb on CuS film successfully, and xanthate copper was formed, making malachite surface

Hydrophobic, and the specific reaction process equation are as follows:

$O_2+2H_2O+4e \rightarrow 4OH^-$

$CuS+2\ C_4H_6OCSS^- \rightarrow Cu(C_4H_6OCSS)_2+S+2e$

$2CuS+4C_4H_6OCSS^-+3H_2O \rightarrow 2Cu(C_4H_6OCSS)_2+S_2O_3^{2-}+6H^++8e$

Because of the different surface activities of malachite surface, the optimal sodium sulfide concentrate for the formation of sulfide film is different, at the best concentration of sodium sulfide, only part of malachite surface formed complete sulfuration area, malachite of incomplete sulfuration area and non sulfuration area is difficult to adsorb butyl xanthate, at the same time, the adsorption of butyl xanthate on sulfide film of malachite surface was dynamic adsorption - desorption process, so it is difficult to achieve the complete recovery of malachite using single butyl xanthate.

Owing to the adsorption characteristics of butyl xanthate on malachite surface, the recovery of malachite increased with the increase of butyl xanthate concentration, when the concentration was more than 120 mg/L, the malachite recovery was basically constant, which was only 76.8% at the concentration of 160 mg/L.

2.3 Condition experiments of benzohydroxamic acid concentration

The concentration of pine oil was 72 mg/L, benzohydroxamic acid was used as the collector without sulfuration, and the effect of benzohydroxamic acid concentration on malachite recovery is shown in Fig.4.

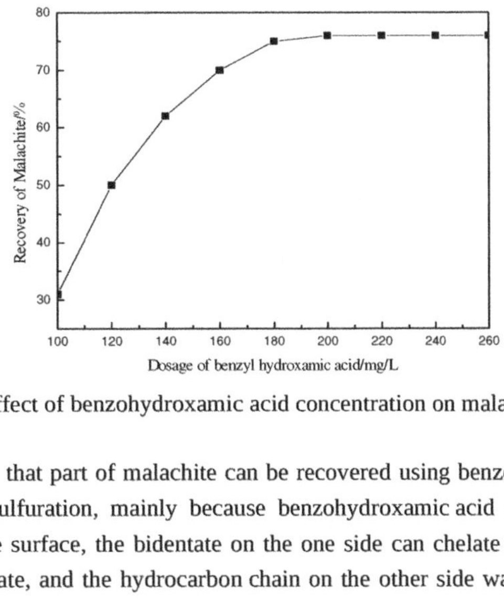

Fig.4 The effect of benzohydroxamic acid concentration on malachite recovery

The results showed that part of malachite can be recovered using benzohydroxamic acid as the collector without sulfuration, mainly because benzohydroxamic acid can occur chemical adsorption on malachite surface, the bidentate on the one side can chelate with Cu^{2+} and form stable penta cyclic chelate, and the hydrocarbon chain on the other side was hydrophobic, thus making malachite floated, the main chemical equation is as follows:

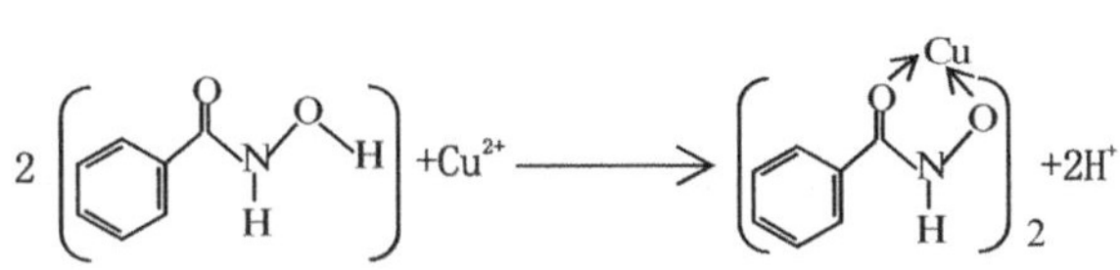

Since the activity of copper ions on malachite surface is different, only part copper ions can form stable chelate with benzohydroxamic acid, so the malachite can not be recovered completely even if the concentration of benzohydroxamic acid has a great increase.

2.5 CONDITION EXPERIMENTS OF DIFFERENT RATIOS OF BENZOHYDROXAMIC ACID TO BUTYL XANTHATE

The concentration of sodium sulfide and pine oil was 300 mg/L and 72 mg/L, respectively. Benzohydroxamic acid and butyl xanthate were used as the combined collector, and the total concentration was 180 mg/L, the effect of the ratio of benzohydroxamic acid to butyl xanthate on malachite recovery is shown in Fig.5.

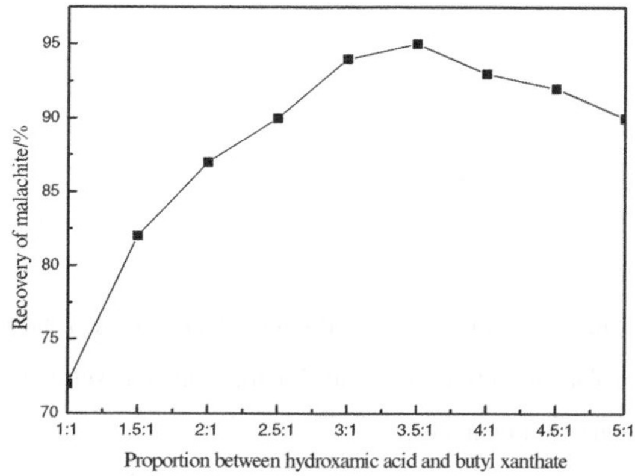

Fig.5 The effect of the ratio of benzohydroxamic acid to butyl xanthate on malachite recovery

When benzohydroxamic acid and butyl xanthate were combined used, the malachite recovery increased gradually with the increase of benzohydroxamic acid ratio, and the malachite recovery was up to 95.6% at the ratio of 3.5:1, while the malachite recovery had a slight decrease when the ratio of benzohydroxamic acid was more than 3.5:1. It is showed that the combined use of benzohydroxamic acid and butyl xanthate can greatly increase the flotation recovery of malachite compared with single benzohydroxamic acid or butyl xanthate, and the mian reasons are as follows:

(1)Synergistic effect between reagents. Benzohydroxamic acid and butyl xanthate have different activities, benzohydroxamic acid with higher activity first forms stable cyclic chelate with Cu^{2+} on malachite surface, van der waals force between the hydrocarbon chains strengthens the alternative adsorption of butyl xanthate on CuS points formed by sulfuration, moreover, the adsorption of butyl xanthate on malachite surface increases with the increase of benzohydroxamic acid adsorption, synergistic effect are created between benzohydroxamic acid and butyl xanthate, increasing the malachite recovery.

(2)The inhomogeneity of malachite surface activity. The bonding orbital is filled with electrons and the antibonding orbital is empty inside malachite, and the crystal is in stable condition. In the crushing and grinding process of large malachite, the specific surface area increases, and more atoms on mineral surface are produced, in their original plate the neighbour atom bonding with these atoms on mineral surface "lost", lattice plane are truncated, which produces excess charge and causes upper state of mineral surface, so it is required the atoms on mineral surface restructure or adsorb foreign molecules or ions in order to reduce the surface energy. Due to the randomness of crushing and grinding process, the surface activity of malachite is different, inevitably making the sulfuration degree of each surface different, and the sulfuration area of malachite surface is divided into: complete sulfuration area, incomplete sulfuration area and non sulfuration area.

(3)The complete sulfuration area on mineral surface can be seen as copper sulfide ore, according to the valence bond theory, the electronegativity of sulfur in copper sulfide is less, and the covalent radius is larger. From the hard and soft acids and bases theory, it can be seen that the electronegativity of sulfur is low, the polarizability is high, and it is easy to be oxidized, the constraints of valence electrons is relatively loose, belonging to the soft base, so it affects the coupling atoms, increases the softness of the linked atom, that is, the copper ion in copper sulfide is soft acid. However, the incomplete sulfuration area and the nonsulfuration area can be considered as copper oxide ore or close to copper oxide ore, while the electronegativity of oxygen in copper oxide is high, the polarizability is low, being difficult to be oxidized, and the constraints of its valence electrons is relatively close, its negative inductive effect makes the copper ions in the copper oxide become a hard acid. According to the hard and soft acids and bases theory, hard acids preferentially coordinate with hard bases, and soft acids preferentially coordinate with soft bases. In full sulfuration area, xanthate type soft base (electronegativity is about 2.7) is easy to adsorb in the complete sulfuration area; while in incomplete sulfuration area and nonsulfuration area, the electronegativity of benzohydroxamic acid is larger (about 3.8), which contains large hardness O and N atoms, so it is easy to chelate with copper ion and adsorb on malachite surface, thus making malachite floated. Therefore, the combined use of benzohydroxamic acid and butyl xanthate can make the collector completely covered on malachite surface, achieving the goal of efficient recovery of malachite.

3 EXPERIMENTAL RESEARCH ON REAL ORE

The copper grade of a copper oxide ore in Congo was 2.35%, and the oxidation rate of copper minerals was 95.4%, the multielement analysis results of the raw ore is shown in table 1, and the phase analysis of copper is shown in table 2.

Table 1 The multielement analysis results of the raw ore/%

Element	Cu	Co	S	As	SiO_2	Al_2O_3	CaO	MgO	Fe_2O_3
Content	2.35	0.017	0.05	0.068	41.52	11.15	8.12	1.66	2.25

Table 2 The phase analysis results of copper/%

Copper phase	Free copper oxide	Combined copper oxide	Primary copper sulfide	Secondary copper sulfide	Total copper
Content	2.45	0.17	0.06	0.007	2.687
Distribution rate	91.18	6.33	2.23	0.26	100.00

The copper oxide of the sample accounted for 97.51%, the binding rate and the oxidation rate were low, and the copper oxide mineral is mainly in the presence of malachite. From single mineral flotation experiments, it can be seen malachite have the highest recovery when benzohydroxamic acid and butyl xanthate are used as the combined collector with the ratio of 3.5:1. In order to verify the feasibility of the combined collector, the comparative tests of different collector were first conducted, the process is shown in Fig.6, and the results are shown in table 3.

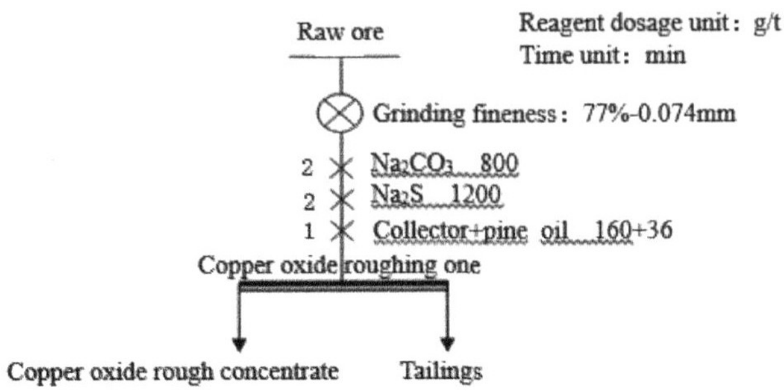

Fig.6 The comparative test process of different collector of a copper oxide ore in Congo

Table 3 The comparative test results of different collector of a copper oxide ore in Congo /%

Products	Yield	Cu grade	Cu recovery	Collectors
Copper oxide rough concentrate	18.06	12.05	80.10	benzohydroxamic acid : butyl xanthate =3.5:1
Tailings	81.94	0.66	19.90	
Raw ore	100.00	2.72	100.00	
Copper oxide rough concentrate	15.31	11.94	67.48	butyl xanthate
Tailings	84.69	1.04	32.52	
Raw ore	100.00	2.71	100.00	
Copper oxide rough concentrate	17.44	11.62	74.73	benzohydroxamic acid
Tailings	82.56	0.83	25.27	
Raw ore	100.00	2.71	100.00	
Copper oxide rough concentrate	18.42	10.03	68.74	pentyl xanthate
Tailings	81.58	1.03	31.26	
Raw ore	100.00	2.68	100.00	

The results showed that the grade and recovery of copper oxide rough concentrate were highest using the combined collector developed by single mineral flotation experiments, and the recovery had a great increase compared with other collectors. So the combined collector had good feasibility to real ore.

According to the process and reagent system of Fig.7, the closed circle experiment was conducted using the combined collector of benzohydroxamic acid and butyl xanthate at the ratio of condition test, and the results are shown in table 4.

Table 4 The closed circle experiment results of a copper oxide ore in Congo /%

Products	Yield	Cu grade	Cu recovery	Malachite content	Malachite recovery
Copper oxide concentrate	7.41	32.12	88.02	52.11	92.67
Tailings	92.59	0.35	11.98	0.33	7.33
Raw ore	100.00	2.70	100.00	4.17	100.00

The results showed that the concentrate with the grade and recovery of 32.12% and 88.02% was obtained using sodium carbonate as regulator, sodium sulfide as vulcanizing agent, and benzohydroxamic acid and butyl xanthate as the collector, the malachite recovery reached to 92.67%. Under the same experimental condition, the comparative test was carried out using butyl xanthate as the collector, the grade of the obtained copper oxide concentrate was 28.04%, and the recovery was 75.61%. Therefore, malachite mineral can be effectively recovered using the combined collector developed by single mineral flotation experiments at a appropriate ratio, and the recovery of copper oxide increased 12%.

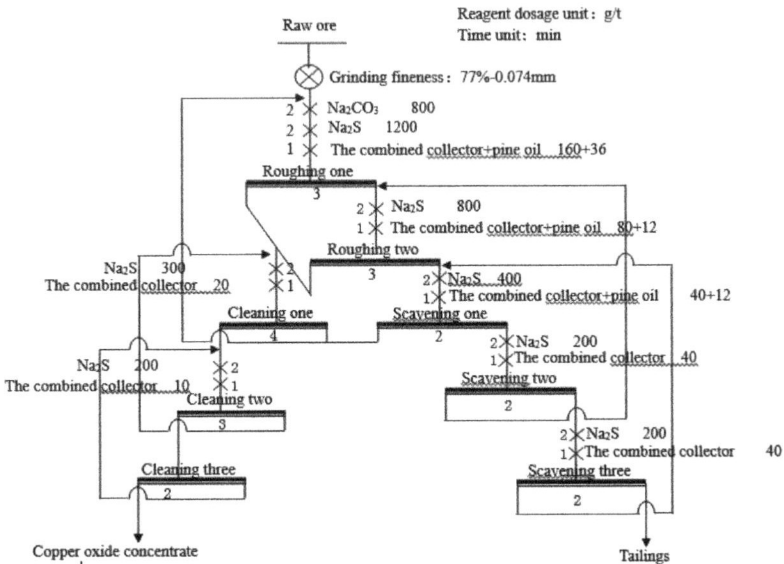

Fig.7 The closed circle experimental process of a copper oxide ore in Congo

4.CONCLUSION

(1)In order to weaken the hydrophilcity of minerals and decrease the collector dosage, it is necessary to vulcanize malachite properly.

(2)The nonuniformity of malachite surface caused the nonuniformity of malachite surface after sulfuration, dividing into complete sulfuration area, incomplete sulfuration area and non sulfuration area, and corresponding to soft acid, solid acid and acid between soft acid and solid acid, respectively. According to the soft and hard copper, the softer alkali like butyl xanthate and the harder alkali like benzohydroxamic acid can be used as the combined collector, and synergistic effect can be created between different collectors.

(3)There is a best ration between collectors, in the range of best ratio, the collector can maximize the collecting capability, which can save reagent and realize efficient collecting. The combined collector of benzohydroxamic acid and butyl xanthate developed in this research can ensure the full play of synergistic effect at the ratio of 3.5:1, and the malachite recovery is more than 95%. Furthermore, the combined collector of benzohydroxamic acid and butyl xanthate has good feasibility to real ore, effectively increasing the recovery of copper oxide ore.

REFERENCES

[1] William X.Chavez，Jr.New Mexico School of mines, The Geochemical settings of copper Oxide ore Deposits: Recognition of Eh-ph Conditions and the Weathering of Copper Mineralization Eleventh Annual V.M. Goldschmidt Conference(2001:3623)

[2] Fu Wenzhang, Gu Jinchuan. Flotation Research on Copper Oxide Ore[J].Multipurpose Utilization of Mineral Resources,1997，（3）：37～42.

[3] Wen Shenglai. Overview on Flotation Method of Low-grade Oxidized ore and Its Bioleaching Technology[J]. Morden Mining，2010(2)：57～59.

[4]Riet B V, Kier L B, Elford H L. Structure-activity relationships of benzohydroxamic acid inhibitors of ribonucleotide reductase.[J]. Journal of Pharmaceutical Sciences, 1980, 69(7):856–857.

[5] Zhou Xia. Studing of Using Foreign Copper Ore Resource in China[J].Nonferrous Mines，2003，12：43～45.

[6] Lu Daogang. The Research Situation and Development Direction of Copper and Cobalt Ore in Katanga Region of Congo[J].World Nonferrous Metal，2009(7)：72～75.

[7] J.Lenormed，T.Salman and R.H.Yoon.Hydoxamate Flotation of Malachite Can.Met.Q.18（1979）125～129

[8] Mao Y B, Wen S M, Fang J J. Studying on Oxide Copper Ore Flotation with Different Collectors[J]. Advanced Materials Research, 2014, 1010-1012:1626-1629.

[9] Lekang O I, Stevik T K, Bomo A M. Evaluation of different combined collectors used in longlines for blue mussel farming[J]. Aquacultural Engineering, 2003, 27(2):89–104.

INVESTIGATION OF THE MICROSTRUCTURAL EVOLUTION BETWEEN PELLET AND SINTER UNDER THE CONDITIONS OF AN OXYGEN BLAST FURNACE

Wentao GUO, Qingguo XUE, Long CHEN, Yingli LIU, Xuefeng SHE, Jingsong WANG*
State Key Laboratory of Advanced Metallurgy, University of Science and Technology Beijing, Beijing 100083 PR China.

ABSRACT

Oxygen blast furnace is a new iron-making technology for carbon saving and emission reduction. The softening process of burden was investigated under the conditions of oxygen blast furnace (OBF) and traditional blast furnace (TBF) using the softening and melting device. The results showed that the slag matrix and myrmekitic metallic iron formed at the edge of the TBF, and the slag matrix and island wustite were in the center. However, the slag matrix and myrmekitic metallic iron formed both at the edge and the center in OBF. Compared to that in TBF, the distribution of the slag and metallic iron showed that the slag was winded or separated in the metallic iron phase in OBF after the melting of burden, increasing the slag-iron separation temperature and the interaction starting temperature and improving the softening and melting properties.

1 INTRODUCTION

At present, the energy consumption of iron making in blast furnace accounts for nearly 70% of the entire steel-making process, and under the circumstances of energy conservation and emissions reduction, this process requires higher efficiency and energy conservation technology. Top gas circulation - full oxygen blast furnace (full oxygen blast furnace) can significantly cut the coke rate, reducing CO_2 emissions and having highly efficient production.[1-3] In recent years, more than_10 processes have been proposed by scholars at home and abroad to answer two key questions about upper cold and down heat.[4-10]

The biggest advantages of OBF is that the reduction potential was improved substantially, and the indirect reduction degree reached 90%,[11] which means that the direct reduction degree reduced significantly. It benefits the reduction of heat output in the high-temperature region and the carbon emission. Meanwhile, the improvement of indirect reduction will certainly influence the size and position of cohesive zone, which will ultimately affect the production efficiency and energy utilize situation. The softening and melting experiment under the condition of OBF proved that the softening interval will broaden, but the melting interval will narrow down, the temperament of dripping will improve, and the permeability index will significantly reduce.[12] when the reduction degree increased, the metal iron phase area increased simultaneously, and the melting point of slag phase, the viscosity , the interfacial tension also changed, which leaded the change of microstructure of burden during the softening and melting process, and the softening and melting performance was improved. C. E. Loo et al.[13] The interaction between sinter and lump could restrain the untimely deformation of the burden layer and improve the softening and melting performance of mixed burden. Research about the change in ore phase structure during softening and melting, the evolution rule of burden properties, and the interaction between different ferric burdens is still relatively scarce. The revelation of the OBF softening and melting mechanism has significance for the determination of OBF process parameters. Using a self-designed experimental apparatus for load softening, this research studied the microstructural evolution of pellets and sinter pre-reduced in the OBF and T-BF gas atmosphere. The interaction effect between different ores in the mixed burden was also investigated. Comparing the microstructure of the OBF case with that of the T-BF case, the microstructural evolution mechanism of ferric burden in the OBF atmosphere was studied, and the softening and melting behavior of ferric burden in the OBF atmosphere was revealed.

2. EXPERIMENT

2.1. EXPERIMENT MATERIALS

The chemical composition of the pellets and sinter used in this research is showed in the **Table 1**. The granular size is 14-15 mm.

Table 1. Chemical compositions of sinter and pellet. (wt%)

	TFe	FeO	SiO_2	Al_2O_3	CaO	MgO	R
Pellet	61.32	3.27	8.83	1.18	0.84	0.7	0.01
Sinter	54.44	8.69	5.75	2.6	10.76	2.48	1.87

2.2. EXPERIMENTAL METHOD

The flow charts of OBF and T-BF process were shown in the **Fig 1**. Among them, the mixed ore was mixed base in binary basicity as 1.3. The experimental atmospheres were from the computing results of blast furnace mathematical model. Drying 500 g samples at 100 degrees Celsius for 3 hours, then placed the samples into a high temperature alloy reaction tube sizes Φ75 mm×800 mm.

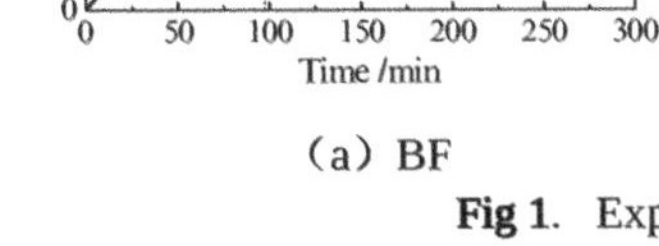

(a) BF

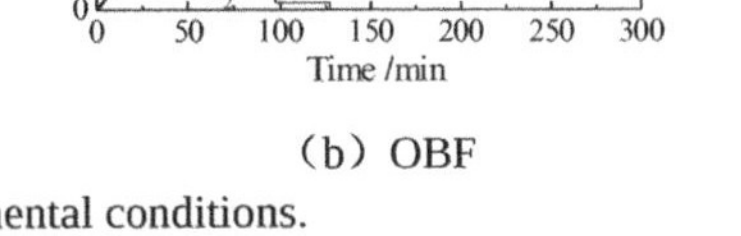

(b) OBF

Fig 1. Experimental conditions.

The experimental apparatus for softening and melting under load is shown schematically in **Fig 2**; it mainly consists of an electric furnace, graphite crucible, silicon carbon tube, displacement meter, differential pressure gauge and auxiliary equipment. The ferrous burden was charged into the graphite crucible, and over and under which coke (20±0.1 g) was placed. Several holes existed in the bottom of the graphite crucible, contributing to the dripping of slag and iron, and the top of the graphite crucible was connected to the graphite pusher. The height of sample in the reaction tube was 65±5 mm. A load of 1 kg/cm^2 was added in the experimental process. The heating process was protected by N_2 before 473 K, and the reductive gas mixture with a flow of 12 L/min was introduced into the reaction tube. When the first iron dripped, the reductive gas mixture was switched to N_2, and the sample was cooled to room temperature. The vertical shrinkage, pressure drop and temperature of the burden layer were all recorded automatically by a computer.

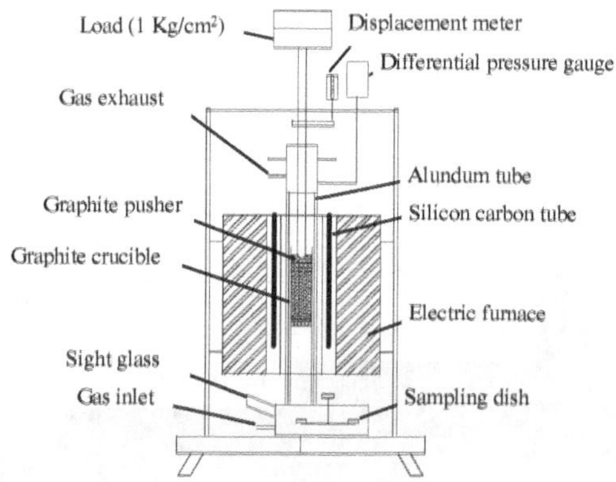

Fig.2 Experimental apparatus.

$T_{10\%}$ and $T_{40\%}$ were defined as the softening start temperature and softening finish temperature, respectively. The softening interval was defined by $\Delta T_B=(T_{40\%}-T_{10\%})$. T_m is the softening finishing temperature, at which the pressure drop of the burden layer reached 490 Pa .[11)] T_m is also defined as the melting starting temperature; Td is the dripping temperature, which is recorded manually when the first iron drips. For research of evolution of the burden microstructure, the experiments were interrupted when the Temperature reached $T_{10\%}$, $T_{40\%}$, or T_m, respectively, and cooled to room temperature with a 12 L/min N_2 gas flow inlet. Then, the samples were inlaid with ethoxyline resin and cut along the direction of the load press. The microstructure was obtained by SEM, and the distribution of elements was obtained by EDS.

3. EXPERIMENT RESULTS

3.1 THE EVOLUTION OF THE BURDEN BED MACROSTRUCTURE DURING THE SOFTENING AND MELTING PROCESS

The evolution of the burden bed macrostructure during the softening and melting process under T-BF and OBF is shown in **Fig 3**. In the OBF case, the reduction degree of the burden particle increased because of the greatly improved reduction potential, the wustite cores in the burden particles were almost distinguished, and the non-deformability was strengthened. The T10% is 1167 °C, which is 29 °C higher than that in the T-BF case. With the increasing temperature, the reduction degree increased continually, and compared to the TBF case, the shrinking speed of the burden bed slowed down. The T10% is 1333 °C, higher than that in the TBF case by 66 °C. The ΔT_B is 166 °C, which is 37 °C higher than that in the T-BF case. Because of the significant reduction of the wustite amount, the proportion of slag phase in OBF case is decreased and the melting point is lowered. The gas permeability of burden bed is also improved; the T_m and T_d are 1415 °C and 1430 °C, which are higher than that those in the T-BF case, at 92 °C and 16 °C, respectively. The softening and melting behavior of burden in the OBF case decreases the position of the cohesive zone in the blast furnace and reduces the gas resistance in the burden bed.

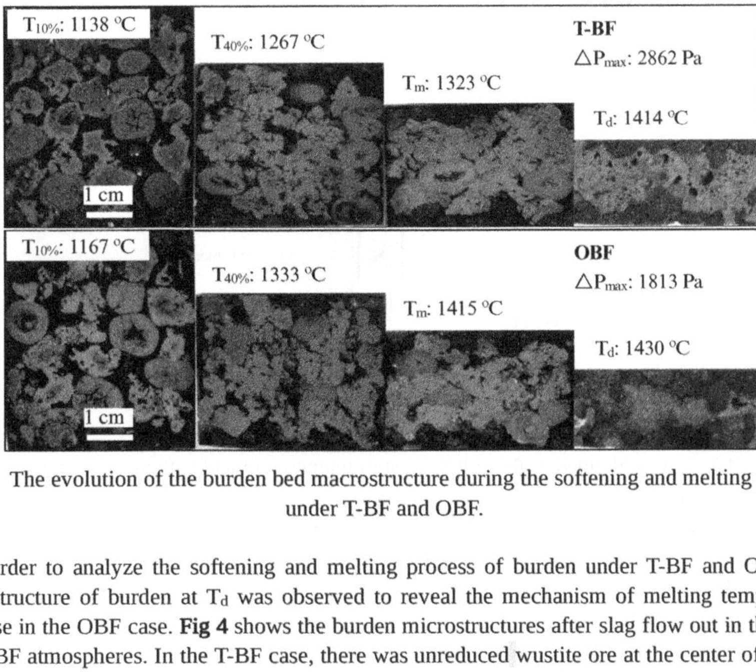

Fig.3 The evolution of the burden bed macrostructure during the softening and melting process under T-BF and OBF.

In order to analyze the softening and melting process of burden under T-BF and OBF, the microstructure of burden at T_d was observed to reveal the mechanism of melting temperature increase in the OBF case. **Fig 4** shows the burden microstructures after slag flow out in the T-BF and OBF atmospheres. In the T-BF case, there was unreduced wustite ore at the center of burden at first, but as the temperature increased, wustite ore began melting to continuously wash the external metallic iron, and finally, low-melting-point slag broke through the metallic iron due to the load to fill the pore spaces between burdens. As shown in **Fig 4**(a), low-melting-point Fe_2SiO_4 phase, formed by wustite, existed among two burden particles, and the gas permeability of the burden bed would significantly decrease due to the outflow of low-melting-point slag phase. In the OBF case, due to wustite ore disappearing at the center of burden, an abundant liquid phase flow out would not exist during the melting process, as only little melting slag oozed from the burden margin. In addition, the increase in the slag melting point would lead to T_d increase, which would decrease the position of the cohesive zone and decrease the thickness. Thus, the ΔP_{max} in the OBF case was lower.

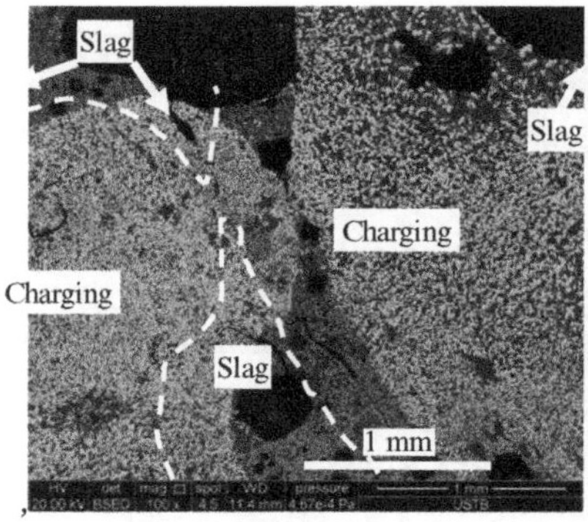

(a)T-BF

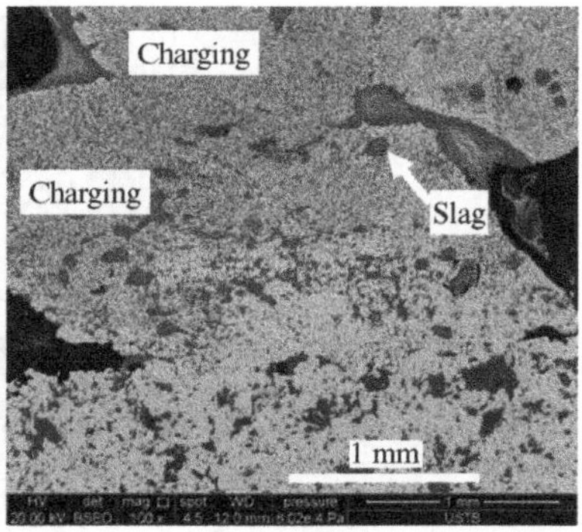

(b)OBF

Fig.4 The burden microstructures after slag flow out in the T-BF and OBF atmospheres.

3.2 THE EVOLUTION OF BURDEN MICROSTRUCTURE DURING THE SOFTENING AND MELTING PROCESS

The microstructural evolution of the interface between the sinter and pellet is shown in **Fig 5**. In the T-BF case, a 10% shrinkage rate was reached at 1138 °C. At this temperature, the original interface still can be identified clearly, and there is an aggregation of metal iron in the pellet side. A temperature of 1267 °C corresponds to a 40% shrinkage ratio, where it is difficult to distinguish the original interface. A bulky vermicular iron phase formed in the pellet side, while a relatively minor iron phase formed in the sinter side. The T_m corresponds to 1323 °C. At this temperature, the metal iron phases contacted each other and formed a structure with the sinter features of the slag phase that was winded or cut by the metal iron phase. It can be determined from the microstructure of the interface that the interaction effect was completely performed. In the OBF case, a 10% shrinkage ratio was reached at 1167 °C, and the interface structure was very similar to that in the T-BF case. At 1333 °C, the temperature corresponded to a 40% shrinkage ratio, and the initial interface still can be distinguished because the vast metal iron phase existing on both sides of the interface impeded the interaction between slag phases. The T_m corresponds to 1415 °C, when some of the iron phases of the interface area have connected and were translated to lamellar form at this time. It can be derived that the interaction was fully performed because the initial interface is already indistinguishable.

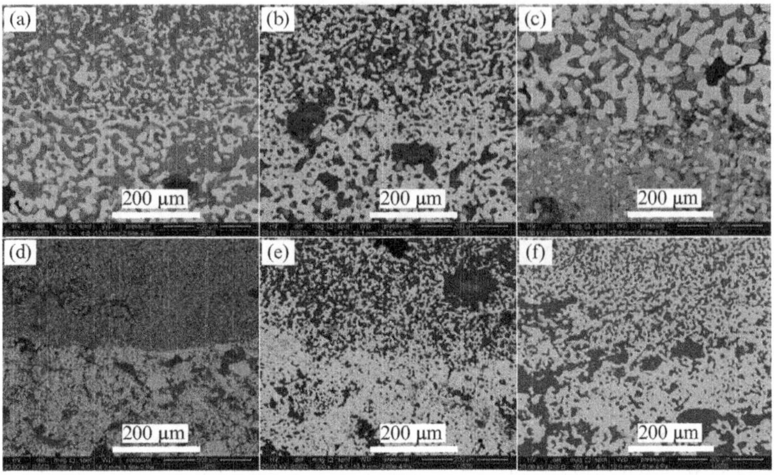

(a)T-BF: $T_{10\%}$; (b) T-BF: $T_{40\%}$; (c) T-BF: T_m; (d) OBF: $T_{10\%}$; (e) OBF: $T_{40\%}$; (f) OBF: T_m

Fig.5 The microstructural evolution of the interface between the sinter and pellet during the softening and melting process under T-BF and OBF.

By observing the interfacial microstructure of the pellet-sinter mixed burden, it can be concluded that the starting temperature of the interaction effect in OBF is higher than that in T-BF. The research of P. F. Nogueira[14,15] revealed that the interaction between different burdens changed the interface basicity Ca. The distribution of Ca and Si in the direction perpendicular to the interface was observed by the EDS line-scanning method. The results of the ratio change rule are shown in **Fig 6**. In the T-BF case, there is a mutation in Ca in the case of 10% shrinkage, which means that the interaction effect has not occurred yet. At a 40% shrinkage ratio, the Ca variation within certain limits of both sides of the interface shows a gradual trend. In the transitional zone from sinter to pellet, the Ca presents a descending trend, which indicates the appearance of the interaction effect. When the shrinkage ratio reached 60%, the Ca tends toward a stable value, which means that the interaction effect has already fully developed. Compared to the above situation, the sample with a 40% shrinkage ratio in the OBF case corresponding to 1415 °C shows that the mutation in Ca still exists in the interface, meaning that the interaction effect has not yet occurred. When the shrinkage ratio reached 60% at 1430 °C, the Ca tended toward a stable value, meaning that the interaction effect had already fully developed. Thus, it can be concluded that the decisive factor for the softening and melting behavior is the Ca change caused by the interpenetration of the slag of different burdens. The change in Ca will certainly lead to a change in the melting point, viscosity, and interfacial tension of slag.

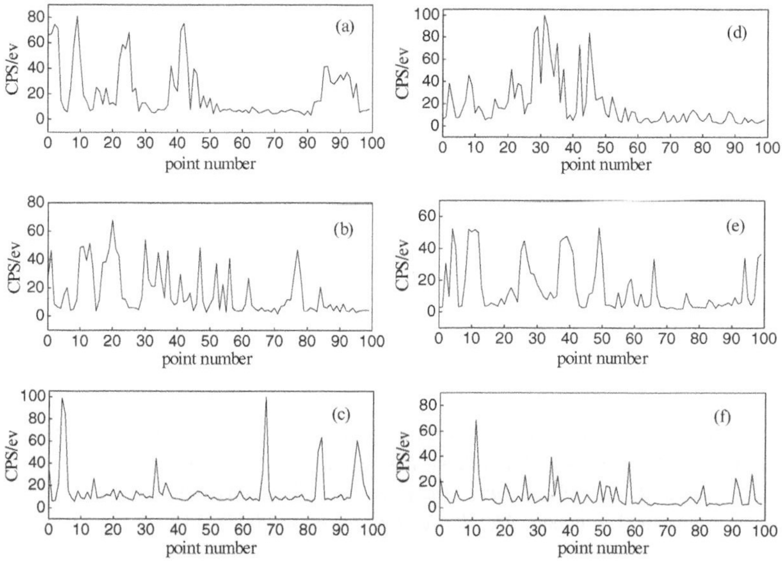

(a)T-BF: $T_{10\%}$; (b) T-BF: $T_{40\%}$; (c) T-BF: T_m; (d) OBF: $T_{10\%}$; (e) OBF: $T_{40\%}$; (f) OBF: T_m

Fig.6 The distribution evolution of Ca in the interface between the pellet and sinter during the softening and melting process under T-BF or OBF conditions.

4. CONCLUSIONS

Compared to the T-BF case, the furnace bed structure during the softening and melting process under the OBF atmosphere is greatly changed, the proportion of wustite in the burden particle is reduced and the amount of iron phase is increased. The non-deformability of the burden particle is strengthened, and the softening and melting properties of the burden are improved.

In the T-BF case, the iron shell of the burden particle is fractured first, followed by the flow of the liquid slag phase. In the OBF case, the liquid slag phase with a higher melting point is exuded from the iron phases, and the melting temperature is obviously increased.

The proportion of wustite in the OBF case is decreased, while the iron phase is increased, which will inhibit the interaction behavior between the pellet and sinter. In the T-BF case, the interaction has already occurred at 1267 °C when the softening process is terminated, while in the OBF case, there is no distinct indication of the interaction effect when the softening process is terminated at 1333 °C.

ACKNOWLEDGEMENTS

The authors gratefully acknowledge the Natural Science Foundation of China and Baosteel under Grant of China (No. 51134008) and National Key Technology R&D Program of China (No. 2011BAC01B02)

REFERENCES

1. K. D. Xu, *Iron and Steel.* 45, 1 (2010).
2. T. Ariyama, R. Murai and J. Ishii J, *ISIJ int.* 45, 1371 (2005).
3. W. Liu, D. King and J. Liu, *JOM.* 61, 36 (2009)
4. Y. H. Han, J. S. Wang and Y. Z. Li, *J Univ Sci Technol Beijing.* 33, 1280 (2011)
5. G. Danloy, J. V. Stel and P. Schmöle, Heat and mass balances in the ULCOS Blast Furnace, *Proceedings of the 4th Ulcos seminar* (Essen, Germany, 2008).
6. F. Fink, *Steel Times.* 224, 398 (1996).
7. W. K. Lu and R. V. Kumar, *Iron and Steel Society of AIME Transactions.* 5, 25 (1984).
8. M. S. Qin, Z. K. Gao and G. L. Wang, *Ironmaking Steelmaking.* 15, 287 (1988).
9. Y. Ohno, H. Hotta and M. Matsuura, *Tetsu to Hagane.* 75, 1278 (1989).
10. P. M. Guo, J. J. Gao and P. Zhao, *J Univ Sci Technol Beijing.* 33, 2 (2012).
11. R. Z. Lan, J. S. Wang, Y. H. Han, X. F. She, L. T. Wao, Q. G. Xue: *J Iron Steel Res Int.*, 19(2012), 13.
12. X. W. An, J. S. Wang, R. Z. Lan, Y. H. Han, Q. G. Xue: *J Iron Steel Res Int.*, 20(2013), 11.
13. C. E. Loo, L. T. Matthews, D. P. O'dea: *ISIJ Inte.*, 51(2011), 930.
14. P. F. Nogueira, R. J. Fruehan: *Metall.Mater.Trans. B*, 35(2004), 829.
15. P. F. Nogueira, R. J. Fruehan: *Metall.Mater.Trans. B*, 36(2005), 583.

NOVEL ENGINEERED CEMENTITIOUS MATERIALS BY USING CLASS C FLY ASH AS A CEMENTITIOUS PHASE

M. F. Riyad[1], M. Fuka[1], R. Lofthus[1], Q. Li[2], N. M. Patel[3], and S. Gupta[1]

[1]Department of Mechanical Engineering, University of North Dakota, ND 58202
[2]USG Corporation, Libertyville, IL 60048
[3]Energy and Environmental Research Center, University of North Dakota, ND 58201

ABSTRACT

This paper reports the synthesis and characterization of novel composites fabricated by using Class C Fly Ash (CFA) as the main cementitious constituent. This study demonstrates the favorable impacts of $Ca(OH)_2$ additions, high temperature curing, and/or carbonation on enhancing the compressive strength of CFA based composites. For example, the CFA matrix had a compressive strength of ~30 MPa after curing under ambient conditions for 28 days. The heat treatment of CFA samples at 70°C for 3h and subsequent carbonation at 87°C for 3 h enhanced the compressive strength to ~66 MPa. The concomitant effect of the addition of 2.8 wt% $Ca(OH)_2$ in the FA matrix, and the heat treatment of CFA samples at 70°C for 3h and subsequent carbonation at 87°C for 3 h further increased the compressive strength to ~87 MPa. The curing period of the CFA based composites can also be drastically reduced by high temperature curing and/or carbonation. For example, conventional curing takes 28 days under ambient conditions, whereas heat treatment and/or carbonation can reduce the curing time to 3-6 h. From thermogravimetric analysis (TGA) and SEM study of the fracture surfaces, the predominant mechanism for strength development was proposed to be enhanced hydration of the heat treated samples, and $CaCO_3$ formation at the phase boundaries of the carbonated samples.

1. INTRODUCTION

Fly Ash (FA) is the principal by-product during the combustion of pulverized coal in thermal power plants [1-5]. FA is generally grey in color, abrasive, mostly alkaline, and refractory in nature [6, 7]. The chemical composition of FA has high percentage of silica (60–65%), alumina (25–30%), magnetite, Fe_2O_3 (6–15%) [6]. The Fly Ash can be classified into two classes – class F fly ash (FFA) and CFA (CFA) [8]. The CFA refer to those having more than 8% CaO and are usually from lignite and subbituminous coal combustion. The FFA are from bituminous and anthracite coals and have less than 8% CaO [5].

Some of the reasons for reutilizing FA are, (a) to minimize disposal costs, (b) decreased area reserved for disposal, thus enabling other uses of the land and decreasing disposal permitting requirements, (c) financial returns from the sale of the by-product or at least can offset of the processing and disposal costs, and (d) the by-products can replace some scarce or expensive natural resources. Therefore, it is critical to find sustainable solution for its utilization rather than land disposal to minimize the environmental challenges associated with its disposal [2, 6, 7].

FA has already partly replaced Portland Cement (PC) in hardened concrete [9-11]. FA usage in structural concrete as a replacement of PC is around 30% of the binder. The resulting concrete has similar or enhanced performance compared to that of Portland cement concrete of equivalent 28 day strength is possible [11]. However, in order to promote wide scale usage of FA, it is critical to develop cementitous materials mainly composed of FA.

It is well known FA can also be activated by alkali to form hardened solid by geopolymerization [12-20]. The alkali activation of waste materials is a chemical process that transforms glassy structures (partially or totally amorphous and/or metastable) into very compact well-cemented composites [12]. In general, two types of cementitious materials can be produced by alkali activation: (a) Si and Ca based, and (b) Si and Al based. The alkali activation of metakaolin or FFA is a typical example of the second situation [13]. Geopolymers has also the potential to replace Ordinary Portland Cement (OPC) [14]. Interestingly, CFA has been used during most of the research studies [15-17]. A few research studies have focused on synthesizing geopolymer from CFA [18, 19]. Nevertheless, in all the studies, high concentration of alkali solution is used to activate the mix.

However, the usage of large amounts of alkali activators raises several concerns due to the, (a) the production of large amounts of alkali is an energy intensive process and can indirectly lead to increased CO_2 emissions, and (b) free alkali and silicate can have deleterious effects on concrete in poorly formulated systems. Therefore, it is critical to develop fundamental understanding whether the FA based cementitious materials can be fabricated with low or no alkali solutions [20].

Recently, Riyad et al. [21] proposed a *"Green Cement Paradigm (GCPa)"*. The main components of this paradigm are: (a) use environmentally friendly precursors and aggregates, (b) minimize or eliminate the usage of alkali or other harmful chemicals as an activator, (c) use of environmentally friendly practices for example, low w/c (w-water, c- cement) ratio etc., and (d) create a cementitious network composed of hydraulic and carbonate bonds for faster ASTM (American Society of Testing and Materials) certification according to industry standards. In order to validate the GCPa, CFA was chosen as a precursor. It was also shown that CFA can have cementitious behavior if the cold pressed compacts are cured with tap water or low alkali solutions [21, 22]. In this paper, we report for the first time, the development of CFA based composites by engineering compositions, and performing different conditioning processes, for example, heat treatment and/or carbonation. We will also report a comprehensive study of mechanical behavior, microstructure analysis, and phase characterization of these composites.

2. EXPERIMENTAL

CFA powder was procured from Headwaters Resources, Sioux Falls, SD. Different compositions were then designed by adding controlled amount of $Ca(OH)_2$ (Part number 21181, Sigma-Aldrich Corp. St. Louis, MO, USA) in CFA (Table 1). All the compositions were mixed with 3 g of water in a ball mill (8000 M, SPEX SamplePrep, LLC, Metuchen, NJ) for 2 minutes. They were then cold pressed in a cylindrical steel die (~25.4 mm diameter, MTI Corp., Richmond, CA) by using a Carver Laboratory press (Model 3853, Carver Inc., Wabash, IN) at ~86 MPa.

The samples were fabricated by using four different processing conditions: (a) the samples were cured in a closed for chamber for 28 days (Condition A), (b) the composition A-2 (Table 1) was heat treated at 50°C, or 70°C, or 90°C, and all the other compositions listed in Table 1 were heat treated at 70°C in a closed chamber in a heating oven for 3 h (Condition B), (c) the samples were carbonated by following the procedure outlined in the next paragraph (Condition C), and (d) the samples were heat treated samples at 70°C by following Condition B, and then they were carbonated by following the procedure outlined in the next paragraph (Condition D).

Table 1: Different compositions of CFA and $Ca(OH)_2$

Composition	Processing	Class C FA (g)	$Ca(OH)_2$ (g)	Water (g)
A-2	Cold Pressed	18	0	3
B-2		17.5	0.5	
C-2		17	1	
D-2		14.4	3.6	
E-2		10.8	7.2	

Before carbonation, all the samples were dried at 100 °C for 24 h in an oven. The dried samples were then placed inside an autoclave chamber (Model 25 X-1, Wisconsin Aluminum Foundry Co., INC., Manitowol, WI). Initially, the chamber was partially filled with 3 L of water, thereafter, the chamber was closed. The chamber was pressurized to ~35 kPa by flowing CO_2 gas (UN 1013, Praxair Inc., Danbury, CT) inside the chamber. The CO_2 gas was then purged from the chamber. This cycle was repeated twice, and the chamber was again pressurized to ~35 kPa. The temperature of the chamber was raised to 87°C, and the samples were carbonated for 3 h. After carbonation, the samples were taken out of the chamber, and then dried at 100°C for 24 h in a drying oven. All the compressive strength measurements were performed on dried samples.

The compressive strengths of the samples were measured by a mechanical testing system (AG-IS, Shimadzu Scientific Corporation, Columbia, MD) at a deflection rate of 1 mm/min. For each composition, the compressive strengths were measured for a set of 3 samples. Secondary electron (SE) images were obtained using a Scanning Electron Microscope (SEM) (JEOL JSM-6490LV, JEOL USA, Inc., Peabody, Massachusetts.) Samples were coated with Au/Pd by using a sputter coater (Balzers SCD 030, BAL-TEC RMC, Tucson AZ USA), and were then mounted on aluminum mounts. XRD patterns were collected from 20-60° (2θ) by using a X-Ray Diffractometer (XRD) (Ultima IV, Rigaku Americas, The Woodlands, TX). Crystalline phase were identified by computer search-match procedures, which employ the ICDD Powder Diffraction file, using MDI Jade 9.0 Software. For thermogravimetric analysis (TGA) (SDT Q600, TA Instruments, New Castle, DE), the fractured pieces of the samples were collected, and crushed into fine powders. A small amount of powdered sample (~75 mg) was then placed inside the instrument. Initially, the furnace was heated at 10°C/min to 1000°C from RT, and then it was cooled to RT. For comparison, TGA analysis was also performed on CFA powders hydrated in water for 24 h and then dried at 100°C for 24 h.

3. RESULTS AND DISCUSSION

3.1 EFFECT OF TEMPERATURE ON COMPRESSIVE STRENGTH

Figure 1 shows the effect of temperature on the mechanical strength of composition A-2 (Table 1) cured by conditions A and B. The compressive strength of the sample was ~30 MPa after curing at RT for 28 days (Condition A). Comparatively, the compressive strength of samples cured by condition B at 50°C and 70°C were similar. Interestingly, slightly higher compressive strength of ~45 MPa was observed for samples cured at 90°C (Condition B). The potential mechanism will be discussed in Section 3.5.

3.2 EFFECT OF $Ca(OH)_2$ ADDITIONS ON COMPRESSIVE STRENGTH

Figure 2 shows the effect of $Ca(OH)_2$ additions (Table 1) on the compressive strength of the samples fabricated by Condition A. Initially, the green strength of all compositions was ~5 MPa, however, a significant enhancement in compressive strength was observed after the addition of $Ca(OH)_2$ in the matrix. For comparison, the samples composed of only CFA (Composition A-2, Table 1) showed a compressive strength of ~30 MPa after curing for 28 days. After the addition of 2.8 wt% $Ca(OH)_2$ (Composition B-2, Table 1), the compressive strength increased to ~22 MPa and ~45 MPa after curing for 7 and 28 days, respectively. Similarly, after the addition of 20 wt% $Ca(OH)_2$ (Composition D-2, Table 1) and 40 wt% $Ca(OH)_2$ (Composition E-2, Table 1), the compressive strength increased to ~38 MPa for both compositions after curing for 7 days, and to ~51 MPa and ~47 MPa after curing for 28 days, respectively. It is interesting to note that the addition of only 2.8 wt% $Ca(OH)_2$ can enhance the compressive of CFA matrix by 50%. The potential mechanism will be discussed in Section 3.5.

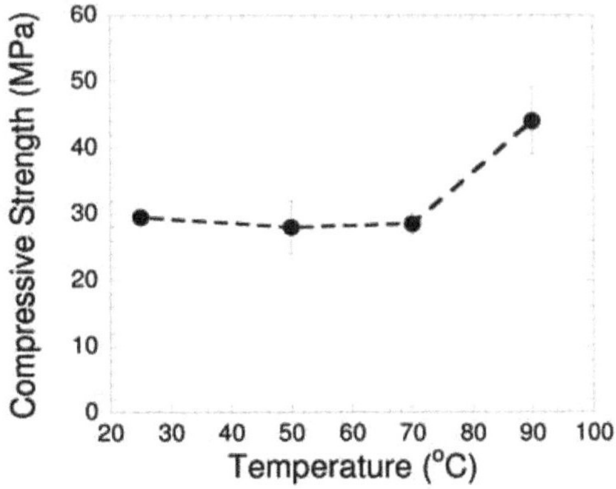

Figure 1: The plot shows the effect of curing temperature (Condition B) on the mechanical performance of A-2 composition (Table 1). For comparison, the samples studied at RT were cured for 28 days (Condition A).

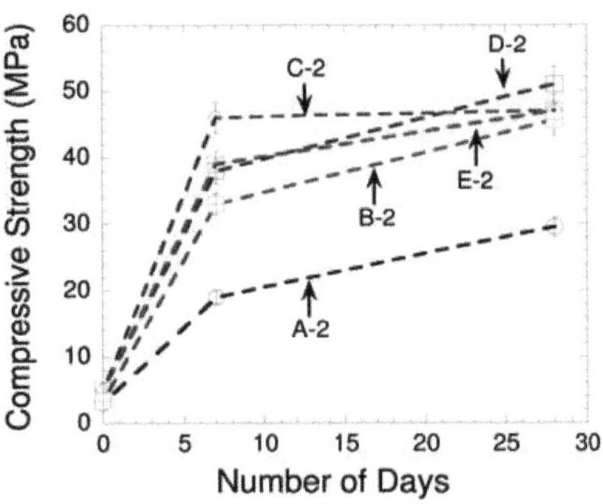

Figure 2: Effect of $Ca(OH)_2$ additions (Table 1) on the compressive strength of samples cured by following Condition A.

3.3 EFFECT OF HEAT TREATMENT AND/OR CARBONATION ON THE COMPRESSIVE STRENGTH

Figure 3 shows the comparison of the compressive strength of the CFA samples treated by Conditions A-D (Table 1). As compared to condition A, all the other curing conditions resulted in the enhancement of the compressive strength; however condition D had the most significant effect on the compressive strength. For example, Composition A-2 (Table 1) had a similar compressive strength of ~30 MPa after curing by either Condition A or B. The carbonation treatment by either Condition C or D enhanced the compressive strength to ~38 and ~66 MPa, respectively. After adding 2.8 wt% $Ca(OH)_2$ in the FA matrix (Composition B-2, Table 1), the compressive strength increased marginally to ~31 MPa and ~45 MPa by curing by Conditions A and B, respectively. However, after the carbonation treatments by Conditions C and D, the compressive strength increased to ~63 MPa and ~87 MPa, respectively.

Comparatively, after the addition of 40 wt% $Ca(OH)_2$ (Composition E-2, Table 1), the compressive strength was similar to Composition B-2 (Table 1) after the samples were cured by either Conditions A, or B, or C. However, the compressive strength increased to ~94 MPa after the samples were cured by condition D. This study also shows that the addition of as low 2.8 wt% $Ca(OH)_2$ in the FA matrix can effectively enhance the strength of CFA samples. The potential mechanism will be discussed in Section 3.5.

3.4 MICROSTRUCTURE AND PHASE ANALYSIS

Figure 4 shows TGA analysis of the CFA based compositions. The CFA powder showed only a small weight loss of ~0.5% during the entire heating cycle. In comparison, all the other compositions showed a weight loss of 2-4 wt% until 200° C (Stage I). The weight loss during Stage I was due to the loss of capillary water and/or decomposition of C-S-H (Calcium Silicate Hydrate) phases from the samples [21, 23]. The ~2 wt% weight loss between 200-600°C is due to the decomposition of $Ca(OH)_2$ phase (Stage II) [23]. A significant amount of weight loss also occurred in the temperature range of 600 - 800°C in the carbonated CFA (Condition C), and heat treated and carbonated CFA (Conditions D) samples (Stage III). This weight loss can be attributed to the decomposition of $CaCO_3$ from the cement matrix [23].

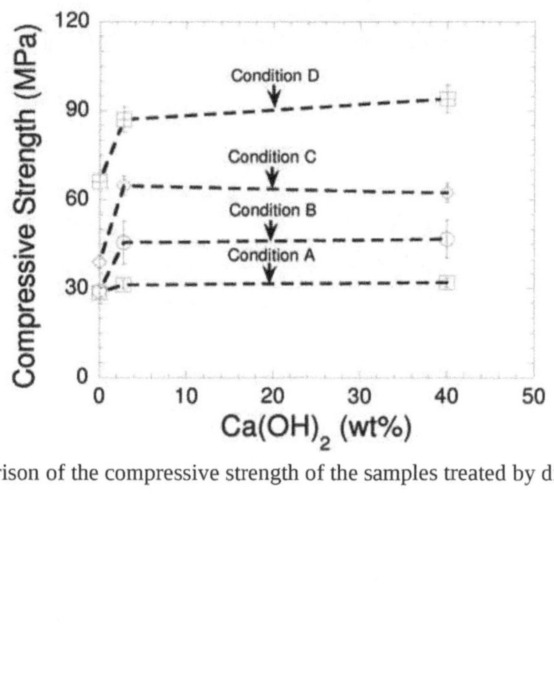

Figure 3: Comparison of the compressive strength of the samples treated by different conditions A-D.

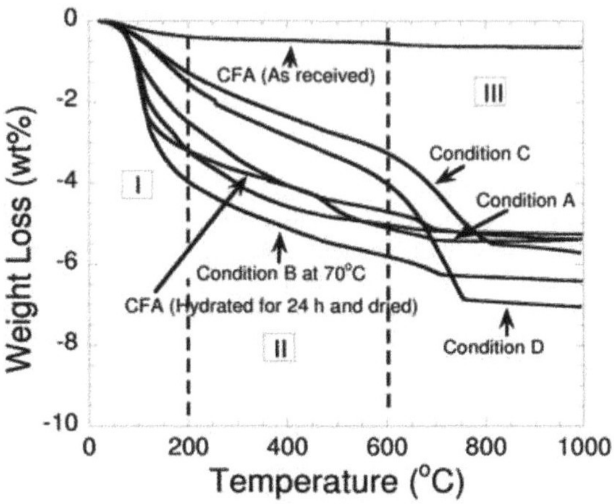

Figure 4: TGA analysis of the A-2 composition (Table 1) cured by different conditions.

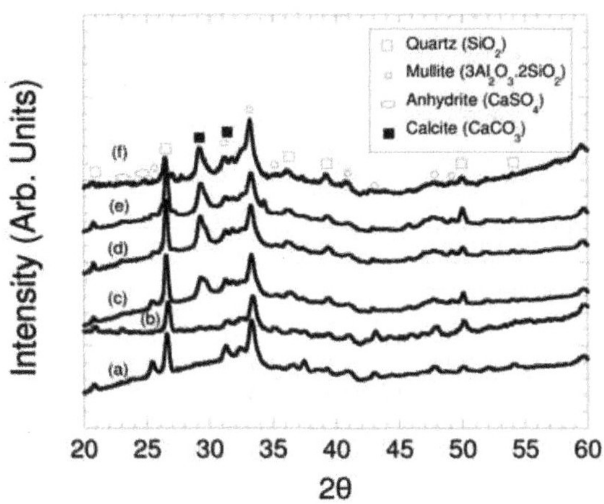

Figure 5: XRD profiles of, (a) as-received CFA, and after, (b) heat treatment at 90°C (Condition B), (c) carbonated (Condition C), (d) heat treatment and carbonated (Condition D) of A-2 composition (Table 1), and carbonation of composition B-2 (Table 1) by using Condition, (e) C, and (f) D.

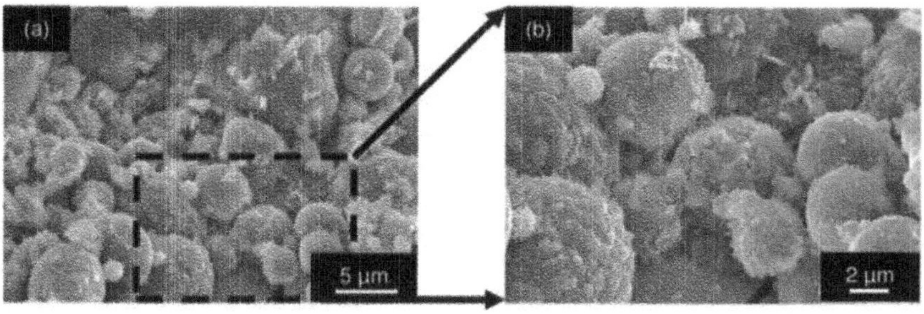

Figure 6: SE SEM micrographs of carbonated CFA (Composition A-2, Table 1) by Condition C at, (a) low, and (b) high magnification.

The XRD of as received CFA (Fig. 5) showed Mullite ($3Al_2O_3.2SiO_2$), Quartz (SiO_2), and anhydrite ($CaSO_4$) are the major crystalline phases. The CFA samples fabricated by condition B also showed similar results which indicate that the hydrated phases are most probably amorphous or cannot be detected by XRD. Comparatively, XRD patterns of all the carbonated samples showed calcite phase ($CaCO_3$) along with other crystalline phases described above. This result shows that the $Ca(OH)_2$ rich phases present in CFA has reacted with the CO_2 during the carbonation experiments.

Figure 6 shows the SEM micrograph of the carbonated CFA sample at low and high magnification, respectively. Calcite ($CaCO_3$) was observed in the XRD analysis (Fig. 5) of all the carbonated samples. Figure 6 (b) clearly shows the deposition of calcite particles around the bigger spherical particles CFA particles. Therefore, we can conclude that $Ca(OH)_2$ present in the CFA has reacted with the CO_2 to form $CaCO_3$ during the carbonation process.

3.5 POTENTIAL MECHANISM

Rovaník [24] reported that the elevated temperature curing can enhance the compressive strength of metakaolin-based geopolymer. Bakharev et al. [25] also reported that the properties of metakaolin-based geopolymeric materials are dependent on curing conditions. More particularly, this study showed that mechanical performance of CFA is enhanced by high temperature curing (Fig. 1). For example, the samples cured at 70°C for 3 h (Condition B) had comparable strength to samples cured for 28 days in ambient conditions. TGA results showed that the samples treated at 70°C (Condition B) showed similar hydration as the samples cured at RT for 28 days (Condition A) (Fig. 4). Based on the current evidence, the mechanical performances of CFA based compositions are enhanced at high temperature due to faster hydration reaction. Nevertheless, authors agree that more studies are needed to understand the detailed mechanism of hydration reactions and concomitant strength development.

Lime-pozzolan mortars have been used by mankind for centuries [26]. Anitohos et al. [27] reported that addition of quicklime in CFA accelerated the degree of reaction of reaction throughout the curing period. The quicklime addition lead to the creation of CH bonds among ash particles, and lead to the increased solubility of SiO_2 leading to a greater release of soluble SiO_2 into the hydrating matrix. Shi [28] also used quicklime to activate the action of natural pozzolans. Shi [28] further reported that for a given CaO content, the quicklime had a higher lime consumption rate than hydrated lime before 28 days, but no difference could be observed thereafter. In addition. lime-pozzolan cements made with quicklime gave higher strengths than the cement made with hydrated lime. In this study, we have also observed strength enhancement in all the compositions containing $Ca(OH)_2$ as an additive as compared to only CFA based compositions. It would be interesting to explore cold pressed quick lime-CFA composition in future.

Haq et al. [29] reported that the carbonation of FA based Geopolymers resulted in improved mechanical properties due to uniformity, presence of strong silica network, and higher geopolymerization. In our studies, we also observed carbonation significantly enhanced the compressive strength (Fig. 3). Furthermore, the combined effect of high temperature curing and carbonation (Condition D) gave the best possible result. More studies are needed to optimize the $Ca(OH)_2$ content and curing conditions to further explore CFA based cementitious compositions.

4. CONCLUSIONS

CFA can be used solely as a cementitious phase. The mechanical performance of CFA compacts can be enhanced by using $Ca(OH)_2$ as a particulate additive. For example, the addition of 2.8 wt% $Ca(OH)_2$ in the CFA matrix can enhance the compressive strength by 50% by curing under ambient conditions at RT (Condition A). In addition, high temperature curing and/or carbonation can also significantly enhance the compressive strength of the samples. For example, the CFA matrix (Composition A-2, Table 1) had a similar compressive strength of ~30 MPa after curing by either Condition A or B. The carbonation treatment by either Condition C or D further enhanced the compressive strength to ~38 and ~66 MPa, respectively. The concomitant effect of the addition of 2.8 wt% $Ca(OH)_2$ in the FA matrix (Composition B-2, Table 1) and curing by Conditions C or D can increase the compressive strength to ~63 MPa and ~87 MPa, respectively. The curing period of the CFA based cementitous compositions can be drastically reduced by high temperature curing and/or carbonation to 3-6 h as compared to curing for 28 days required for conventional curing under ambient conditions.

5. ACKNOWLEDGEMENT

One of the author (SG) would like to acknowledge the University of North Dakota startup and seed funding for support. This work was also supported in part by the National Science Foundation (NSF) through the University of Minnesota MRSEC under Award Number DMR-0819885. NDSU Electron Microscopy Center core facility is also acknowledged for the microscopy. This material is also based upon work supported by the NSF under Grant No. 0619098, and 1229417. Any opinions, findings, and conclusions or recommendations expressed in this material are those of the author(s) and do not necessarily reflect the views of the National Science Foundation.

6. REFERENCES

1. "Characterization of the coal fly ash for the purpose of improvement of industrial on-line measurement of unburned carbon content", K. Styszko-Grochowiak, J. Golaś, H. Jankowski, S. Koziński, Fuel **83**, 1847 – 1853 (2004).
2. "Petrographic examination of coal-combustion fly ash", J. C. Hower, Int. J. Coal. Geol. **92**, 90 - 97(2012).
3. World Coal Association, 2011. Coal statistics. http://www.worldcoal.org/resources/ coal-statistics/.
4. "Revised 2009 Coal Combustion Product (CCP) Production & Use Survey Report", American Coal Ash Association, 2011 (http://www.acaa-usa.org/associations/8003/ files/2009_Production_and_Use_Survey_Revised_100511.pdf).
5. "Application of solid ash based catalysts in heterogeneous catalysis", S. Wang, Environ. Sci. Technol. **42**, 7055–63 (2008).
6. "A review on the utilization of fly ash", Prog. Energy Combust. Sci., M. Ahmaruzzaman **36** (3), 327-363 (2010).
7. "Mercury flow via coal and coal utilization by-products: A global perspective", A. B. Mukherjee, R. Zevenhoven, P. Bhattacharya, K. S. Sajwan, R. Kikuchi, Resour. Conserv. Recy. **52**, 571–591 (2008).
8. "ASTM C618-12a: Standard Specification for Coal Fly Ash and Raw or Calcined Natural

Pozzolan for Use in Concrete ASTM International", http://www.astm.org/Standards/C618.htm.
9. "A review of the multi-component utilisation of coal fly ash", R. S. Blissett and N. A. Rowson, Fuel **97**, 1-23 (2012).
10. "Supplementary cementitious materials", B. Lothenbach, K. Scrivener, R.D. Hooton, Cem. Conc. Res. **41**, 1244–1256 (2011).
11. "Towards maximising the use of fly ash as a binder", M.J. McCarthy, R.K. Dhir, Fuel **78** (1999) 121–132.
12. "Alkali-activated fly ashes: A cement for the future", A. Palomo, M.W. Grutzeck, M.T. Blanco, Cem. Conc. Res. **29**, 1323–1329 (1999).
13. "Characterisation of fly ashes. Potential reactivity as alkaline cements", A. Fernandez-Jimenez, A. Palomo, Fuel **82**, 2259–2265 (2003).
14. "Geopolymer technology: the current state of the art", P. Duxson, A. Fernandez-Jimenez, J. L. Provis, G. C. Lukey, A. Palomo, J. S. J. van Deventer, J. Mater. Sci **42**, 2917–2933 (2007).
15. "NaOH-activated ground fly ash geopolymer cured at ambient temperature", K. Somna, C. Jaturapitakkul, P. Kajitvichyanukul, and P. Chindaprasirt, Fuel **90**, 2118-2124 (2011).
16. "The mechanical properties of fly ash-based geopolymer concrete with alkaline activators", G. S. Ryu, Y. B. Lee, K. T. Koh, Y. S. Chung, Constr. Build. Mater. **47**, 409–418 (2013).
17. "Composition and microstructure of alkali activated fly ash binder: Effect of the activator", A. Fernandez-Jimenez, A. Palomo, Cem. Conc. Res. **35**, 1984 – 1992 (2005).
18. "Alkali-activated complex binders from class C fly ash and Ca-containing admixtures" X. Guo, H. Shi, L. Chen, W. A. Dick, J. Hazard. Mater. **173**, 480–486 (2010).
19. "Compressive strength and microstructural characteristics of class C fly ash geopolymer", X. Guo, H. Shi, and W. A. Dick, Cem. Concr. Compos. **32(2)**, 142-147 (2010).
20. "The role of inorganic polymer technology in the development of 'green concrete", P. Duxson, J. L. Provis, G. C. Lukey, J. S.J. van Deventer, Cem. Conc. Res. **37**, 1590–1597 (2007).
21. "On the Usage of Class C Fly Ash as a Sole Reactive Precursor for Cementitious Phase", M. F. Riyad and S. Gupta, J. Chem. Biol. Interfaces 2, 55-60 (2014).
22. S. Gupta, & M. F. Riyad, "Low-Alkali Fly Ash Cement and Method of Making" (US Patent Application Submitted).
23. "Evaluation of the Hydration of Portland Cement Containing Various Carbonates by Means of Thermal Analysis", R. Gabrovšek, T. Vuk, V. Kaučič, Acta Chim. Slov., **53**, 159–165 (2006).
24. "Effect of curing temperature on the development of hard structure of metakaolin-based geopolymer ", P. Rovnaník, Construction and Building Materials, **24**, 1176-1183 (2010).
25. "Geopolymeric materials prepared using Class F fly ash and elevated temperature curing", T. Bakharev, Cement and Concrete Research, **35**, 1224-1232 (2005).
26. "Prehistory of concrete", R. Malinowski, and Y.Garfinkel, Concrete Int., **13**, 62–68(1991).
27. "Influence of quicklime addition on the mechanical properties and hydration degree of blended cements containing different fly ashes", S. K. Antiohos, A. Papageorgiou, V. G. Papadakis, and S. Tsimas, Construction and Building Materials, **22**, 1191-1200 (2008).
28. "Studies on Several Factors Affecting Hydration and Properties of Lime-Pozzolan Cements", C. Shi, Journal of Materials in Civil Engineering, **13**, 441-445 (2001).
29. "In-situ carbonation of alkali activated fly ash geopolymer", E. U. Haq, S. K. Padmanabhan, and A. Licciulli, Construction and Building Materials, **66**, 781-786 (2014).

EFFECTS OF COMPOSITION CHANGES ON THE SINTERING PROPERTIES OF NOVEL STEEL SLAG CERAMICS

Lihua Zhao [1,2], Yu Li [1,2], Feng Jiang[1,2], and Daqiang Cang [1,2]

[1] State Key Laboratory of Advanced Metallurgy
[2]School of Metallurgy and Ecological Engineering, University of Science and Technology Beijing
Beijing, China
Corresponding author: Yu Li leeuu00@sina.com

ABSTRACT

Pyroxene ceramics are designed based on $CaO-MgO-SiO_2-FeO$ system, composing of pyroxene group minerals as main crystal phases. Steel slag, which is rich in component of CaO and FeO(or Fe_2O_3), could be used as a kind of main raw materials in such novel ceramics and the prepared ceramics have good mechanical properties. In this paper, steel slag in combination with quartz, talcum, clay and feldspar was converted to novel ceramics by traditional preparing process. Effect of composition changes on the properties of the production was studied by scanning electron microscope (SEM) and X-ray diffraction (XRD) techniques coupled with the tests of physical and mechanical properties of the samples. The results revealed that the sintering temperature and flexural strength decreased with increasing calcium and iron and decreasing magnesium content. Part of the iron oxide or RO involved in crystal phase formation, while the rest were unchanged in the sample.

INTRODUCTION

Around one hundred million tons steel slag is produced every year in China, but the utilization ratio of steel slag is lower than 30%[1]. The ways of utilization of steel slag usually include internal recycling within the steel industry, such as reuse as raw materials in sintering, and external recycling in other industries, such as being used as raw materials of cement, concrete and road bed, being applied to agriculture as soil amendment as well as being used in wastewater treatment by virtue of its absorptivity[2-8]. However, none of these means could help to improve the utilization level of steel slag both in quantity and quality at present. In recent years there have been many reports on producing ceramics from steel slag. HE *et al*[9]. produced glass-ceramics by using steel slag as main raw materials; Daqiang Cang *et al*[10]. produced novel ceramics with high flexural strength of 143MPa and low water absorption of 0.02% from Ca-rich and Fe-rich steel slag and the main crystalline phases were detected as diopside, augite and some other pyroxene minerals. The $CaO-MgO-SiO_2-FeO$ system ceramics are a compromising route for the commercial use of steel slag due to the fact that traditional ceramics can't reuse large amounts of iron-rich and calcium-rich slag as main raw materials for the strictly limitation on less composition of

CaO and Fe_2O_3 than about 3% and 1% , respectively.

The structural formula of pyroxene is $M_2M_1T_2O_6$, where T stands for Si^{4+}, while there are many alternatives for M_1 and M_2[11]. In the CaO-MgO-SiO_2-FeO system ceramics, M_1 could be iron or magnesium and M_2 could be iron, calcium or magnesium. Xianbin Ai *et al*[12]. have studied how MgO in raw materials influences the sintering properties as well as the characteristic of the ceramics; Lihua Zhao *et al*[10]. have studied how the changes of raw materials influences the crystal phases in the ceramics; but there haven't been any reports on how the fluctuation of elements influences the sintering properties and characteristics of the ceramics. In fact, it is pretty important to know the regular pattern because different elements may form different crystal phases in composition or distribution in ceramics, which then influences the characteristics of the products. Therefore this paper focuses on influences of the elements composition changes on the sintering properties.

2 EXPERIMENT

The steel slag was obtained from SHANDONG steel and iron plant (China), and other materials were from SHANDONG Provence. All materials were dried at 105°C for 24h and then crashed.

A total of six mixtures were prepared on the same ratio of (CaO+MgO+FeO)/SiO_2 closed to 1 to guarantee the composition of as-mixed samples in the range of pyroxene component, in the proportions summarized in Table 1.

300g of the test samples were milled for 30min with 300ml water, using an alumina ball mill and spheres, at 300 cycles/min. The slurry were sieved through an 180μm sieve and then dried at 105°C for 24h. 6-10.wt% water was added in the dried powder and then pressed in a laboratory uniaxial press at 80MPa into rectangular (50mm*100mm*10mm) specimens.

Specimens were sintered in an electric muffle in air atmosphere in the thermal interval 900-1200°C. The heating rate was 7°C/min and the holding period was 60min. Finally, the fired samples were subjected to physical and mechanical tests such as linear shrinkage, water absorption and flexural strength. The crystallized samples obtained from the heat treatment were then studied by XRD using a Rigaku DMax-RB powder diffractometer. The microstructure characterization was performed in the scanning electron microscope (SEM) in MLA-250(FEI) operating at 25 KV. Right before SEM test, the samples were etched with 0.5vol. % HF for 1.5 minutes at room temperature and coated with carbon to be analyzed.

Table 1 The Main Chemical Composition of the Six Mixtures

Sample No.	CaO	SiO_2	Fe_2O_3	Al_2O_3	MgO	$(CaO+MgO+FeO)/SiO_2$
J1	21.78	46.57	7.55	6.82	12.55	1.03
J2	22.68	45.78	7.97	7.58	10.84	1.02
J3	23.63	44.89	8.37	7.96	9.74	1.03
J4	25.00	43.65	8.99	8.91	7.48	1.03
J5	27.32	41.53	10.01	10.25	4.12	1.03
J6	27.75	41.18	10.23	10.82	2.96	1.02

3 RESULTS

3.1PHYSICAL AND MECHANICAL PROPERTIES

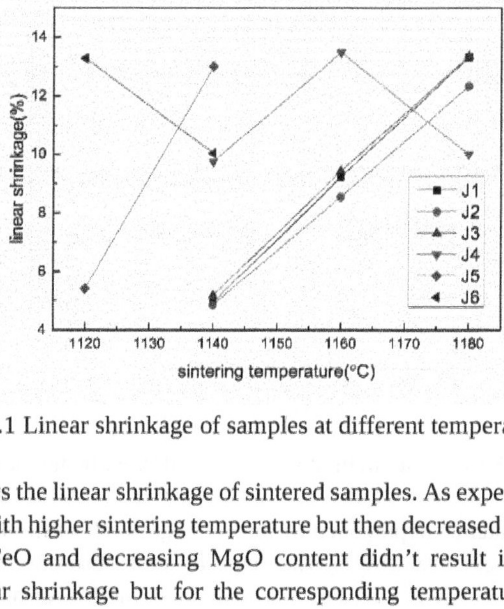

Fig.1 Linear shrinkage of samples at different temperatures

Figure 1 shows the linear shrinkage of sintered samples. As expected, firing shrinkage values increased with higher sintering temperature but then decreased because of over-fired. Increasing CaO, FeO and decreasing MgO content didn't result in the change of the maximum of linear shrinkage but for the corresponding temperature of the maximum values. The maximum values of linear shrinkage is up to nearly 14%.

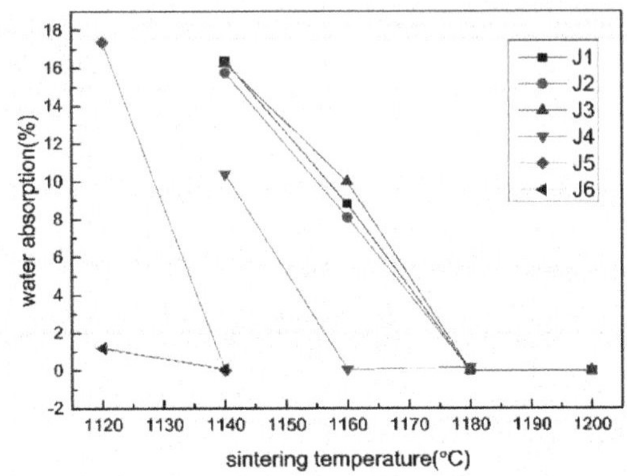

Fig.2 Water absorption of samples at different temperatures

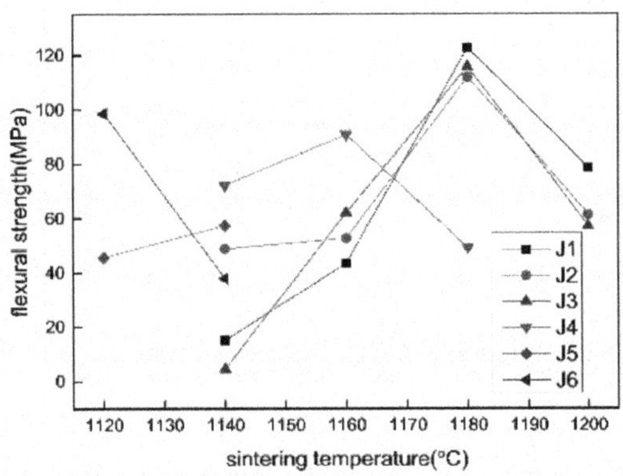

Fig.3 Flexural strength of samples at different temperatures

The results of water absorption and flexural strength versus sintering temperature are shown in Fig.2 and Fig.3, respectively. The trends of water absorption are consistent with the flexural strength. Figure 3 indicates that flexural strength decreased with increasing calcium and iron and decreasing magnesium content, although the flexural strength of J1-J3 didn't change significantly. The sintering temperature of the maximum of flexural strength decreased with increasing calcium and iron and decreasing magnesium content. Table 2 shows the maximum flexural strength and the corresponding sintering temperature of the six samples. Sintering temperatures decreased and flexural strength decreased first and then increased with the increasing calcium and iron and decreasing magnesium content. This may due to the differences of low-melting-point and the types of the crystals.

Table 2 The Maximum Flexural Strength and the Corresponding Sintering Temperature of the Six Samples

Sample No.	J1	J2	J3	J4	J5	J6
Sintering temperature (°C)	1180	1180	1180	1160	1140	1120
Flexural strength (MPa)	122.67	111.87	115.67	90.6	57.34	98.59

3.2 X-RAY DIFFRACTION AND SCANNING ELECTRON MICROSCOPY

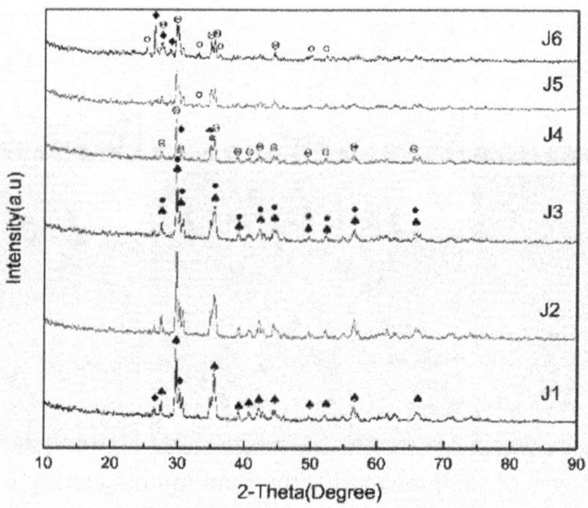

Fig.4 XRD patterns of the samples fired at their optimum sintering temperatures (•—Diopside, °—Andradite, *—Almandine, ♠—Augite, ♦—Quartz, ♣—Dmisteinbergite, ϴ—Diopside ferrian)

XRD analysis of the six fired samples at their optimum sintering temperatures (shown in table 2) are given in Fig.4. XRD results pointed out the diopside and augite as the major phases in sample J1-J3. Phases developed in J4 are diopside ferrian, almandine and quartz. Diopside ferrian, andradite and quartz are the major phases in J5, while the dominated phases in J6 are diopside ferrian, quartz and wollastonite. This suggests that the main phases tends to change from pyroxene or augite to iron diopside, andradite and quartz as the increase of calcium and iron content and reduction of magnesium.

SEM images of six samples fired at their optimum sintering temperature (shown in Table 2) are shown in Fig.5. Back scattered electron detector (BESD) was used in this research. SEM and BESD showed little difference among J1-J3 from the appearance; J4 has larger crystal grains and pore size, this might be responsible for its low flexural strength; pores in J5 connect to form channels, and the crystal grains in J5 have irregular shapes and are distributed disorderly; J6 is relatively denser than J5. According to the XRD results, it is the type along with the shape of crystals that determines the mechanical strength of the samples. It is also notable that the white dots and lumps in all samples have been identified by EDS as iron oxides or FeO-MgO (i.e. RO phase) which came from steel slag. This means that not all the iron have been converted to form crystals. That is to say there's still some iron left in the form of RO or free iron.

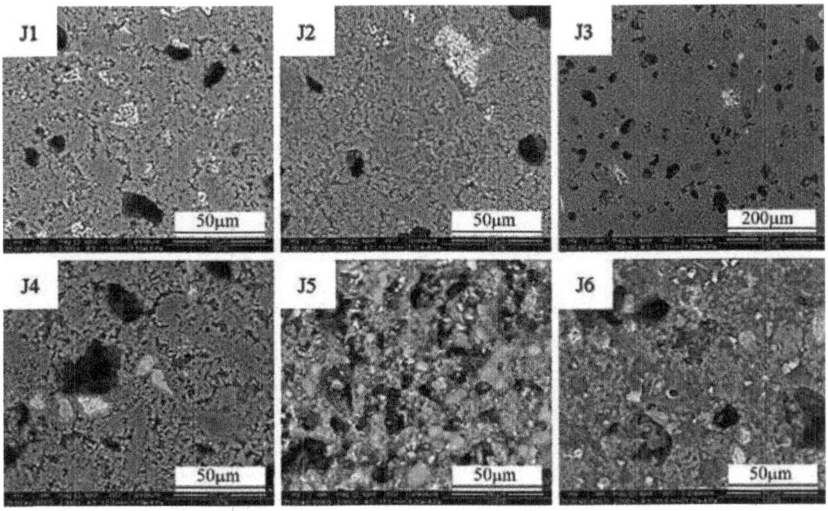

Fig.5 SEM images of samples fired at their optimum sintering temperatures

CONCLUSIONS

Under the premise of $(CaO+MgO+FeO)/SiO_2=1$ the optimum sintering temperature of samples decreased, while the flexural strength firstly decreased and then increased with the increase of iron and calcium content and the reduction of magnesium content. The main crystal phases transformed from pyroxene and diopside to hedenbergite, andradite and quartz.

ACKNOWLEDGEMENTS

This work has been financially supported by the National Natural Science Foundation of China (NSFC, NO. 51474027) and Beijing Higher Education Young Elite Teacher Project (NO.YETP0348)

REFERENCE

[1]China Resources Comprehensive Utilization annual report. (National Development and Reform Commission, Beijing, 2014).
[2]Baciocchi, R., Costa, G., Di Bartolomeo, E., Polettini, A. & Pomi, R. Carbonation of

stainless steel slag as a process for CO2 storage and slag valorization. *Waste and biomass valorization* 1, 467-477 (2010).

[3]Barca, C., Gérente, C., Meyer, D., Chazarenc, F. & Andrès, Y. Phosphate removal from synthetic and real wastewater using steel slags produced in Europe. *Water research* **46**, 2376-2384 (2012).

[4]Cao, W. & Yang, Q. Properties of a Carbonated Steel Slag-Slaked Lime Mixture. *Journal of Materials in Civil Engineering*, 04014115, doi:10.1061/(asce)mt.1943-5533.0001049 (2014).

[5]Das, B., Prakash, S., Reddy, P. S. R. & Misra, V. N. An overview of utilization of slag and sludge from steel industries. *Resour Conserv Recy* **50**, 40-57, doi:DOI 10.1016/j.resconrec.2006.05.008 (2007).

[6]Lee, M. S. *et al.* Evaluating the efficiency and temporal variation of pilot-scale constructed wetlands and steel slag phosphorus removing filters for treating dairy wastewater. *water research* **44**, 4077-4086 (2010).

[7]Oluwasola, E. A., Hainin, M. R. & Aziz, M. M. A. Characteristics and Utilization of Steel Slag in Road Construction. *Jurnal Teknologi* **70** (2014).

[8]Yu, J. & Wang, K. Study on characteristics of steel slag for CO2 capture. *Energy & Fuels* **25**, 5483-5492 (2011).

[9]He, F., Fang, Y., Xie, J. & Xie, J. Fabrication and characterization of glass–ceramics materials developed from steel slag waste. *Materials & Design* **42**, 198-203, doi:10.1016/j.matdes.2012.05.033 (2012).

[10]Zhao, L., Li, Y., Zhou, Y. & Cang, D. Preparation of novel ceramics with high CaO content from steel slag. *Materials & Design* **64**, 608-613, doi:10.1016/j.matdes.2014.08.015 (2014).

[11]Cameron, M. & Papike, J. J. Structural and chemical variations in pyroxenes. *Am. Mineral.* **66**, 1-50 (1981).

[12]Ai, X. B., Li, Y., Gu, X. M. & Cang, D. Q. Development of ceramic based on steel slag with different magnesium content. *Advances in Applied Ceramics* **112**, 213-218, doi:10.1179/1743676113y.0000000078 (2013).

EFFICIENCY GAINS IN POWERTRAIN COMPONENTS BY MOLYBDENUM-ALLOYED SPECIAL STEELS

Hardy Mohrbacher
NiobelCon bvba
Schilde, Belgium

ABSTRACT

Powertrain components play a crucial role in the efficiency and performance of vehicles and machines. These components should not only last under the prevailing severe loading conditions but also consume as little as possible energy by themselves during production and operation. Typically these components are being fabricated from special steels involving additional heat treatments. Increased demands by either downsizing in the vehicle industry or high and erratic loading as occurring in heavy equipment such as windmills have been pushing existing alloys to their limits. The powertrain of a windmill accounts for approximately 25 percent of the total equipment cost and is a major cause for downtime. Recent developments demonstrated that by dedicated alloying of molybdenum to existing special steels considerable performance improvements and efficiency gains could be achieved. The paper explains the underlying metallurgical reasons and demonstrates examples of gear and bearing steels.

INTRODUCTION

Increasingly stringent demands by authorities to reduce vehicle exhaust emissions have forced the automotive industry to take a variety of measures for lowering fuel consumption (Table 1). Much in focus has been the reduction of vehicle weight by using high strength steel (HSS) and therefore lower gage or by incorporating low-density materials such as aluminum, magnesium or carbon fiber reinforced plastics. Steel clearly has the smallest carbon footprint in generating these materials and is the only means of reducing weight and cost simultaneously. The intensive use of such advanced materials, often being applied in a mix, has led to weight savings of several ten kilograms in smaller cars to several hundred kilograms in larger vehicles. An often quoted value states that vehicle weight reduction of 100 kg results in an average reduction of fuel consumption by 0.5 liters per 100 km. Weight reduction is particularly relevant in stop-and-go traffic where the vehicle mass has to be accelerated frequently. However, aerodynamics (drag coefficient) becomes the more relevant influencing factor during cruising at more constant speeds.

Increasing efforts are being done for optimizing the powertrain. A major trend is that of downsizing the engine and incorporating one or several turbochargers. The characteristic of such engines is an almost instantly high torque at low revolutions (Figure 1) putting rather sudden and high loads on the gearbox. Gearboxes have also evolved to 8 or even 9 transmission ratios allowing operating the engine in the optimum rpm range under nearly all conditions. Another mega trend in the automotive industry is the partial or full electrification of the powertrain. However, despite of officially low CO_2 emission values emanating from artificial driving cycles, the real CO_2 emission value of an electric vehicle much depends on the energy mix fed into the electric grid. Effectively low CO_2 values can only be achieved when the electric energy used for battery charging mainly originates from wind, solar or hydro power plants. Any vehicle, irrespective of the applied propulsion method, involves rotating parts that are prone to frictional power losses. Reducing friction in bearings, gears and drive shafts can achieve further fuel savings in the order of a few percent.

Lightweight strategies increase the contact stresses above P_{0max} of 2.14 GPa (or FZG load stage 14) due to the reduction of component sizes. Such high contact pressure requires advanced technical solutions particularly in terms of materials and treatments. Steels with improved mechanical properties offering a reduced coefficient of friction, resistance to high contact pressure and potential to apply a low friction surface coating represents a viable alternative to conventional gear and bearing steels in this respect.

The present paper focuses on optimizing special steel alloys used in powertrain applications. A modification of the well-established gear steel 20MnCr5 capable of accepting substantially higher loads will be presented. The steel also has the potential for depositing a low-friction coating after carburizing treatment. Furthermore, the slip-rolling resistance, friction and wear behavior of steels such as 36NiCrMoV1-5-7, Cronidur 30 (AMS 5898), 20MnCr5 (SAE 4820 or 5120) and 100Cr6H (SAE E52100) was analyzed with regard to the application in high-pressure contacts.

Table 1: Typical measures of reducing fuel consumption of cars.

Technology	Weight	Fuel Economy	Cost	Cost per 1% FE improvement
Light weighting (HSS)	-140kg	5%	-€100	-€20
Turbo charged gasoline engine (down-sized)	-20kg	10%	€200	€20
Turbo charged Diesel engine (vs. gasoline)	+100kg	30%	€1000	€33
Advanced stop-start system	+5kg	5%	€200	€40
Light weighting (Al/Mg)	-280kg	10%	€750-1250	€75-125

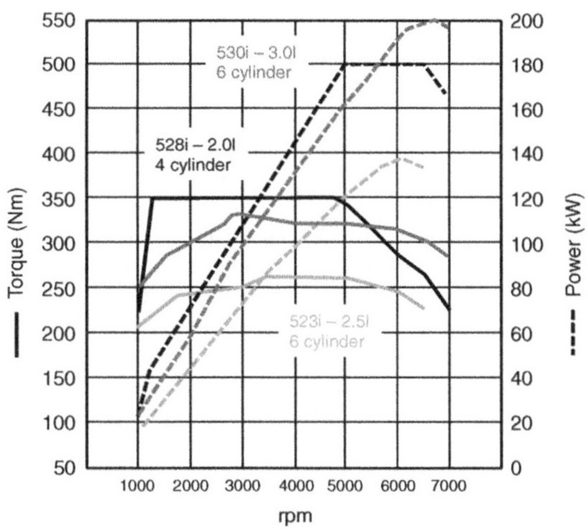

Figure 1: Power and torque characteristics of naturally aspirated 3.0l, 2.5l) and downsized turbocharged engines (2.0l) (data by BMW AG).

HIGH PERFORMANCE GEAR STEELS

The general requirements for high performance gear components are a hard case providing wear resistance and a tough core preventing brittle failure under high impact loads [1]. Accordingly, various alloy concepts and thermo-mechanical treatments have been developed to achieve this property combination. Alloy concepts for gear applications significantly vary in different markets due to historical drivers (e.g. automotive, machine building, military), practical experiences as well as local availability of alloying elements (Table 2).

Table 2: Common carburizing steel grades for gear in various geographical markets.

Steel grade	Standard		C	Si	Mn	P	S	Cr	Mo	Ni	Region
20MnCr5	EN 10084 (1.7147)	min.	0.17	-	1.10	-	-	1.00	-	-	Germany
		max.	0.22	0.40	1.40	0.035	0.035	1.30			
18CrNiMo7-6	EN 10084 (1.6587)	min.	0.15	-	0.50	-	-	1.50	0.25	1.40	
		max.	0.21	0.40	0.90	0.025	0.035	1.80	0.35	1.70	
15CrNi6	EN 10084 (1.5919)	min.	0.14	-	0.40	-	-	1.40	-	1.40	France
		max.	0.19	0.40	0.60	0.035	0.035	1.70		1.70	
17NiCrMo6-5	EN 10084 (1.6566)	min.	0.14	-	0.60	-	-	0.80	0.15	1.20	Italy, France
		max.	0.20	0.40	0.90	0.025	0.035	1.10	0.25	1.50	
SAE 8620	SAE J1249	min.	0.18	0.15	0.70	-	-	0.40	0.15	0.40	North America
		max.	0.23	0.35	0.90	0.030	0.040	0.60	0.25	0.70	
SAE 9310	SAE J1249	min.	0.08	0.15	0.45	-	-	1.00	0.08	3.00	
		max.	0.13	0.35	0.65	0.025	0.040	1.40	0.15	3.50	
20CrMnTi	GB T 3077 - 1999	min.	0.17	0.17	0.80	-	-	1.00	0.00	-	China
		max.	0.23	0.37	1.10	0.035	0.035	1.30	0.15	0.30	
20CrMnMo	GB T 3077 - 1999	min.	0.17	0.17	0.90	-	-	1.10	0.20	-	
		max.	0.23	0.37	1.20	0.025	0.035	1.40	0.30	0.30	
SCM420	JIS	min.	0.18	0.15	0.60	-	-	0.90	0.15	-	Japan

Carburizing steel must comply with the following key requirements in view of component properties and durability:

Chemical composition / hardenability

- Homogeneity / microscopic and macroscopic cleanness
- Mechanical properties (strength, fatigue strength and toughness)
- Wear resistance, rolling strength and vibration resistance
- High and uniform dimensional stability [2].

A performance map (Figure 2) relates an easily measurable property like in this case surface hardness to a complex system property such as the tooth root endurance strength. The fact that for a given surface hardness a rather wide range of tooth root endurance strength levels can be obtained suggests that alloy composition, microstructure and thermo-mechanical treatment have an extremely high impact on the actual gear performance. Another system property of high importance for gear durability is the resistance to gear tooth flank micro pitting. High contact pressure, the status of lubrication, material properties, microstructure and chemical composition influence this system property. For case carburizing steels three quality levels can be distinguished: grade ML stands for the minimum requirement; grade MQ represents requirements which can be met by experienced manufacturers at moderate cost; grade ME represents requirements which must be aimed at when higher allowable stresses are desirable.

The approach practiced is to design alloying concepts taking the entire processing route into consideration including modified or innovative heat treatments. A fundamental way of dealing with these demands is to adjust the chemical composition of carburizing steels. In this respect one can principally define two approaches. An economically driven approach aims achieving a defined performance spectrum with a cost reduced alloy concept whereas a performance driven approach targets superior properties at equal or moderately increased cost. The current work considers both approaches focusing on modified molybdenum-based alloy concepts including niobium microalloying. Thereby innovative heat treatment conditions have also been tested.

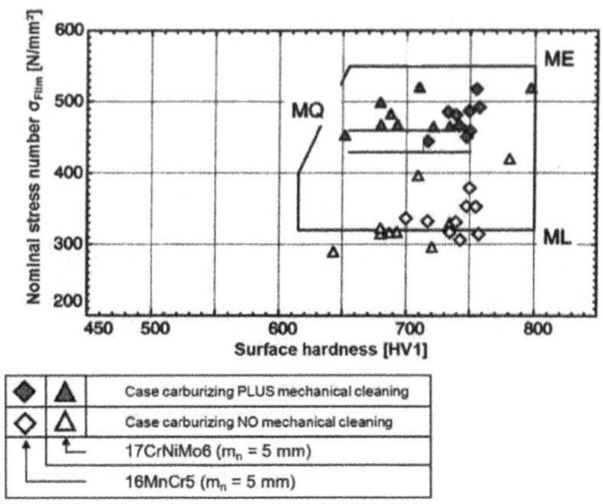

Figure 2: Tooth root load carrying capacity range of state-of-the-art case hardened gears and indication of quality levels (ME, MQ, ML) [3].

Optimization concept for carburizing steels

As performance reference, grade 18CrNiMo7-6 (1.6587) has been selected since this grade currently is being widely used for demanding gear applications in Europe. The task was to modify the main alloying elements in a way to achieve either the same performance at lower alloy cost or better performance at similar alloy cost. The following approach is considered to be relevant in this respect:

Improving hardenability

- Increasing core tensile strength and toughness
- Increasing fatigue strength in both core and case
- Reducing quench distortion and thus residual stresses
- Improving microstructural stability to withstand elevated temperatures during service.

Considering the stress state at the gear tooth root [4], the most critical area appears to be below the surface, whereas the surface-near region is less critical. In case hardenability is insufficient and core strength is too low, the applied stress can locally exceed the material strength. This will result in poor tooth root fatigue endurance. Residual tensile stress due to quench distortion can add to the applied stress and worsen the situation, while shot peening introducing compressive residual stress has a beneficial effect. Degradation of the microstructure during service due to tempering effects (hot running) leads to softening which is detrimental. This also applies for retained austenite present in the outer quench layer.

Additionally, a typical surface near defect in a carburized layer, the intergranular oxidation layer has to be avoided because it acts as a fatigue fracture initiation site and reduces the fatigue strength of the tooth. The soft zone caused by intergranular oxidation results in surface softening in the carburized layer [5]. Elimination of surface structure anomalies is thus an important aspect in the development of high fatigue strength gears. Improving toughness can effectively be achieved by raising the tempering temperature. However, this requires reinforcing the tempering resistance in order not to lose strength. A fundamental way to deal with these issues is to adjust the chemical composition of the carburizing steel. Accordingly, the chemical composition of carburizing steels can be further developed to achieve the above goals using the following guidelines:

Prevent intergranular oxidation → reduce Si, Mn, and Cr.

- Prevent MnS inclusions → reduce S, limit Mn
- Prevent TiN inclusions → control Ti / N wt.% ratio close to 3.4
- Improve hardenability → increase Mo.
- Improve toughness → increase Ni and Mo.
- Refine and homogenize grain size → balance Nb, Ti, Al and N microalloying addition.
- Strengthen grain boundaries → reduce P and S, add Mo and Nb.

Increasing base hardness and core hardenability

Several alloying elements, besides carbon, contribute to hardenability as such: molybdenum, manganese, chromium, nickel as well as boron microalloying. For cost reduction reasons alloys using higher manganese and chromium additions, eventually combined with boron microalloying have been favored for many gear applications. However, such cost reduced alloy concepts, although providing good hardenability, have limitation in terms of toughness and tempering resistance. Besides, the prevention of intergranular oxidation requires Mn, Cr and also Si levels to be reduced. In the other extreme alloy producers have developed richly alloyed steels for those applications where transmission failure causes high replacement and outage costs. An example is 15NiMoCr10-4 (C:0.15%, Si:1.1%, Cr:1%, Mo:2% and Ni:2.5%), which is used in high end applications e.g. in aerospace or Formula-1 gear. However such steel requires special melting technology and is not widely available. Comparing this steel to another high-Ni steel (14NiCrMo13-4) the increase of the molybdenum content from 0.25% to 2.0% brings about a significant improvement of hardenability, surface hardness and also tempering resistance [6] (Figure 3). The high tempering resistance of the material bears two important advantages. Firstly, it allows performing duplex treatments, i.e., the case hardened surface obtains a second treatment such as PVD coating or plasma nitriding (PN) to further increase surface hardness. These treatments are usually performed in a temperature window of 300 to 500°C. It is thus prerequisite that the hardness obtained in the underlying material after quenching from the carburizing temperature is not degraded by the second heat cycle. Secondly, many conventional case carburizing steel grades are restricted to a maximum operating temperature of 120 to 160°C. A steel grade with high tempering resistance can be operated hotter without degrading. Elevated operating temperatures may occur for instance by frictional heating when the transmission experiences lubrication problems.

Good tempering resistance in a typical gear steel base alloy can also achieved with lower molybdenum additions as indicated in Figure 4. Already a Mo addition of 0.5 to 0.7% provides good resistance against softening for tempering parameters up to around $16x10^3$. Resistance against softening under a tempering parameter of $16x10^3$ means that a secondary treatment at a temperature of 450°C for up to 10 hours should be possible. This condition is typical for plasma nitriding. Microalloying of Nb further enhances the tempering resistance, obviously as a synergy effect with Mo. Molybdenum and niobium have to some extend similar metallurgical effects. Both exert strong solute drag on grain boundaries as well as dislocations [7] and also lower the activity of carbon [8]. These fundamental effects are noticed by delayed recovery or recrystallization as well as a reduced rate of pearlite growth thus increased hardenability. The solubility of both elements in austenite is however very different. Molybdenum has a good solubility [9] whereas that of niobium is low [10]. Therefore niobium precipitates as NbC particles at rather high temperature. Manganese, chromium and particularly molybdenum increase the solubility of niobium in austenite [11]. Accordingly, more niobium will be in solution after quenching from austenitizing temperature, which is then available for fine precipitation during tempering treatment acting against softening.

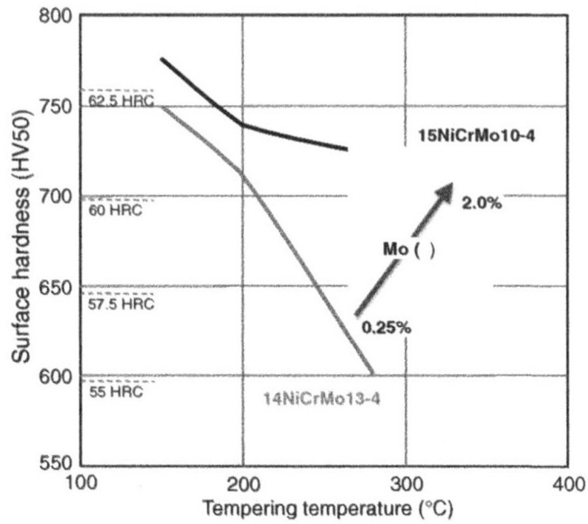

Figure 3: Surface hardness of the carburized layer in relation to tempering temperature and effect of increased molybdenum content.

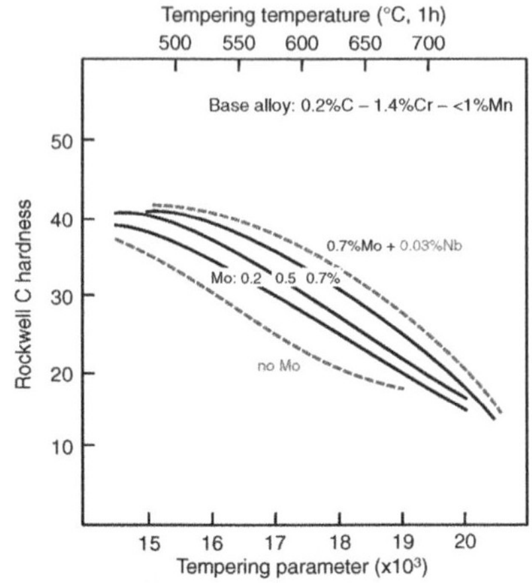

Figure 4: Effect of tempering resistance in function of the molybdenum content and synergy effect with niobium (HP = T x (20 + log t) x 10^{-3} [T in K, t in h]).

Adding niobium in combination with titanium to carburizing steel has a marked effect on the grain size distribution [12-19] as shown in Figure 5 for grade 18CrNiMo7-6. Not only is the grain size generally much finer but also the scatter range is narrower. The microalloyed variant safely avoids prohibited grain sizes despite the high carburizing temperature (1030°C) and the long duration (25 hours). Similar good results of grain size stability have been obtained with modified variants of 25MoCr4 and 20CrMo5 grades Figure 6. The Nb and Ti dual microalloyed 25MoCr4 variant reveals resistance to coarsening up to 1050°C carburizing temperature whereas the Nb-only microalloyed 20CrMo5 variant is stable up to 1000°C. The latter alloy indicates that for very high carburizing temperatures indeed addition of multiple microalloy elements appears to increase the temperature stability of pinning precipitates. However, at standard carburizing conditions below 1000°C also the Nb-only microalloyed concept exhibits very fine austenite grain size with rather narrow scattering.

Developed gear steel for high temperature carburizing and subsequent surface treatment

Based on the individual and synergetic effects of alloying elements described before, the intended processing route and the desired property profile a modified alloy concept has been designed (Table 3) for a full scale production trial including gear running tests. The developed alloy design can be considered as a modified 20MnCr5 grade. It is aiming for higher performance than that of 18CrNiMo7-6 at similar alloy cost. The content of carbon is increased for higher maximum hardness while Mo and Ni are added for increased hardenability and tempering resistance. Niobium microalloying is applied for austenite grain size control. The achieved mechanical properties of the developed case-carburizing steel obtained after heat treatment indeed corresponds to the postulated expectations (Figure 10 and Table 3). The toughness of the developed steel is lower than that of 18CrNiMo7-6 due to the reduced nickel alloy content, yet remains still on a level clearly above the minimum requirement.

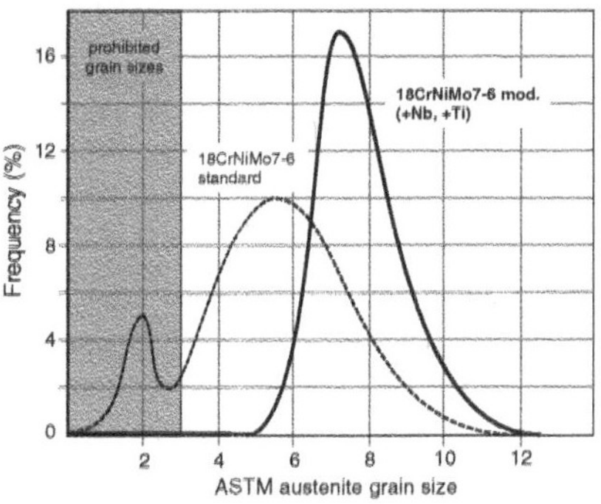

Figure 5: Grain size distribution of a Nb+Ti microalloyed 18CrNiMo7-6 (1.6587) heat compared to a conventional analysis
(carburizing conditions: 1030°C / 25 h).

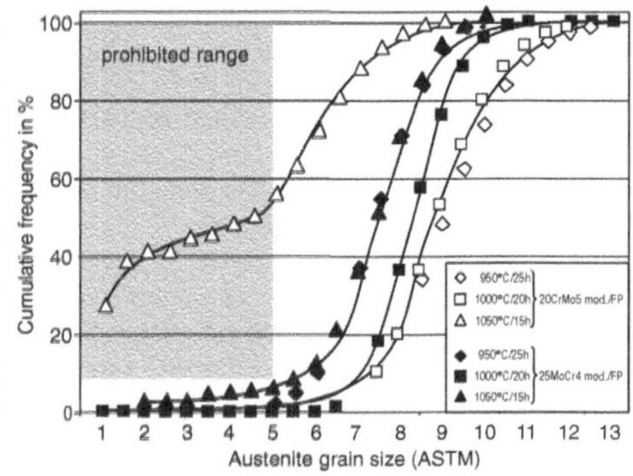

Figure 6: Grain size stability under various carburizing conditions for Nb microalloyed 20CrMo5 (1.7264) and 25CrMo4 (1.7325).

Table 3: Chemical composition of the modified carburizing steel 20MnCr5.

Alloying element	**C**	**Si**	**Mn**	**Cr**	**Mo**	**Ni**	**Nb**
20MnCr5 mod.	0.26	0.12	1.46	1.23	0.54	0.91	0.03

Table 4: Mechanical properties of developed case-carburizing steel (hardened at 880°C/2h + oil/180°C/2h) vs. reference grade 18CrNiMo7-6.

Property	**20MnCr5 mod.**	**18CrNiMo7-6**
Tensile strength, R_m (MPa)	1758	1182
Impact energy, A_v (J)	47	80
Rotating fatigue limit $\sigma_{(50\%)}$@N=10^7 (MPa)	722	510
Hardenability@11 mm (HRc)	51	41
Hardenability@25 mm (HRc)	50	36

Table 5: Hardness characteristics after various heat treatments.

Treatment after carburizing at 1030°C	**20MnCr5 mod.**		
	Surface (HV1)	**Core (HV10)**	**550HV1 depth (mm)**
As quenched	769	544	2.04
Freezing -70°C/2h - Tempering 200°C/2h	717	505	1.53
Plasma nitriding at 400°C	994	432	0.8
Plasma nitriding at 440°C	1009	422	0.35

The heat treatment behavior of the developed alloy has been tested by a carburizing process operated at 1030°C to a nominal case depth range of 0.95 to 1.2 mm. For determining the depth of the case layer a limit hardness of 550HV was defined according to ISO 6336-5. The targeted surface hardness was set to 680-700 HV. Additionally a secondary plasma nitriding treatment has been also performed at 400 and 440°C, respectively. Table 4 summarizes the hardness data for the various pilot heat treatments. Due to its increased tempering resistance the developed steel shows very high surface hardness of around 1000HV after plasma nitriding whereas the core hardness is being reduced. Nevertheless a core hardness of more than 400HV is still a high value. With some further optimization this steel has potential of fulfilling the case depth requirements at secondary treatment temperatures up to 440°C.

Operational performance of the developed carburizing steel was tested and benchmarked at FZG TU Munich, Germany. The tooth root load-carrying capacity was tested in a pulsator rig. Investigations on the flank load carrying capacity were performed by running tests on a back-to-back gear test rig according to DIN ISO 14635-1 as described in [20, 21]. The test gears for these investigations were case hardened after the gear milling. Subsequent to case carburizing the test gears were mechanically cleaned by shot blasting. The flanks as well as the tooth roots of the test gears for the investigations on the tooth root bending strength were not ground. For the running tests gear wheels with a module of 5 mm and a gear ratio of 17/18 were used. This test gear is typical for the examination of the pitting load capacity.

A performance benchmark for the gear tooth root fatigue strength (DIN3990) of the developed steel against established case carburizing alloys is shown in Figure 12. In this graphic the grey shaded are indicates the typical performance range of state-of-the-art carburizing grades established in Europe. Additionally some non-European grades are indicated. The developed steel is situated above the quality level ME and thus performing better than established steel grades.

In order to determine the flank pitting load capacity repeated gear running tests in the high cycle as well as low cycle fatigue range were carried out as described in [21]. The test rig was driven with a constant speed of 3000 rpm and all test runs were performed under oil spray lubrication (approx. 2 l/min into the tooth mesh) with FVA 3 + 4% Anglamol 99 (S-P-additive), a mineral oil of viscosity class ISO VG 100, and an oil temperature of 60°C. Figure 14 represents a benchmark comparison of the obtained data against established case carburizing grades. The developed steel (20MnCr5 mod.) exhibits a very high pitting endurance limit, outperforming the leading established alloys of quality level ME.

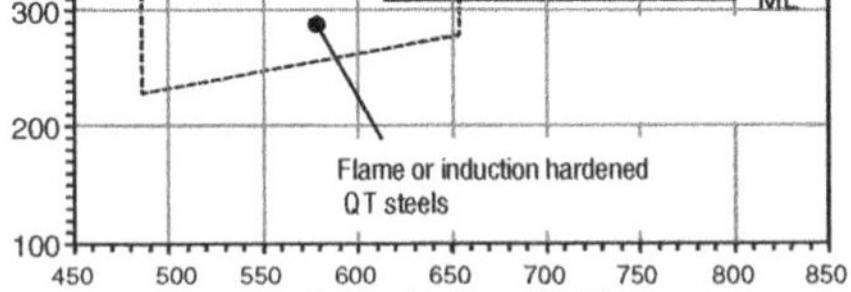

Figure 7: Benchmarking of gear tooth root fatigue of developed case-carburizing steel vs. established alloys (DIN3990).

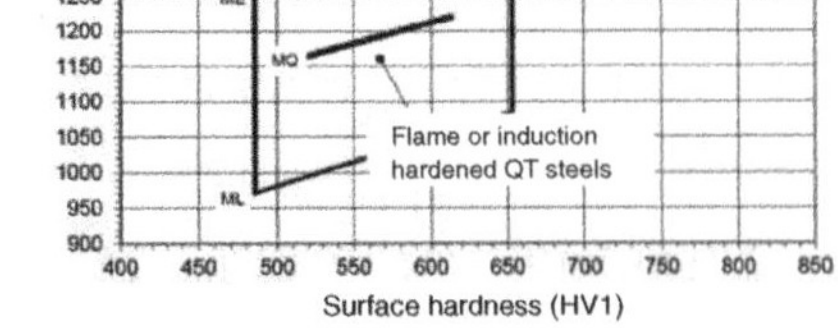

Figure 8: Benchmarking of gear tooth flank micropitting of developed case-carburizing steels vs. established alloys (DIN3990).

The current results suggest that the developed steel 20MnCr5 mod. has the potential of providing an economically viable solution for highly loaded gear in heavy machinery and vehicles. Its use in vehicle transmission could enable downsizing of components reducing weight. In larger transmissions such as used in trucks and heavy machinery its use can help avoiding unexpected failure and extending warranty periods.

STEELS WITH IMPROVED SLIP ROLLING RESISTANCE

Energy dissipated in gears and bearings by rolling friction gives rise to power loss and hence contributes to reduced efficiency as well as increased emissions. The energy dissipation is related to micro-slip and friction at the contact interface, inelastic properties of the material and roughness of the rolling surfaces. Friction can be lowered by lubrication as well as by low-friction coatings on the surfaces of the mating rolling bodies. With increasing contact pressure due to lightweight design of bearing elements inelastic deformation of the material becomes larger leading to hysteresis losses. Furthermore, surface degradation related to rolling fatigue increases roughness and can cause hard low-friction coatings can to spall off. Furthermore, increased torques or contact pressures raises oil film temperatures on the gear tooth flanks towards 200°C, leading to in-situ tempering and thus hardness loss of many state-of-the-art steels.

In previous investigations on the slip-rolling resistance of thin film coatings [22-25], the substrates were mainly made of the well-known bearing steels 100Cr6H (SAE E52100) and Cronidur 30 (AMS 5898, 1.4108). Cronidur 30 is a martensitic cold working steel melted under pressurized nitrogen. The use of pressurized electro slag re-melting in combination with sophisticated rolling technology, extremely high cleanliness and a homogeneous microstructure can achieve high durability values at a maximum hardness of 60 HRC. An important aspect of this steel grade is its high temper resistance up to 500°C making it also compatible to typical thin film coating processes. Both steels 100Cr6H and Cronidur 30 are widely used in the bearing industry. However, nitrogen alloyed steels, like Cronidur 30 are not widely available and are expensive. The steel grade 36NiCrMoV1-5-7 can be a lower cost alternative having good availability. Other advantages of this steel are:

- •High tempering resistance allowing the application of thin film coatings
- •Uncomplicated heat treatment
- •Reduced friction as uncoated steel under mixed/boundary conditions
- •High load carrying capacity
- •Low wear as uncoated steel
- •Tribological compatibility with state-of-the-art lubricants

The three steels, of which the chemical composition is given in Table 6, have been benchmarked by slip rolling testing using the Optimol 2-disc tribometer equipment and procedures described in [22]. The mechanical properties of these steels after heat treatment are summarized in Table 7. The hardness of grade 36NiCrMoV1-5-7 is clearly lower than that of 100Cr6H and Cronidur 30 yet its toughness is much better. Furthermore, its tempering resistance is particularly good allowing operations up to 510°C without significant hardness loss.

Table 6: Chemical composition of steel grades tested for slip rolling resistance.

Alloying element	C	Si	Mn	P	S	Cr	Mo	Ni	N	V
100Cr6H	1.0	0.25	0.35			1.50				
Cronidur 30	0.25-0.35	<1.0	<1.0			14.0-16.0	0.85-1.1	<0.5	0.3-0.5	
36NiCrMoV1-5-7	0.30	0.25	0.2	0.002	0.002	1.50	0.8	3.0		0.3

Table 7: Mechanical properties (heat treated condition) of steel grades tested for slip rolling resistance.

Steel grade	**100Cr6H (SAE E52100)**	**Cronidur 30 (AMS 5898)**	**36NiCrMoV1-5-7**
Hardness [HRC]	65.8	62.2	~ 54
Ultimate Strength R_m [MPa]	~ 2300	~ 2300	1474
Fracture Toughness K_{IC} [MPa$\sqrt{m}$]	~ 16.5	~ 21	120
Charpy Toughness KV [J]	-	-	30
Maximum ServiceTemperature [°C]	150	475	510

Friction behavior under slip rolling conditions

Slip rolling tests were executed over 10 million cycles (132 km) at three different contact pressures with self-mating tribo-couples having the geometry shown in Figure 9. BMW factory fill engine oil SAE 0W-30 VP1 was used as lubricant at an operating temperature of 120°C. The slip rate was adjusted to 10 percent corresponding to a sliding speed of 0.08 m/s. For the steel grade 36NiCrMoV1-5-7 five runs were repeated (min.-max. range indicated in Figure 9) and compared to statistical reference data of the two other steel grades.

Generally, all data (Figure 9) indicate a decreasing evolution of friction over the duration of the test. That means power losses reduce over run time. It is also apparent that steel grade 36NiCrMoV1-5-7 reveals a lower coefficient of friction (COF) over the entire run time at all three contact pressures. Most distinguished is the difference in COF at the mean contact pressure of 1.5 GPa where the COF of 36NiCrMoV1-5-7 is approximately 30 percent lower than for the established steel grades. Under this condition the COF of the 36NiCrMoV1-5-7 self-mating tribo-couple approaches the low level that is characteristic for steel coated with low-friction DLC thin films. With increasing contact pressure, the COF of the three steel grades assimilate. Those of grades 100Cr6H and Cronidur 30 slightly reduce whereas the COF of grade 36NiCrMoV1-5-7 slightly increases. It is thus evident that grade 36NiCrMoV1-5-7 has the potential of reducing friction even without low-friction coating at contact pressures above FZG 14.

Wear behavior under slip rolling conditions

Wear resistance in slip rolling tribo-contacts is an essential aspect as neither raceway nor ball should show significant damage over lifetime. As an evaluation criterion in the current tests it was specified that the surface damage should be not larger than 1 mm^2 over at least 10 million load cycles. Evaluation of the wear behavior indicated that grade 36NiCrMoV1-5-7 clearly has the potential of fulfilling the requirements for slip rolling wear resistance under all three contact pressure levels (Figure 10). The wear rate of grade 36NiCrMoV1-5-7 despite its comparably lower hardness is comparable to that of 100Cr6H and Cronidur 30. However, the wear rate of DLC coated steels is one order of magnitude lower. However, some individual samples of grade 36NiCrMoV1-5-7 showed premature failure (Figure 11). More detailed analysis could identify that the origin of premature failure was related to sub-surface non-metallic inclusions in the steel such as calcium sulfide (CaS) as well as aluminum oxide (Al2O3) and magnesium oxide (MgO) inclusions. Non-metallic inclusions acts as crack initiators, especially when they are locally agglomerated directly beneath the cyclically stressed surface. Thus, techniques such as vacuum induction melting (VIM) and vacuum arc re-melting (VAR) are necessary to improve steel cleanliness avoiding avoid rolling contact fatigue. Furthermore, refinement and homogenization of the microstructure is advisable for optimum toughness and fatigue resistance. This can be achieved by niobium microalloying, which has not yet been applied in the current commercially available grade 36NiCrMoV1-5-7.

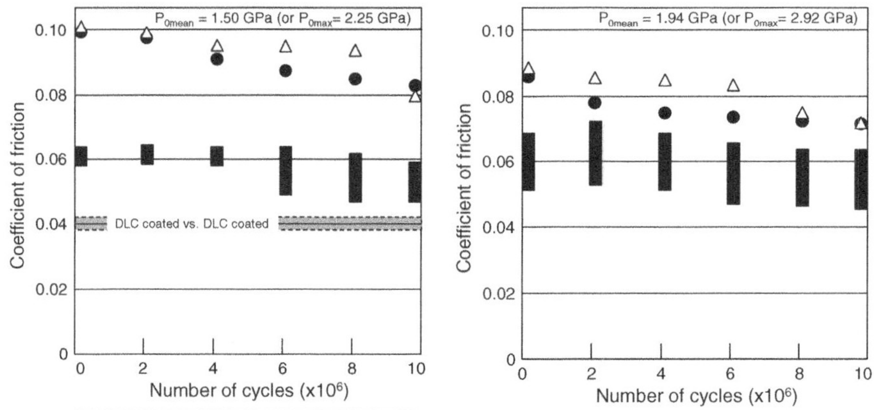

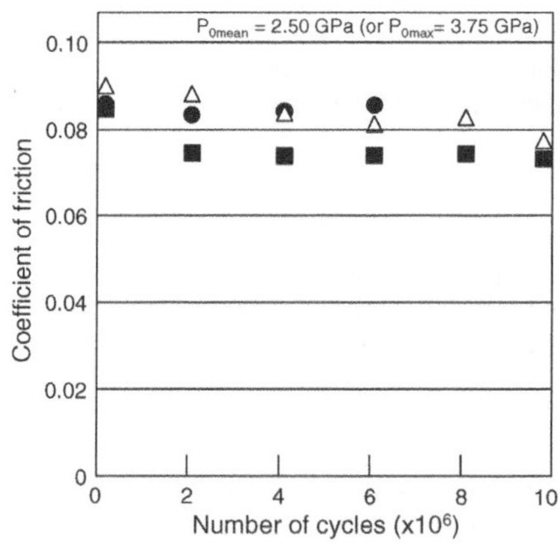

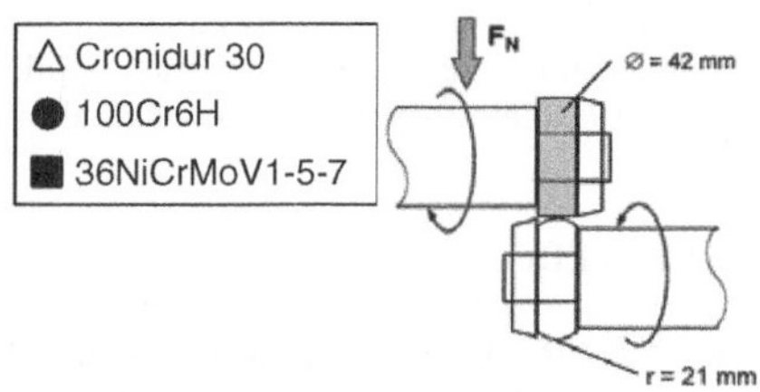

Figure 9: Evolution of the coefficient of friction over 10 million load cycles at different contact pressures. Self-mating tribo partners of cylindrical vs. round-shaped disc. Operating temperature: 120°C. Lubrication: BMW factory fill engine oil SAE 0W-30 VP1.

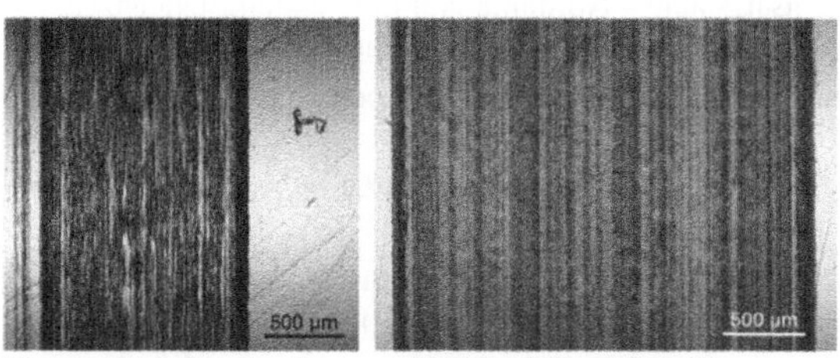

Figure 10: Wear tracks on cylindrical disc after 10 million load cycles under 1.5 GPa (left) and 2.5 GPa (right) mean contact pressure.

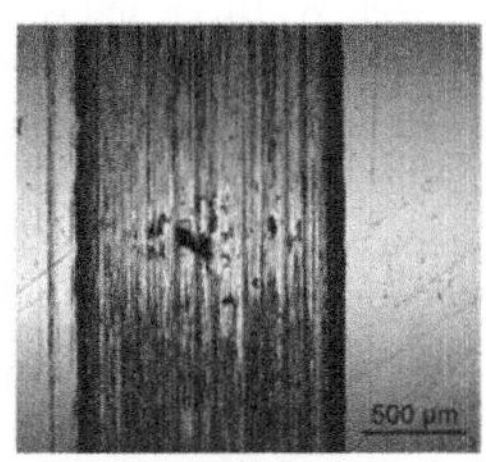

Figure 11: Pre-mature damage on cylindrical disc due to sub-surface non-metallic inclusions.

A recent study [26] demonstrated the significance of frictional losses in the vehicle powertrain. A total reduction of friction by 30 percent could reduce fuel consumption by around 12 percent. Grade 36NiCrMoV1-5-7 provides such a reduction in tribo-contacts subjected to 1.5 GPa mean pressure. Alternatively, DLC coated steels offer even lower friction. However, sufficient tempering resistance of the steel is prerequisite for the coating deposition. Both molybdenum alloyed steels, Cronidur 30 as well as 36NiCrMoV1-5-7 provide the required tempering resistance for such surface treatment.

CONCLUSIONS

It has been demonstrated that significant improvements of steel properties relevant to powertrain applications are possible without applying expensive metallurgy or sophisticated treatments. In gear applications the optimized steel 22MnCr5 mod. showed much higher pitting resistance than the existing state-of-the-art steels. In bearing applications the existing steel 36NiCrMoV1-5-7 revealed a surprisingly low coefficient of friction under slip rolling conditions. In both steels molybdenum is a key alloying element. Molybdenum not only provides excellent hardenability but also severely raises the tempering resistance. The latter is crucial when operating temperature is high or when an additional surface treatment such as plasma nitriding or thin film deposition is desired. Niobium as a microalloying element controls and homogenizes the steel microstructure thus improving toughness and fatigue resistance. It also enhances the molybdenum-based tempering resistance. Detailed understanding of molybdenum-niobium based alloying concepts gained in the presented studies allows further optimizing and tailoring properties of special steel grades for powertrain applications. The envisaged benefits in terms of better efficiencies and reliability in powertrain applications as well as simplified production routes in making these steels will clearly outweigh alloy costs associated with molybdenum and niobium additions.

ACKNOWLEDGEMENT

The experimental benchmarking tests on both molybdenum-alloyed steels were financially supported by the International Molybdenum Association, London (www.imoa.info). The Gear Research Centre (FZG) of the Technical University Munich and the Federal Institute for Materials Research and Testing (BAM), Berlin are gratefully acknowledged for executing the experimental tests. Dr. Frank Hippenstiel of BGH Siegen kindly provided the tested steel materials.

REFERENCES

[1] F. Hippenstiel, K.-P. Johann and R. Caspari: Tailor Made Carburizing Steels for Use in Power Generation Plants, European Conference on Heat Treatment 2009, Strasbourg (France).

[2] H. Spitzer, W. Bleck and R. Flesch: Einsatzstähle – Normung und Entwicklungstendenzen, ATTT/AWT-Tagung Einsatzhärtung, 29./30. April 1998, Aachen, Tagungsband, p. 11-20.

[3] N. Bretl, S. Schurer, T. Tobie, K. Stahl and B.-R. Höhn, Thermal Processing for Gear Solutions, Fall/Winter 2014, p. 52.

[4] A. Stenico, Werkstoffmechanische Untersuchungen zur Zahnfußtragfähigkeit einsatzgehärteter Zahnräder, Ph.D. thesis, TU Munich (Germany), Sep. 2007.

[5] M. Uno, M. Hirai and F. Nakasato, Effect of Alloying Elements on the Properties of Cr-free Carburizing Steels, The Sumitomo Search No. 39, September 1989, p. 33.

[6] E. Dehner and F. Weber, GearTechnology (www.geartechnology.com), July 2007, p. 42.

[7] F. Togashi and T. Nishizawa, J. Japan Inst. Metals, Vol. 40, No. 1 (1976), p. 12.

[8] T. Tanaka and T. Enami, Tetsu-to-Hagané, 58 (1972), p. 1775.

[9] E.J. Pavlina, J.G. Speer and C.J. Van Tyne, Scripta Materialia 66 (2012) p. 243.

[10] 25.V.K. Lakshmanan and J.S. Kirkaldy, Metallurgical Transactions A, Vo. 15A, 1984, p. 541.

[11] S. Koyama, T. Ishii and K. Narita, Jap. Inst, of Metal J., 35, (1971) p. 1089.

[12] B. Huchtemann and V. Schüler, Thyssen Technische Berichte, Heft 1/93 (1993), p. 97.

[13]M. Kubota and T. Ochi, Development of Anti-Coarsening Steel for Carburizing, Materials Science Forum Vols. 539-543 (2007), p. 4855.

[14]M.J. Leap and E.L. Brown, Mater. Sci. Technol., Vol.18, 2002, p. 945.

[15]T. Kimura and Y. Kurebayashi, Niobium in Microalloyed Engineering Steels, Wire Rods and Case Carburized Products, Proc. of the Int. Symp. Niobium 2001, TMS (2001), p. 801.

[16]F. Hippenstiel, Tailored Solutions in Microalloyed Engineering Steels for the Power Transmission Industry, Materials Science Forum Vols. 539-543 (2007), p. 4131.

[17]K. A. Alogab, D. K. Matlock, J. G. Speer and H. J. Kleebe, ISIJ International, Vol. 47 (2007), No. 2, p. 307.

[18]K. A. Alogab, D. K. Matlock, J. G. Speer and H. J. Kleebe, ISIJ International, Vol. 47 (2007), No. 7, p. 1034.

[19]K. Klenke, R. Kohlmann, P. Reinhold and W. Schweinebraten, Improved performance by high temperature carburizing shown by the example of VW4521+Nb, Proc. Steels in Cars and Truck 2008, Verlag Stahleisen Düsseldorf (2008), p. 173.

[20]H. Mohrbacher, Proc. Int. Conf on Advances in Metallurgy of Long and Forged Products, AIST 2015, p.59.

[21]B.-R. Höhn, K. Stahl, J. Schudy, T. Tobie, B. Zornek, Gear Technology, June/July 2012, p. 60

[22]Woydt, M., Scholz, C., Manier, C.-A., Brückner, A., Weihnacht, V., Slip-rolling resistance of ta-C and a-C coatings up to 3,000 MPa of maximum Hertzian contact pressure, Materialwissenschaft und Werkstofftechnik, Volume 43, Issue 12, pp. 1019 – 1028, 2012.

[23]Manier, C.-A., Slip-rolling resistance of novel Zr(C,N) thin film coatings under high Hertzian contact pressures, BAM-Ph.D. series, Volume (Band) 60, Berlin, 2010.

[24]Manier, C.-A., Ziegele H., et al.: Zirconium-based coatings in highly stressed rolling contacts as alternative solution to DLC and ta-C coatings, Wear, Vol. 269, Issues 11 – 12, pp. 770 – 781, 2010.

[25]Manier, C.-A., Theiler, G., Spaltmann, D., Woydt, M., Ziegele, H.: Benchmark of thin film coatings for lubricated slip-rolling contacts, Wear, Vol. 268, Issues 11 – 12, pp. 1442 – 1454, 2010.

[26]J. Beulshausen, J. Geiger, S. Pischinger, B.-R. Höhn, autotechreview, Volume 3, Issue 1, Januar y 2014, p. 30.

NIOBIUM CARBIDE – AN INNOVATIVE AND SUSTAINABLE HIGH-PERFORMANCE MATERIAL FOR TOOLING, FRICTION AND WEAR APPLICATIONS

Hardy Mohrbacher
NiobelCon bvba
Schilde, Belgium

Mathias Woydt
Federal Institute for Materials Research and Testing (BAM)
Berlin, Germany

Jef Vleugels, Shuigen Huang
KU Leuven – Department of Materials Engineering (MTM)
Leuven, Belgium

ABSTRACT

Transition metals IVB to VIB in the periodic table of elements are strong carbide formers. Most prominent is tungsten carbide, which is typically consolidated with cobalt binder and is used for tooling, mineral processing and wear applications. Surprisingly little information was available about niobium carbide (NbC). Recently, pure NbC was synthesized from Nb_2O_5 and its properties were characterized in detail. Furthermore NbC hardmetal grades were synthesized using various metallic binders. Based on its tribological characteristics NbC appears to have great potential in tooling, friction and wear applications. Niobium carbide is largely available, efficient to produce, provides comparably low friction in many relevant tribo-contacts and shows low wear. NbC and Nb_2O_5 have so far no REACH classification related to human toxicology. Neither are these listed as substances of very high concern contrary to WO_3 and Co_3O_4. The paper demonstrates the key characteristics of NbC and discusses its sustainability and reliable value chain.

INTRODUCTION

Carbides of the transition metals represented by groups IVB to VIB in the periodic table of elements are typically used as hard phases in technical alloys. The hardness of these transition metal carbides is superior to that of most minerals (Figure 1). Metals with high melting point such as cobalt, nickel or molybdenum are serving as ductile binder for hardmetals and cermets. Casting alloys containing dispersed hard phase particles provide the platform for a wide variety of metal-matrix composites.

In the arena of hardmetals, tungsten carbide has been the dominating hard phase since decades. Other transition metal carbides such as those of chromium, vanadium and titanium are also well established. Surprising little attention has been paid to niobium carbide, however. Nevertheless, niobium carbide has been occasionally used as minority phase in hardmetals, castings and tool steels. In these applications it serves either as grain refiner or as hard phase helping to enhance wear resistance, limits grain growth and enhances hot hardness. While some data on the fundamental properties of pure niobium carbide have been reported [1] reliable information on its tribological performance was not readily available. Recently it was shown however, that pure NbC as well as metal bonded NbC have a pronounced wear resistance under dry sliding versus other monolithic ceramics and carbides [2].

Hardmetals based on tungsten carbide are mainly produced by a powder metallurgy (PM) based sintering process due to the high solubility of the hard phase in alloys and in their melts as well as the mismatch in density between molten binder alloys and hard phase.

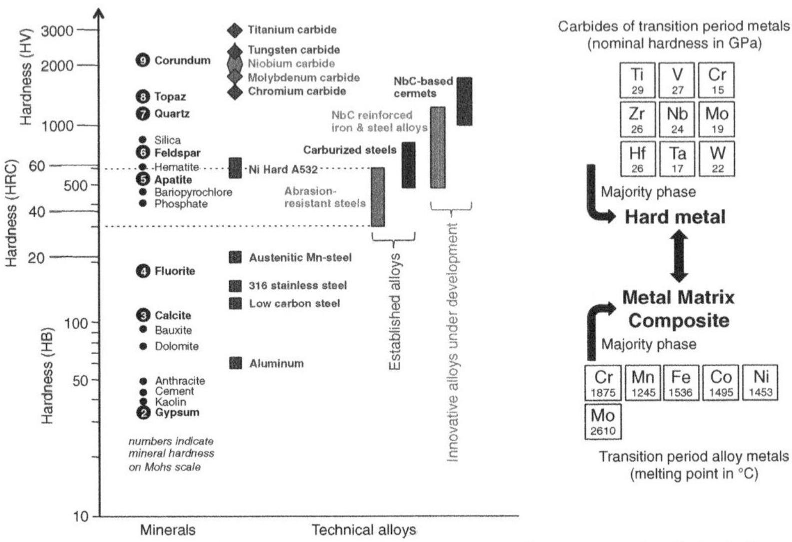

Figure 1: Hardness comparison between mineral species and technical alloys; definition of hard metal and MMC materials.

Cobalt and nickel primarily serve as binder phase. The powder metallurgical route, however, limits the size of components, whereas castings would allow the fabrication of much bigger components. Alternatively hardmetals can be deposited as clad layers or thermally sprayed coatings onto a softer substrate material. Two well-known properties of NbC are its similar density to and its low solubility in iron. The density of NbC is very close to that of iron so that the mismatch in densities, depending on the carbide stoichiometry, and iron alloy composition is less than 10 percent. Its solubility in iron-based alloys as well as in nickel and cobalt is particularly small. Utilizing these two properties makes alternative production routes feasible for producing components as compared to PM-based processing routes.

Material health and safety issues with established hardmetals recently caused significant concerns. The European "Registration, Evaluation, Authorization and Restriction of Chemical substances program", also known under the acronym REACH [3], currently classifies cobalt as very toxic for the human health. Furthermore tungsten carbide-cobalt hard metal dust was shown to be more toxic in combination than either pure cobalt or tungsten carbide alone [4]. So far, Ni powders do not have the same hazardous classification as Co powders. Niobium, on the contrary, is known to be one of the most biocompatible metal. No particular hazards or critical notifications for its oxide (Nb_2O_5) and carbide (NbC) were filed in the framework of REACH (Table 1). Biocompatibility evaluation using cell viability tests indicated no significant difference of NbC over a wide range of stoichiometry as compared to the established human body implant alloy Ti-6Al-4V [5].

The explored global tungsten reserves are estimated to around 3.10 million tons metal content or 6.384 million tons of tungsten as "reserves and resources" [6]. The reserves of the currently operating niobium mines, including the known deposits and secondary resources, largely exceed those of tungsten. In addition, significant new deposits have been discovered. Approximately 90 percent of the refined niobium (FeNb) is used as microalloy addition in the world steel industry. The current production capacity of FeNb and purified Nb_2O_5 is far bigger than its consumption. Both products can be precursor of niobium carbide.

Table 1: Summary of health & safety classification and labeling (extract)

Substance	[CAS] {EINECS}	Hazard Class and Classification Code			
		Mutagenic category 2	Carcerogenic category 1A	Reproductive toxicity category 1B	Carcerogenic category 2
Co_3O_4	[1308-06-1] {215-157-2}	Yes (H341, suspected of causing genetic defects)	Yes (H350, may cause cancer when inhaled)	Yes (H360, may damage fertility or the unborn child)	—
WO_3	[1314-35-8] {215-231-4}	—	—	—	Yes (H351, suspected of causing cancer)
Nb_2O_5	[1313-96-8] {215-213-6}	—	—	—	—

Yet, the use of niobium carbide as a hard material is much less established than that of tungsten carbide. Nevertheless, its physical and chemical properties indicate an outstanding potential of niobium carbide for applications where currently tungsten carbide is being the standard material.

METHODS OF NIOBIUM CARBIDE SYNTHESIS

Niobium carbide can be synthesized by various practical methods of which two are being considered here. One method is the direct carbo-thermal reduction of pure niobium oxide (Nb_2O_5) at elevated temperature (>800 °C) by the solid-state reaction:

$$Nb_2O_5 + 7 \cdot C \rightarrow 2 \cdot NbC + 5 \cdot CO(g) \quad (1)$$

Reaction (1) does not represent the actual steady state reaction, since the reduction process is controlled by a gas-solid phase reaction (2) involving the so-called Boudouard reaction (3) as an intermediate step:

$$Nb_2O_5 + 9 \cdot CO(g) = 2 \cdot NbC + 7 \cdot CO_2(g) \quad (2)$$

$$7 \cdot CO_2(g) + 7 \cdot C = 14 \cdot CO(g) \quad (3)$$

Thermodynamic data indicate that reaction (2) is feasible at temperatures higher than 1000 °C. However, experiments indicated that for a complete conversion of Nb_2O_5 to oxide-free NbC reaction temperatures of at least 1200 °C must be applied. Since NbC oxidizes rapidly at temperatures above 480 °C the reaction atmosphere must be oxygen-free. An inert and reducing atmosphere of $95\%N_2 + 5\%H_2$ was found to be practicable.

Representative NbC powders synthesized from Nb_2O_5 powder using the carbo-thermal reduction process show primary grain sizes in the sub-micron range (100 – 600 nm) reflecting the original oxide crystallite size (Figure 2). The NbC crystallites are agglomerated to larger particles (Figure 3), which can be broken down by an appropriate milling process after carbo-thermal reduction. XRD analysis indicated that these powders consist of pure niobium carbide, i.e., without traces of residual oxides or free carbon.

An alternative product developed by the Brazilian niobium mining company Companhia Brasileira de Metallurgia e Mineração (CBMM) is an innovative ferroalloy containing a high amount of primary NbC comprising controlled size and compact morphology. The production process is based on the standard alumino-thermic reduction utilized for ferro-niobium production. Yet, by adding a carbon source directly to this process primary precipitation of NbC occurs into an iron matrix instead of forming FeNb intermetallic phases (Figure 4). The carbide particles being formed in this process have sizes in the range of 10 to 50 µm and are of cubic morphology

(Figure 5). At the current development stage, the volume share of NbC phase embedded in an iron matrix is around 60 percent. This ferroalloy can be directly added to an iron melt for producing iron-based alloys containing a lower volume fraction of dispersed carbide particles. On the other hand, the Fe-NbC ferroalloy, in particular after further increasing the carbide volume share, can serve as a source of pure carbide supply (melt grown NbC). In this process, the iron matrix is dissolved from the ferroalloy by a leaching process using either hydrochloric or sulphuric acid (Figure 5).

Figure 2: Morphology of $NbC_{1.0}$ crystallites obtained by carbo-thermal reduction of CBMM´s high purity Nb_2O_5 powder (de-agglomerated)

Figure 3: Agglomerated $NbC_{0.88}$ crystallites after carbo-thermal reduction of CBMM´s high purity Nb_2O_5 (used for “H1”)

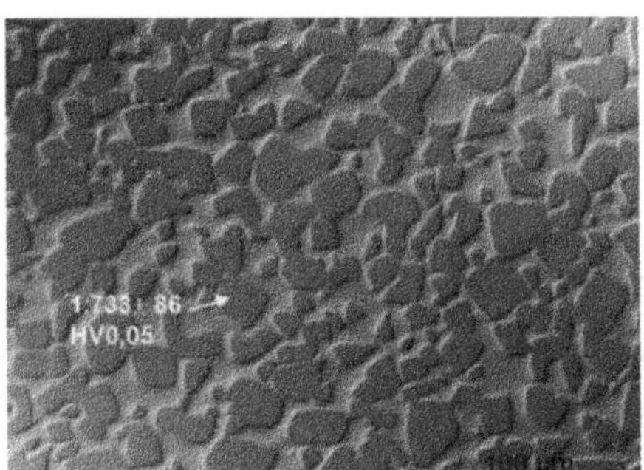

Figure 4: $NbC_{0.94}$ particles formed by primary precipitation in a carbon-alloyed iron melt.

Figure 5: $NbC_{0.95}$ particles precipitated in liquid iron after chemical leaching by sulphuric acid.

BASIC PROPERTIES OF NIOBIUM CARBIDE

Table 2 resumes some of the key physical properties referring to room temperature measurements of stoichiometric niobium carbide in comparison to those of tungsten carbide published in literature [1]. The melting temperature of NbC is particularly high, which is expected to limit thermal softening as well as the tendency to adhesive wear. Data by Kelly et al. [7] indicated that the yield strength of NbC at temperatures above 1000°C is substantially higher than that of WC. The density of NbC is only about half of that of WC. This means that for producing a given component half the mass of NbC is needed, making the use of NbC economically attractive. In dynamic applications where high accelerations and speeds are involved, the reduced mass results in lower inertia and kinetic energy, respectively. Since the density of NbC is very close to that of iron, gravity segregation should be rather low when dispersing solid NbC particles in melts of iron-based alloys. This likewise applies for spin casting processes where higher density particles segregate outwards and lighter density particles segregate inwards. The Young's modulus of NbC is one of the highest amongst the refractory metal carbides, yet lower than for WC. Hence, components made from NbC will show a somewhat higher elastic deflection under mechanical loading compared to WC. On the other hand, contact stresses under a given load will be lower in NbC. Thermal stresses induced between NbC and metallic binder alloys are expected to be lower than for WC. This is reasoned by the larger coefficient of thermal expansion that is closer to that of typical metallic binders as well as to the lower Young's modulus of NbC. Due to its metallic character niobium carbide has a relatively good thermal and electrical conductivity. Its thermal conductivity substantially increases at elevated temperature. Frictional heat generated in tribological contacts can thus be efficiently evacuated. The reasonably good electrical conductivity allows processing of NbC by for instance electrical discharge machining (EDM). It also helps avoiding the build-up of substantial static electricity in tribo-contacts with electrical isolators such as for instance polymer-based composites.

Table 1: Comparison of basic physical properties of niobium and tungsten carbide.

	Melting point (°C)	Density (kg/m³)	Young's modulus (GPa)	Shear modulus (GPa)	Vickers hardness (GPa)	CTE (ppm/K)	Thermal conductivity (W/m.K)	Specific heat (J/gr·K)	Electrical resistivity (μΩ·cm)
NbC	3520	7780	380-480	214	17-22	6.6	14	0.351	35
WC	2870	15630	620-720	262	24-30	5.5	84	0.203	20

One of the most crucial properties with respect to wear resistant applications is the material hardness. From literature data and current experiments it is evident that the hardness of stoichiometric niobium carbide is below that of tungsten carbide (Table 1). Yet, its hardness is still far superior to most minerals and technical alloys (Figure 1) so that sufficient resistance against abrasive wear in tribo-contacts should be given. On the other hand, finishing efforts when manufacturing niobium carbide tools are expected to be lower than those for tungsten carbide tools.

The exact hardness value of these carbides is influenced by several parameters such as the level of porosity, indentation load [8] and temperature [9,10]. An increase in each of these parameters results in a decrease of the measured hardness value. Considering the temperature dependence of hardness, a two-stage behavior can be observed. In the lower temperature range, hardness decreases moderately with increasing temperature. In the higher temperature range this decrease becomes significantly steeper. The transition between the two regimes (drop in hardness) occurs in the temperature interval of 500-800°C for WC [9]. Niobium carbide exhibits this transition at a higher temperature interval around 900°C [11]. This implies a higher hot-hardness of NbC as compared to WC, which is of relevance for metal cutting operations. Recent hot-hardness measurements on NbC- and WC-based materials confirm this trend [12].

Contrary to WC, NbC is forming sub-carbides, like Nb_6C_5 or Nb_4C_3, over a wide range of carbon stoichiometry according to the Nb-C binary phase diagram shown in Figure 6 [13]. Carbon deficiency in the range of NbC to $NbC_{0.8}$ has a significant influence on physical properties of the material. With increasing carbon deficiency in that range elastic moduli [14], lattice constant [15], density [15], heat capacity [16] and magnetic susceptibility [17] decrease. Yet, hardness strongly increases with increasing carbon deficiency in NbC as shown in Figure 7 [18] exceeding the level of WC. In summary, properties of niobium carbide can be tailored by adjusting the carbon deficiency in the compound.

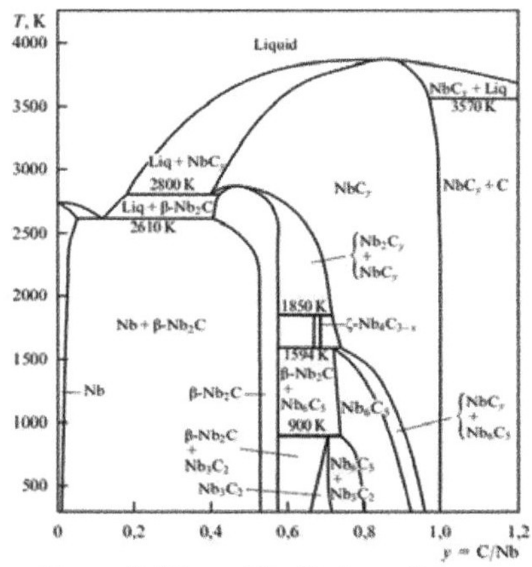

Figure 6: Binary Nb-C phase diagram indicating the existence of various sub-carbide phases [13]

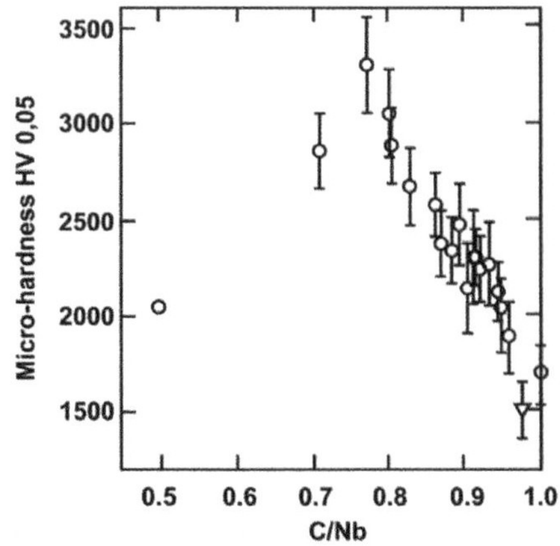

Figure 7: Micro-hardness of niobium carbide phases as function of C/Nb ratio in the compound [18].

The solubility of a carbide compound in a contacting metal at high temperature is a very important property with respect to the processing and wear performance of the carbide in hot tribo-contacts with metal. The solubility of niobium carbide metals such as iron, cobalt or nickel has been reported to be very low in contrast to the rather good solubility of tungsten carbides in these metals [19,20]. The solubility of NbC in austenite is shown in Figure 8 based on an established solubility product [21]. In this plot equilibrium solubility isotherms of niobium and carbon are indicated for various temperatures. Only when the carbon content is extremely low, niobium has a reasonably good solubility. In the area of micro-asperities of hot tribo-contacts, NbC should dissociate into solute niobium and carbon according to the stoichiometric ratio. The intersection of the stoichiometry line with the solubility isotherm defines the maximum solubility of NbC in a tribo-contact at that specific temperature. Carbon already present in an iron-based alloy further limits the solubility of NbC to extremely low values. The solubility of tungsten carbide in austenite is several orders of magnitude larger than that of niobium carbide [22] and responsible for high crater wear. Accordingly, WC-based components bear the risk of efficiently dissolving into iron-based alloys in hot tribo-contacts or during cutting/machining operations whereas NbC-based tools will suffer very limited material loss by dissolution into the chips under the same condition. Consequently a protective coating preventing the dissolution is not necessary on niobium carbide components.

Due to its low solubility, solid NbC particles can be readily dispersed into liquid cast iron preserving its original morphology and without significantly disturbing the carbon balance in the alloy. Because of the close match of densities between liquid iron and NbC the tendency to segregation by either gravity or centrifugal force is relatively small. This allows manufacturing of near net-shape components by either static or spin casting. NbC particles contained in used components can be easily recycled by re-melting corresponding scrap together with fresh alloy.

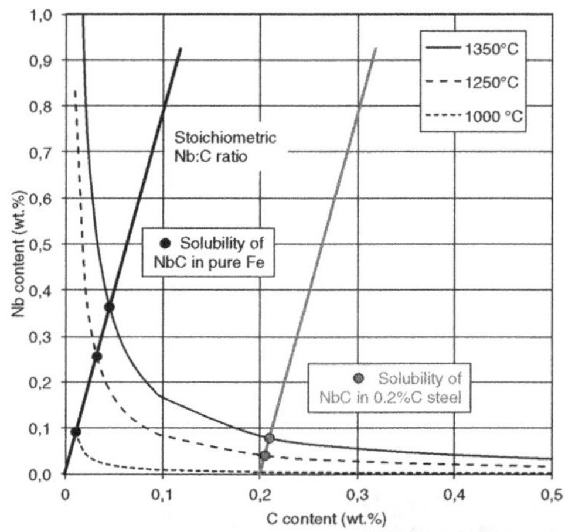

Figure 8: Solubility of NbC in austenite based on the solubility product log[Nb][C]=3.4–7920/T where T is the absolutetemperature;[Nb] and [C] are concentrations in wt.% [21].

METAL BINDERS AND SECONDARY CARBIDE ADDITION

Metal binders have the important function of consolidating the hard carbide phase leading to an acceptable combination of hardness and toughness with regard to the final application. In hardmetals or cermets the metallic binder is the minority phase whereas it becomes the dominating phase in a metal matrix composite (MMC) as indicated in Figure 1. The standard binders for WC are cobalt and nickel, which can also used to consolidate NbC particles. Due to its particularly low solubility in iron, carbon steel, stainless steel or cast iron can also used as binders for NbC. Another possible binder is the intermetallic phase of iron aluminide (Fe_3Al). Iron aluminide is a low cost alternative to stainless steel, also comprising a better ductility, lower density and excellent oxidation resistance up to 1,200°C.

Previous work revealed significant NbC grain growth in NbC-Co based cermets during pressureless sintering for 1 h at 1360°C [23]. NbC grain growth, however, can be effectively suppressed when consolidating NbC-Co cermets in the solid state by hot pressing or spark plasma sintering (SPS) [23,24].

Small additions of NbC to WC-based hard metal result in efficient grain size control. Likewise, the addition of other secondary carbides has the potential of suppressing grain growth in NbC-based hard metal or cermets.

Compared to pure NbC-Ni, limited NbC grain growth and homogeneous grain size distribution could be obtained upon adding secondary phase carbides or Al metal. In that respect WC and Mo_2C were found to significantly enhance hardness and toughness of NbC-Ni cermets. Figure 9 demonstrates several examples of these approaches highlighting the grain size controlling effect on NbC by aluminum additions to iron and nickel binders as well as Cr_3C_2 and especially Mo_2C secondary carbide additions. Accordingly, NbC as hard phase offers a wide variety of possibilities for tailoring hard metal and cermet combinations to the exact needs of an application.

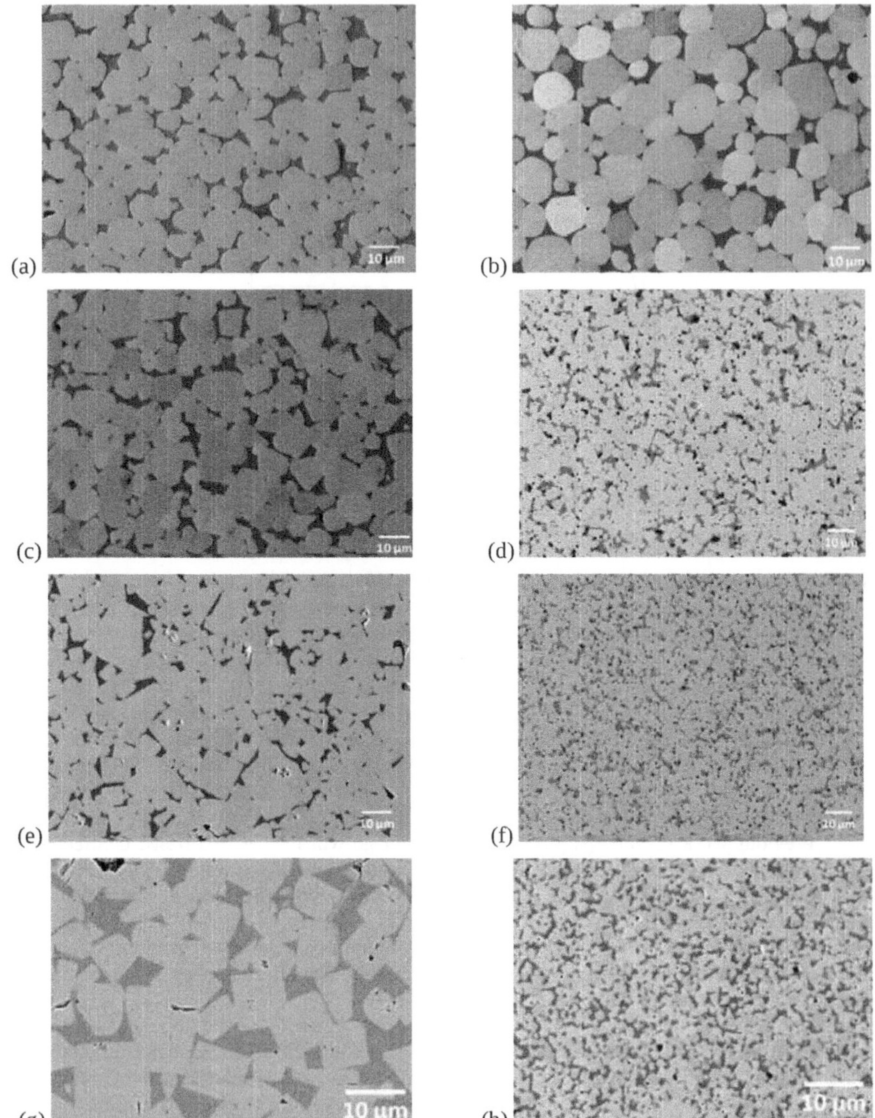

Figure 9: Back scattering electron micrographs of NbC cermets vacuum sintered at 1420°C.
(a): NbC-12 vol% Co → (b): NbC-12 vol% Co + 0.9 vol% Cr_3C_2
(c): NbC-12 vol% Fe → (d): NbC-12 vol% (Fe + 25 at% Al)
(e): NbC-12 vol% Ni → (f): NbC-12 vol% (Ni + 25 at% Al)
(g): NbC-15 vol% Ni → (h): NbC-15 vol% Ni + 5 vol% Mo_2C

The Vickers hardness (HV_{10}) and fracture toughness (K_{IC}) of various NbC cermets prepared under exactly the same sintering conditions are being related to each other in Figure 10. Generally, the hardness of the cermets can be divided into four categories. The Co bonded WC-NbC-VC/Cr_3C_2 cemented carbides have the highest hardness, followed by the Ni bonded NbC-MeC cermets and the NbC-Ni and -Fe cermets which have a lower hardness. The NbC-Co cermets prepared by conventional sintering exhibited the lowest hardness level. Since all grades have a comparable binder content, the difference in hardness of the NbC matrix cermets can be directly attributed to the nature of the binder and the constituent components, i.e., the binder composition and the NbC grain size. Based on the microstructural analysis, the refinement of the NbC grain size in the Al-doped Fe/Ni binder NbC cermets improved the hardness. The hardness of the Fe-Al, Ni-Al, Co and Ni binders are difficult to compare directly since they exhibit different grain sizes. Considering the pure metal binder, the Fe bonded NbC cermets had the highest hardness of 1,294 kg/mm^2, followed by the Ni and Co bonded cermets. For cobalt bonded WC-NbC-VC/Cr_3C_2 cemented carbides, hardness rises with increasing content of grain growth inhibitor.

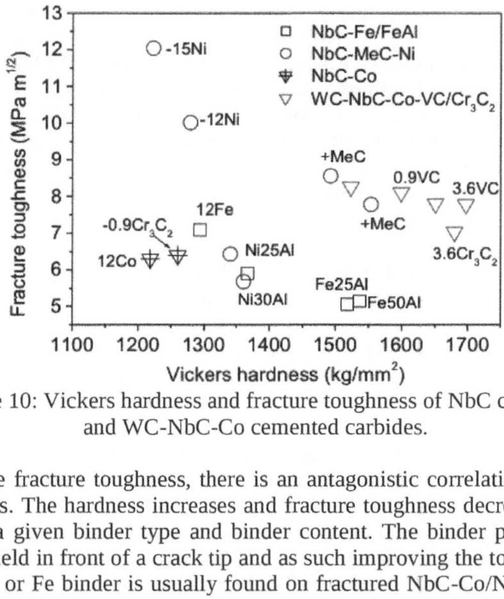

Figure 10: Vickers hardness and fracture toughness of NbC cermets and WC-NbC-Co cemented carbides.

Regarding the fracture toughness, there is an antagonistic correlation to the hardness of NbC matrix materials. The hardness increases and fracture toughness decreases with decreasing NbC grain size for a given binder type and binder content. The binder plays a crucial role in shielding the stress field in front of a crack tip and as such improving the toughness [25]. Ductile failure of the Co, Ni or Fe binder is usually found on fractured NbC-Co/Ni/Fe interfaces. Since intermetallic phases such as Fe-Al and Ni-Al have an intrinsically low room temperature ductility, the Al doped cermets have accordingly a lower fracture toughness. For a given low hardness value around 1,250 kg/mm^2, the Ni bonded NbC cermets had the highest toughness of 12 MPa m$^{1/2}$, whereas the Fe and Co bonded cermets exhibited toughness values of only 6-7 MPa m$^{1/2}$. When hardness was increased to 1,500 kg/mm^2, the transition metal carbide (MeC) added NbC-Ni cermets still had the highest toughness of 8.5 MPa m$^{1/2}$, whereas the corresponding NbC-12 vol% (Fe-25 at% Al) cermet had a toughness of only 5 MPa m$^{1/2}$. Analysis of the microstructures and mechanical properties (Figure 9) reveals that the addition of Al and secondary carbides have a great impact on the final NbC grain size and microstructure as well as the concomitant increased hardness and reduced toughness.

TRIBOLOGICAL BEHAVIOR OF NbC-BASED HARD MATERIALS

Friction and wear are system dependent attributes of interacting surfaces. Therefore the tribological performance of carbides can only be reliably judged using specific tests simulating a defined tribo-contact condition under a wide range of operating conditions. Such simulations have been performed for many hard materials. Yet, for niobium carbide no reliable tribological data could be found in literature. Consequently an extensive test program in a wide range of operating conditions has been set-up to benchmark pure niobium carbide as well as NbC-based hardmetals and MMCs against established other hard materials.

The tribometers used for unidirectional sliding are proprietary developments of BAM and the details are disclosed elsewhere [26]. They comply with ASTM G99 (DIN 50324) and with DIN EN 1071-13:2010. The wear volumes of stationary and rotating specimen were calculated from stylus profilometry and the wear scar diameters were determined by using ASTM D7755-11 or DIN 51834-3. The wear rate k_v is defined as the ratio of volumetric wear to the product of normal load F_N and the sliding distance *s*. The coefficient of friction (CoF) and the total linear wear of both tribo-partners were recorded continuously. One test per combination of parameters was performed, because the testing philosophy at BAM is to screen over a wide range of operating conditions rather than doing repeated tests. Repetitions can be done, however, for specific test conditions. The tribological data of WC-, Cr_3C_2- and (Ti,Mo)(C,N)-based materials elaborated under the same test conditions were available from references [27,29,30] for benchmarking.

The frictional level of different NbC grades and the Fe_3Al-NbC MMC displayed in Figure 11 compares well with different WC-based or Cr_2C_3–based hardmetals or monolithic alumina [27]. At room temperature (RT), the friction of binderless HP-NbC1 increased with increasing sliding speed having values of around 0.3 at low sliding speed and reaching a level of 0.8 at high sliding speed similar to monolithic alumina, whereas all metal bonded NbCs presented an opposite trend. Friction decreased with increasing sliding speed. It thus qualifies NbC for traction and frictional applications such as clutches or brake rotors rather than for low friction bearings (Figure 11). Metal bonded NbC grades showed an opposite trend to pure HP-NbC as the coefficient of friction decreased with increasing sliding speed at RT (Figure 11) [31]. At low sliding speed the value can reach unity while at high sliding speed it drops to a level of 0.4. This behavior is quite similar to that of Cr_2C_3–based hardmetals. The friction behavior of the Fe_3Al-NbC MMC is very similar to that of the NbC-based hardmetals.

At 400°C friction of all NbC and hard metal grades decreased with increasing sliding speed but were on average lower as compared to WC grades. The friction of NbC grades at high sliding speeds was lowest at 400°C. Low friction at high sliding speed is a favorable property for cutting tools reducing cutting forces thus achieving a given cutting performance at reduced machine power.

In comparison to other hardmetals, binderless HP-NbC comprised a particularly high wear resistance especially at RT, which is more or less independent from sliding velocity (Figure 12) [27]. The wear resistance of HP-NbC as well as of Co- and Fe_3Al-bonded NbC hardmetals is highest at RT and high sliding velocities.

At RT the metal-bonded NbC grades displayed a rather constant evolution of the wear rate with increasing sliding speed (Figure 12) [2]. The RT wear rates of the MMC Fe_3Al-$NbC_{0,94}$ are slightly higher than those of the NbC-bearing hardmetals and for all NbC grades the wear rates decreased with increasing sliding speed, what is beneficial for cutting applications. It is also apparent, that metallurgically synthesized or sintered NbC bearing materials tribologically compete with WC and Cr_3C_2 bearing hardmetals or cermets. At 400°C, the cobalt bonded NbCs presented lower wear rates than the NbCs bonded with Fe_3Al or having such a matrix, because of the well-known effect of Co_3O_4 formed by tribo-oxidation lowering the wear rate. It was recently shown, that reduced $Nb_{12}O_{29}$ contributes to the wear resistance of NbC [32], especially under

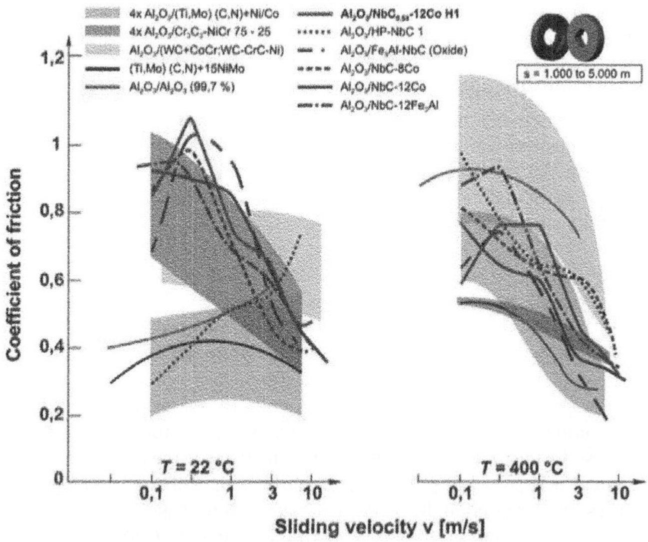

Figure 11: Coefficient of friction of NbC-based hardmetals and Fe_3Al-NbC (MMC) compared to different ceramics and hardmetals under dry friction at RT and 400°C

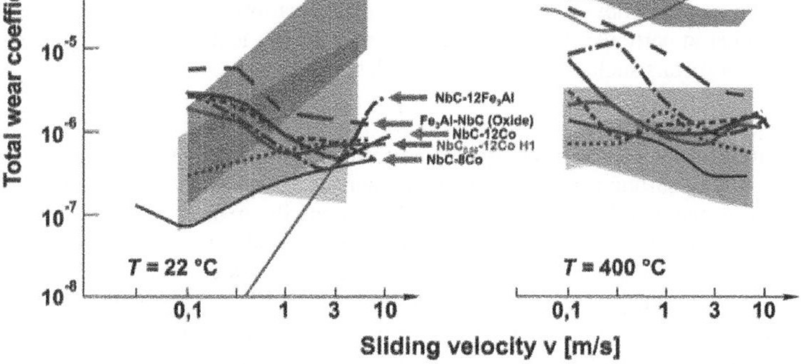

Figure 12: Total wear coefficients of NbC-based hardmetals and Fe_3Al-NbC (MMC) compared to different ceramics and hardmetals under dry friction at RT and 400°C

high sliding speeds at RT, where the wear rate of $Nb_{12}O_{29}$ at 7 m/s of k_v= 4.9 10^{-6} $mm^3/N{\cdot}m$ is close to that of NbC-based cermets.

SUSTAINABILITY CONSIDERATIONS

The data and knowledge on niobium carbide generated during recent years have indicated that this innovative but not yet very well known material has a large spectrum of possible applications. Niobium carbide is in many properties competitive to the established tungsten carbide and in some particular features clearly superior to it. More important, niobium carbide based materials offer sustainable solutions with regard to several aspects as follows:

Economic sustainability

Due to its comparably low density a given component manufactured from NbC has approximately half the mass in comparison to WC. That mass reduction is larger than the price difference between NbC and WC resulting in an effective cost saving when using NbC.

Niobium based raw materials (oxide, ferroalloys) are not noted on the London Metal Exchange and are not exposed to noteworthy price volatility. This price stability facilitates raw material budget planning and allows reliable cost calculation for finished products.

Niobium based raw materials mainly originate from mines in Brazil and Canada. The supply is stable and not exposed to political uncertainties or embargos. The available capacities and known reserves are large enough to cover expected future demands for NbC without constraints. The supply chain situation, the historical governance indicators and the explored global availability including the long-term geological availability (mined deposits and confirmed reserves) do not justify the inclusion of niobium in the list of critical raw materials.

Environmental sustainability

Mining operations at the premier niobium producer CBMM in Brazil, which stands for approximately 80 percent of the global niobium supply, fulfill the highest environmental standards. The mine was ISO 14001 certified in 1997 covering mining operations as well as production of all relevant niobium-based alloys and oxides.

No particular hazards or critical notifications for its oxide (Nb_2O_5) and carbide (NbC) were filed in the framework of REACH. Niobium carbide over a wide range of stoichiometric compositions appeared to have a biocompatibility comparable to Ti-6Al-4V, a widely used material for human body implants.

Niobium carbide allows the application of a large variety of binder metals. When the use of cobalt or nickel in particular application causes concerns, it is possible to deviate to uncritical iron-based binder metals such as for instance stainless steel.

Technical sustainability

Niobium carbide can be easily generated from standard grade niobium oxide by one-step direct reduction with carbon using the same equipment as for tungsten carbide production. Hence, no specific investment will be necessary at established powder manufacturers.

The low solubility of niobium carbide in binder metals and particularly iron facilitates manufacturing. In applications where iron-based materials have to be machined, NbC-based tools do not require a protective coating to diminish crater wear. This reduces cost and carbon footprint as compared to manufacturing WC-based tools, which do require a coating.

The close match of densities between NbC and iron-based alloys as wells as its low solubility allow efficient production of NbC reinforced materials by casting. Used components of that material can be easily recycled by re-melting. Thus, wear resistant components in any size and shape with a very high content of hard phase can be economically produced.

CONCLUSIONS

Research and development of niobium carbide-based hard materials started only recently, yet showing already a very high potential for many technical applications. Industrial pilots of these new materials are in progress, especially in areas where established tungsten carbide-based materials are causing economic, environmental or technical concerns. First results of NbC reinforced castings demonstrated already substantial lifetime improvements for components used in the mining industry. Recent benchmark trials with cobalt-bonded NbC hardmetals indicated favorable performance in high-speed machining of steel components.

REFERENCES

[1] H. O. Pierson, Handbook of Refractory Carbides and Nitrides : properties, characteristics, processing, and applications, Noyes Publications (1996), ISBN 0-8155-1392-5.

[2] M. Woydt and H. Mohrbacher, The tribological and mechanical properties of niobium carbides (NbC) bonded with cobalt or Fe_3Al, WEAR, 321 (2014), 1-7.

[3] European Commission, REACH Review 2012, Document 52013DC0049.

[4] N.N., Cobalt in Hardmetals and Cobalt Sulfate, Gallium Arsenide, Indium Phosphide and Vanadium Pentoxide, WHO-IARC Monographs on the Evaluation of Carcinogenic Risks to Humans, Vol. 86, 2006, ISBN 92 832 1286 X.

[5] M. Braic, V. Braic, M. Balaceanu, A. Vladescu, C. Zoita, I. Titorencu, V. Jinga and F. Miculescu, Preparation and characterization of biocompatible NbC coatings, Thin Solid Films , 04/2011; 519(12):4064-4068.

[6] M. Liedtke and M. Schmidt Rohstoffrisikobewertung – Wolfram DERA Rohstoffinformationen Nr. 19, Ausgabe: April 2014, ISSN: 2193-5319

[7] A. Kelly et al., J. Amer. Ceram. Soc., 1967, Vol. 50, No. 5, p. 253-256.

[8] A. Nino et al., Materials Transactions, Vol. 51, No. 9 (2010) pp. 1621 to 1626.

[9] M.-Y. Lee, High temperature hardness of tungsten carbide, Metallurgical Transactions A, Vol. 14A, 1983, p. 1625-1629.

[10] G. Asai, D.L. Paulson and H.J. Kelly, Properties of cast Columbium Carbide-Cast alloys, RI 7479, Bureau of Mines Report of Investigations, February 1971.

[11] Y. Kumashiro at al., J. of Materials Science (Letters), 15, 1980, p. 1321-1323.

[12] Dr. J. Nunn, private communication, National Physical Laboratory, 2015.

[13] H. Wiesenberger, W. Lengauer and P. Ettmayer, Acta Mater., Vol. 46 (2), 1998, pp. 651-666.

[14] A.S. Kurlov, A.I. Gusev, Int. J. of Refractory Metals and Hard Materials 46 (2014) 125–136.

[15] C.P. Kempter, E.K. Storms and R.J. Fries, J. of Chemical Physics 33, 1873 (1960).

[16] A. A. Rempel et al., Heat capacity of niobium carbide, phys. stat. sol. (a) 113, 353 (1989).

[17] A. Gusev and A. Rempel, Order and disorder of strongly non-stoichiometric compounds, 2001, ISBN 3-540-41817-2.

[18] L. Ramqvist, Jernkontorets annaler (Jernkont. Ann.) 152, (1968) 465.

[19] R. Edwards and T. Raine; "The solid solubility of some stable carbides in cobalt nickel and iron at 1250"; Pulvermetallurgie, Springerverlag, (1952), 232-242.

[20] J. P. Guha and D. Kolar, "Systems of Niobium Monocarbide with Transition Metals"; Journal of the Less-Common Metals, 29 (1972) p. 33-40.

[21] V. K. Lakshmanan and J. S. Kirkaldy, Solubility Product for Niobium Carbide in Austenite, Metallurgical Transactions A, Vol. 15A (1984) 541-544.

[22] E. J. Pavlina, J. G. Speer and C. J. Van Tyne, Equilibrium solubility products of molybdenum carbide and tungsten carbide in iron, Scripta Materialia 66 (2012) 243–246.

[23]S.G. Huang, L. Li, O. Van der Biest and J. Vleugels. Influence of WC addition on the microstructure and mechanical properties of NbC-Co cermets, J. of Alloys and Compounds, 2007; 430:158-164.
[24]S.G. Huang, O. Van der Biest, L. Li, J. Vleugels. Properties of NbC-Co cermets obtained by spark plasma sintering, Materials Letters 2007;61:574-577.
[25]M. Kotoul, On the shielding effec25t of a multiligament zone of a crack in WC-Co, Acta Mater 45, 1997;45:3363-3376.
[26]M. Woydt and K.-H. Habig, High temperature tribology of ceramics, Tribology International 22 (1989) 75-88.
[27]M. Woydt, Tribologie des céramiques (Tribology of ceramics), Encyclopédie Techniques de l'Ingénieur, Paris, March 2013, Tri 4 550, ISBN 978-2-85059-311-6
[28]M. Woydt, Application des données tribologiques des matériaux (The use of tribological quantities of materials), Encyclopédie "Techniques de l'Ingénieur, Paris, June 2012, BM 7 006v2, ISSN 2114-2300.
[29]L.-M. Berger, C. C. Stahr, S. Saaro, S. Thiele, M. Woydt and N. Kelling, Dry sliding up to 7.5 m/s and 800°C of thermally sprayed coatings of the TiO_2–Cr_2O_3 system and (Ti,Mo)(C,N)-Ni(Co), WEAR 267 (2009) 954-964.
[30]L.-M. Berger, S. Saaro, M. Woydt, WC-(W,Cr),C-Ni – the unknown hardmetal coating, Therm. Spray Bull. 1/08 (2008) 37-40.
[31]M. Woydt and H. Mohrbacher, Friction and wear of binderless niobium carbide, WEAR 306 (2013) 126–130.
[32]S.G. Huang, J. Vleugels, H. Mohrbacher and M. Woydt, Densification and tribological profile of niobium oxide, submitted to WEAR

Materials for Nuclear Applications and Extreme Environments

MICROSTRUCTURE OF YTTRIA DOPED CERIA AS A FUNCTION OF OXALATE CO-PRECIPITATION SYNTHESIS CONDITIONS

Brissonneau Laurent*, Mathieu Aurore, Tormos Brigitte, Martin-Garin Anna

CEA, DEN, DTN/SMTA/LIPC Cadarache F-13108 Saint-Paul-lez-Durance, France

*corresponding author : laurent.brissonneau@cea.fr

ABSTRACT

In sodium fast reactors (SFR), dissolved oxygen in sodium can be monitored via potentiometric sensors with an yttria doped thoria electrolyte. Yttria doped ceria (YDC) was chosen as a surrogate material to validate the elaboration process of such sensors. The material must exhibit high density and a fine grain microstructure to be resistant to the corrosion by liquid sodium and thermal shocks. Thus, the oxalic coprecipitation route was chosen to avoid milling steps that could bring impurity incorporation which is suspected to induce grain boundary corrosion in sodium. Powder and sintered pellets were analyzed by BET, XRD and SEM. The results show that synthesis conditions are of primary importance on the process yield, the oxalate powder microstructure and, eventually, on the ceramic density and microstructure. The impurity content was limited by controlling the synthesis, calcination and sintering steps.

INTRODUCTION

In sodium cooled fast reactors, the oxygen content in the sodium must be kept under a specific value (usually few ppm), as it controls the corrosion rate of the stainless steel pins [1]. The oxygen monitoring by potentiometric sensors can contribute to this control function, and are found to be very interesting because they are specific towards oxygen (when plugging indicators do not discriminate the insoluble impurities). In this context, the use of potentiometric sensors appears of great interest as it would bring a redundant, diversified and specific monitoring of the oxygen dissolved in sodium. In such sensors, the difference of oxygen activity in the sodium and in a reference induces a voltage drop according to Nernst's law. The reference and the sodium are separated by a solid electrolyte, which can be brazed on a metallic tube. The tightness between the outside and the inside of the sodium pipe is then easily performed on this metallic tube (the alternative being a long ceramic tube where tightness is performed via cooled sodium).The quality of the electrolyte material is critical to insure the adequate operating of the sensors in the reactor environment [2]. It should be compatible with sodium at high temperature (about 673K -400°C), mechanically resistant and a purely ionic conductor. Many oxygen sensors were developed in the 1970's and 1980's, but they suffer of their poor robustness. Actually, no more commercial oxygen sensors for sodium are available.

Yttria doped thoria was commonly used as the electrolyte material in potentiometric sensors for liquid sodium. Thoria is a stable oxide compatible with sodium and the incorporation of yttrium in its lattice produce oxygen vacancies (Y^{III} in a Th^{IV} lattice), making the ceramic become an ionic conductor. A highly dense fine grains high purity ceramic is generally required to resist to sodium corrosion and mechanical stress [3]. Fine grains dense materials will provide good mechanical properties to the ceramic, that might suffer from thermal shocks, in particular when they are inserted in the sodium. Also, dense material limits the sodium penetration and then its reaction with the electrolyte. The small grains allow the dilution of the impurities at the grain boundaries and then make them less sensitive to sodium attack (in particular for precipitated silica). It was found that oxalic co-precipitation can provide a suitable route for the elaboration of reliable and robust sensors [2], because good yttrium homogenization can be obtained as well as very small grains without any milling process that can induce pollution. Thus it was decided by the CEA to develop a similar process to produce such sensors in order to potentially test them in the ASTRID prototype reactor. The process is first developed on a surrogate material, in order to gain experience and to qualify the equipment, before working with the radioactive thorium material. This paper deals with the results obtained on yttria doped ceria obtained by oxalic co-precipitation. The final microstructure and the impurity incorporation are discussed as a function of the elaboration conditions.

EXPERIMENTAL PART

Oxide Synthesis, Pressing And Sintering

Cerium IV nitrate was used as a precursor in most of the syntheses with the aim of approaching the synthesis conditions with thorium (Th^{IV} is the only oxidation state for Th, when Ce^{III} or Ce^{IV} are both possible for cerium). The cerium nitrate $Ce(NO_3)_4$ is provided as a liquid solution of 1.5M ionic concentration (Alfa Aesar 13093-17-9, 99.99%). The cerium III nitrate $Ce_2(NO_3)_3$ used for the last synthesis (L) is in crystal form (Alfa Aesar 11329, 99.5%) as well as yttrium nitrate $Y_2(NO_3)_3$ (Aldrich 13773-79-8, 99.99%).

The cationic ratio is x=15%Y in all cases, the expected final composition is $Ce_{0,85}Y_{0,15}O_{1,925}$ ($x=n_Y/(n_{Ce}+n_Y)$= 15%, n number of moles). The nitrates are mixed in water; the cationic concentration of the solution is 1 M. The precipitation is obtained by pouring the cationic solution into oxalic acid, generally slightly in excess of stoichiometric proportion, reaction (1). The mixing of the nitrates in the acid solution can be performed drop by drop from a burette or by a rapid pouring. The concentration of the acid solution is slightly lower than the cationic one, as recommended in [4]. The pH value can be set at a higher value by adding ammoniac during the precipitation [4]. The variations in these steps (reactant concentrations, pH value, mixing rate…) are reported in Table I for the different syntheses.

$$2(1\text{-}x)\, Ce(NO_3)_4 + 2x\, Y(NO_3)_3 + 3H_2C_2O_4 \rightarrow Ce_{2-2x}Y_{2x}(C_2O_4)_3 + 6H^+ + (8\text{-}2x)NO_3^- \qquad (1)$$

In this mixed oxalate, both cerium and yttrium are at the oxidation state +III, no charge compensation is required. After few minutes of maturation, the oxalate precipitate is separated by ultracentrifugation, rinsed in water until a neutral pH is obtained, centrifuged, rinsed in ethanol and centrifuged once more. It is dried at 373K (100°C) during at least 4 hours. The oxalate is gently milled in a mortar (alumina, zirconia and agate were used as the mortar material) and calcined in air between 673K (400°C) and 973K (700°C) in alumina crucibles, according to reaction (2):

$$Ce_{2-2x}Y_{2x}(C_2O_4)_3 + (2\text{-}x/2)O_2 \rightarrow 2Ce_{1-x}Y_xO_{2-x/2} + 6CO_2 \qquad (2)$$

From test I the vessel was systematically cleaned in nitric acid and rinsed in ultrapure water. At each steps, the reactive and products are carefully weighed. Samplings for XRD analysis, SEM observations or thermogravimetric analysis are performed before and after calcination.

The oxide powder is gently milled in mortar and uniaxially pressed (manual hydraulic press SPECAC™) at 300MPa or 400MPa in a carbide three-shell mold. Green pellets have a 8-12mm large diameter with a height of about 1 mm. The mold surface is lubricated with few zinc stearate before each test.

The pellets are placed in an alumina boat inside a horizontal tubular furnace Nabertherm™ RHTH 120/300/18. The tube is in C799 alumina. The sintering atmosphere is flowing synthetic air. The synthesis, calcination and sintering conditions are reported in Table I.

Table I: Synthesis conditions, calcination and sintering of the yttrium doped ceria

Test reference	Acid concentration	Rapport. stœchiométrique acide/nitrate	pH	Type of pouring	Oxyde mass (g)	Temperature/time for calcination	Sintering conditions
A	0.15 M	R=1	6	Drop by drop	2 *3	673K / 4h 973K / 4h	1753K, 4 h
C	0.24 M	R=3	6	Drop by drop	2 *3	673K / 4h 973K / 4h	1773K, 4 h
D	1 M	R=3	6	Drop by drop	2 *3	773K / 4h 973K / 4h	1773K, 4 h
E_1	0.3 M	R=1.5	6	Maturation 1 hour	1		
E_2	0.3 M	R=1.5	6	Drop by drop	1		
E_3	0.3 M	R=1.5	3	Drop by drop	1	773K / 4h	1773K, 4 h
E_4	0.3 M	R=1.5	1	Drop by drop	1		
E_5	0.3 M	R=1.5	6	Direct	1		
E_6	0.6 M	R=3	6	Drop by drop	1		
F1	0.2 M	R=1.5	3	direct	3	773K / 4h	1773K, 4 h
F2			1		3		
G	0.2 M	R=1.0	6	direct	3*15	773K / 4h	1773K, 2 h
H	1 M	R=1.0	6	direct	3*15	773K / 4h 973K / 4h	1773K, 2 h
I	0.5 M	R=1.0	6	direct	2*15	773K / 4h	1773K 2 h, 4h
J	0.25 M	R=1.0	6	direct	3*15	773K / 4h	1773K, 4 h
K	0.2 M	R=1.0	5 à 6	direct	26*6	773K / 4h	1773K, 4 h
L	0.2 M	R=1.0	5 à 6	direct	30*12	773K / 4h	1773K 1h, 2h, 4h, 8h

Characterization

Direct observations of the surface of the pellets were usually performed, but some of them were also polished in order to observe the bulk microstructure. In the latter case, a thermal etching (1200°C 2h) is necessary to reveal the microstructure. The metallographic examinations were performed in a LaB_6 ZEISS™ EVO°HD15MA with backscattered electrons. It is equipped with a SDD SX50 energy dispersive spectrometer for chemical analysis. Image analyses were performed with the AnalySIS™ software.

XRD was performed in the θ-θ mode with a BRUKER™ D5000 diffractometer, BET with a Quantachrome Autosorb-1, using nitrogen as absorbed gas. The thermogravimetric analyses are carried out in a horizontal furnace Mettler Toledo™ TGA/SDTA851e thermobalance, gas analysis is performed with a micro chromatograph Agilent™ 2891a. Density determination was carried out with a Mettler-Toledo MS14S (±0,05 mg) equipped with a volumetric mass determination kit MS DNY-43, from a weighing in air (pellets dried at 373K - 100°C) and a weighing in pure ethanol. The accuracy (from measurement scattering) is ±0.04 $g.cm^{-3}$. It was found that pure ethanol satisfactorily penetrates in the ceria porosity. An incomplete penetration would induce an under-estimation of the density (as a part of the open porosity would be accounted as close porosity). As the open porosity is not taken into account in this method, it might lead to an over-estimation of the real density: the relevance of the density determination is then confirmed by the microscopic observation of the surface.

RESULTS

Synthesis Yield

The synthesis yield is obtained by the ratio of the cations in oxide form over the cations in the nitrate solution. The calculated values are reported in Table 2.

The oxalate precipitation was observed immediately after the pouring of the nitrate in the oxalic acid (reaction 2). The maturation time is usually of a few minutes, but for test with incomplete reaction, oxalate precipitation could be observed in the rinsing water several hours after the test (yellow coloration). Complete yields were obtained only when direct pouring was used, but it is not a sufficient condition. Also, it is necessary that the oxalic solution is diluted enough in order to avoid the temperature increase of the solution due to the exothermicity of the reaction (tests H, I and J) and to insure a correct control of the pH. It should be noted that complete yield can be obtained even if the oxalic acid is only slightly over stoichiometric (reaction 1).

Yields above 100% were obtained for batch of high quantity of powder (synthesis K 150g), which for an accurate monitoring was not required. When accurately monitored, the yields are rather close to 100% (rather 102%). A yield above 100% cannot be explained by the presence of impurities (Si, Al) as the last pellets contained few of them. It is rather attributed to water or CO_2 absorption. In fact two mass losses of about 1% are observed in TGA measurements for the oxide powder: one between 343K (70°C) and 393K (120°C), one between 423K (150°C) and 700°C (973K) (the latter being rather linear with time). Each time, CO_2 release is observed, but it accounts for less than about 10% of the total mass loss between 343K (70°C) and 393K (120°C) and 30% of the total mass loss between 423K (150°C) and 700°C (973K). Rather similar results are obtained under air and under Ar/H_2 3% (slightly higher mass loss under reducing atmosphere). Under air, few hydrogen is also released between 443K (170°C) and 643K (370°C). After sintering, the total mass loss is about 4%.

The synthesis with the cerium III nitrate is much easier. The pH is easier to adjust, the precipitation is instantaneous and the oxalate exhibits a white color (found only for the syntheses at pH=1 with Ce^{IV} nitrate). The yellow color of the oxalate obtained with Ce^{IV} at pH=6 could be induced by impurity co-precipitation.

Table II: test results: doping ratio x, reaction yield, oxide density, pellet grain size and pollutants

Test reference	final x (%)	Yield (%)	Density ($g.cm^{-3}$)	Density (% *vs* theoritical)	Grain sise (µm)	Impurity in pellets
A	14.0	70	<5.9			Si>Ca>>Al Rods
C	14.8	69	6.2-6.5	90-95	1..7 ± 0..7	P
D	14.9	76	6.1-6.4	89-93	0.8-1.8	No
E_1	15.0	76.4	6.4	93		
E_2	15.0	80.3	5.7			E_2 : Si>Ca>>Al
E_3	15.0	74.1	6.7 ?	97.5		Rodes + Al_2O_3 grains
E_4	15.0	88.2	6.7 (open)	?		
E_5	15.0	100.2	6.6	96	E_5 : 2 ± 1	E_5 : Al_2O_3 grains
E_6	14.9	56.6	5.9	86		
F	14.9	98.5 93.1	6.8 (open.) 6.3 (open++)	? < 92		No
G	15.1	96.1	6.77±0.02	98.6	bulk 1±0.4 surface 1.4±0.4	bulk Si>Ca>>Al rods + grains in & around porosity
H	15.1	100 66 70	6.77±0.02 6.5 (open++)	98.6	1.3 ± 0.4	P Plates
I	15.0	72-100	6.6	96		P. Si. Ca Dendrites around porosity P Plates
J	15.0	82-101	6.3-6.6 (open++)	92-96		oui
K	14.7-15.3	98-106	6.7	97.5	1.8 ± 0.7	P ; Al_2O_3 grains
L	15.0	97-110	6.76 ± 0.02 6.75 ± 0.02 6.75 ± 0.01 6.76 ± 0.02	98.4 -99.2	0.6-0.75±0.25 0.9-1.0 ± 0.4 1 -1.3 ± 0.4 1.5 - 1.65 ± 0.4	Si. Ca. Al in bulk around porosity

Powder Microstructure

For the oxalate, the hydration number per cerium unit at any pH was calculated to be between 1.5 and 5 from the weight measurements. The XRD peaks can correspond to the cerium oxalate hydrated 5 times (per cerium unit) but for some syntheses, the compounds could not be identified. It is likely that the oxalate can be dehydrated during the more or less long drying time at 100°C. No yttrium oxalate peaks were detected; it is then concluded that the co-precipitation leads to the formation of a mixed oxalate.

As only one Fm3m cubic phase is observed by XRD for the oxide powder, it can be concluded that an yttrium cerium mixed oxide is formed. The oxide crystallite size was estimated from the width of the peaks, using the Scherrer's formula. It varies from ≈10 nm (6-11 nm) for a calcination temperature of 400°C or 500°C, to ≈30 nm (26-32 nm) for a calcination temperature of 973K (700°C).

Some micrographies of the oxide powders are reported in Figure 1. Obviously, the powder is constituted by grains forming plates or parallelepipeds (not rectangular) of few micron size (so containing several crystallites) which are more or less agglomerated (the biggest agglomerates have often common centre). The grains size varies from 1 to 10 µm. The smallest grains, forming few agglomerates, were obtained for the syntheses performed at high pH (3 or 6) and rapid pouring (test E5, F1, G…). However, such microstructure was not obtained when the mixture was inadequate or when the temperature in the reactor rose, leading to bigger size grains and agglomeration (tests H and I).

It should be noted that oxide and oxalate powder grains are isomorphic. The calcination does not lead to the morphologic transformation of the powder.

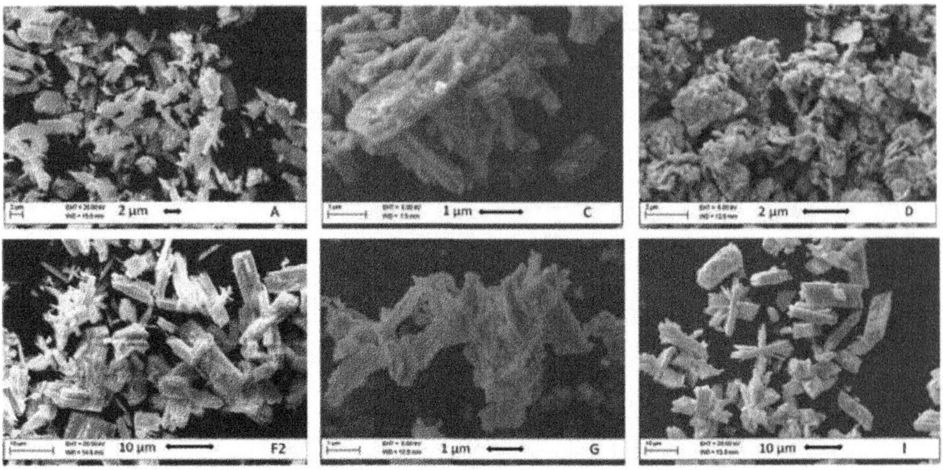

Figure 1 : SEM micrographies of the oxide powder in different synthesis conditions

The BET specific surface of the oxalates is lower than 10 $m^2.g^{-1}$, when it is larger than 20 $m^2.g^{-1}$, up to 85 $m^2.g^{-1}$, for the oxides. As the powder morphology is unchanged, it could be considered that the ratio of the specific surfaces should vary as $(\rho_Y^{2/3}.M_Y^{1/3})/(\rho_A^{2/3}.M_A^{1/3})$, where M holds for the molar mass and ρ for the density of the oxalate A and oxide Y respectively. According to this assumption the ratio of the specific surface of the oxalate over the oxide one should be 1.66, when it is lower than 0.5.
In fact, the calcination leads to the formation of a large porosity inside the grains, without changing their morphology. This surface specific increase is interpreted as the creation of a large mesoporosity induced by the release of the gaseous molecules (CO_2, CO) during the calcination [5].

For two powders calcined and synthesized in similar condition, excepted pH, the powder synthesis at pH=1 exhibits a bigger grains size and a lower specific surface (57 $m^2.g^{-1}$) than the one at pH=3 with a smaller grains size and a bigger specific surface (77 $m^2.g^{-1}$). The specific surface decreases as the calcination temperature is increased (22 and 85 $m^2.g^{-1}$ for the test powder C calcined at 673K (400°C) and 973K (700°C), respectively). This can be interpreted as following: the large mesoporosity created by the rapid release of the gas at 400-500°C gets progressively close or eliminated as the crystallite size increases at 700°C (intergranular sintering).

Pellet Microstructure

The highest pellets densities were obtained only with powders with a grain size lower than 2 µm (see Table II). For the dense pellets, the porosity is generally intergranular (see Figure 2). Some pellets were thought to be dense after the volumetric mass determination by double weighing, but it appears that a large interconnected open porosity can exist (tests E3, H, J).

The average grain size varies between 1 and 2 µm. It is smaller than the powder grain size, which proves that they are constituted of crystallites of smaller size. It was not possible to find a relation between the calcination temperature and the pellet grain size.

The yttrium over cations (x) ratio was checked by EDS. For x=15%, the results are systematically higher in the EDS analyses: between +1% and +30%, but especially on non-polished surface. On polished surface, the difference is less than +10% for x=15%. Then, EDS is not really adequate to check the scattering in yttrium concentration. The tentative to analyze the XRD peaks was not successful. A slight deviation of the cell parameter was observed, but its decrease is much lower than the reported values in the literature [6] (a=0,5410 nm for x=15% to be compared to 0.5406nm as proposed by Bryan Balazs et al., 0.5412nm for x=0%).

Pollution

The main impurities observed on or in the pellets are reported in Table II. This is illustrated in Figure 2, with micrographies of the most frequent type of pollutions.

In the first tests, many pollutants are observed on the surface, forming rods (50µm long, 2µm wide) or dendrites made of grains. They are composed of silicon and calcium oxide mixed with yttrium, aluminum and cerium. In some cases, phosphorus was also observed, and in some tests, alone in plates (sometimes in large quantities, in some particular locations at the surface).

After the test F aluminum, which was until then hardly observed, becomes one of the main detected impurity. Si and Ca are sometimes observed, but in lower quantity than in the previous syntheses. As the china mortar was replaced by an agate one and then by a zirconia or a pure alumina mortar, the milling could be one of the main causes of the large pollutions observed in the first tests.

No impurity could be detected in the powders by SEM, but the detection of impurity is not easy when very small grains are observed.

Sample G was cross section polished and observed by SEM. Some few large spherical or ellipsoidal pores (size between 10 and 100 µm) are observed with impurities at their edges and bottom (Figure 2). Impurity grains are oxides rich in aluminum (with yttrium) or in silicon (with cerium and few calcium). In one case, some SiO_2 was observed inside the porosity. It reacts during the sintering and formed a mixed cerium silicon oxide at the interface. This compound is also visible around the porosity. For the test G, the powder was milled in an agate (SiO_2) mortar.

However, silicon and calcium incorporations, as well as few phosphorus and aluminum are sometimes found in the next tests though: the powder was milled in a zirconia mortar, the vessel was systematically rinsed in nitric acid, ultrapure water was used and the alumina furnace tube was heated under vacuum at 1973K (1700°C) to volatilize residual impurities.

For the test L, the only observed impurities (SiO_2, P) could be attributed, according to their size and location, to residual contamination on the pressing heads of the pressing machine (due to their polishing).

But impurities were also found in the bulk by polished cross section, always at the vicinity of large spherical/ellipsoidal porosities. These impurities formed rods and grains containing Si, Ca and Al (with Y in the rods and Ce in the grains). The crystal growth appears to be oriented, and the apparition of this large porosity could be related to these pollutants.

An ultrapure Ce^{III} nitrate was used for the first of the 30 syntheses of test conditions L. In that case, the oxalate is easier to mill, white color, as well as the oxide pellets after sintering. A lower quality cerium was used for the second part of the syntheses: the oxalate is still white but the pellets are orange color after sintering.

Thus it is likely that there are many origins for the impurities: the systematic washing up of the vessel with nitric acid, rinsing with ultrapure water, milling with mortars more resistant to abrasion clearly improve the purity of the pellets. However, it is likely that the main origin of the impurity (at least Si and Ca) is the aqueous synthesis, at the higher pH, especially when Ce^{IV} nitrate is used. In such conditions, it is likely that the impurities come from the reactants or from the vessel and co-precipitate with the mixed yttrium cerium oxide. More specifically, the phosphorus could come from the yttrium nitrate, as the yttrium is extracted from phosphorus based compounds (monazite).

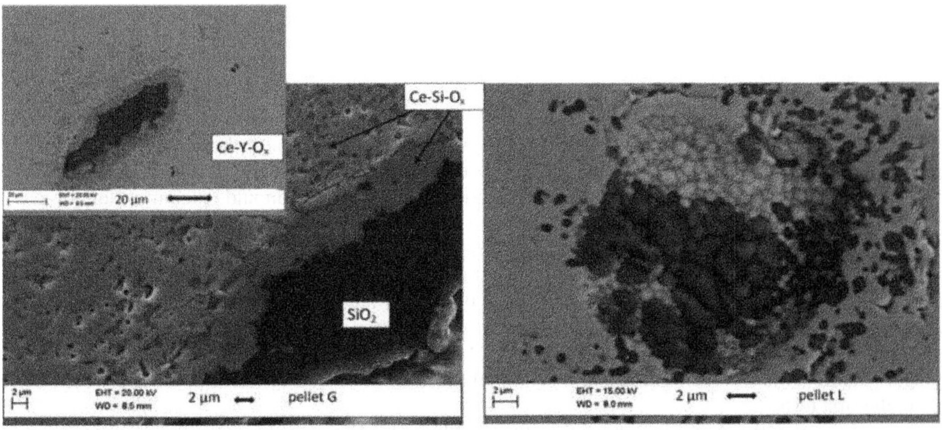

Figure 2 : Typical impurities observed on (left test G) and in (right test C) the pellets.

DISCUSSION

These results show that the aqueous synthesis must be tightly controlled in order to obtain the target microstructure and a complete yield. A small aggregates powder (about one micron) is necessary to obtain a highly dense ceramic. The elaboration of large quantities of a high quality powder is limited in a simple reactor by the control of the synthesis conditions, pH and temperature (for example in test H conditions, three syntheses, three different yields).

In order to obtain small size aggregates, the nucleation of the oxalate particles must be favored at the expense of their growth. In the classical theory of nucleation, the size of the critical nucleus determines the rate of nucleation (the smaller the faster); it is proportional to $\sigma/k.T.\ln S$, with σ the surface energy of the nucleus, k the Boltzmann's constant, T the temperature (in Kelvin) and S the supersaturation defined as the concentration in the solution over the equilibrium concentration. The direct pouring of the cationic solution favors the cation supersaturation in acid and then decreases the size of the critical nucleus. A high pH favors the dissociation reactions of $H_2C_2O_4$ (pK_{A1}=1.2, pK_{A2}=4.3) without being limited by the formation of a strong acid (HNO_3, displacement of reaction 2 to the right side). From pK_A values, it can be calculated that at pH=1, the ratio $[C_2O_4^{2-}]/[H_2C_2O_4]$ is $10^{-3,5}$, when it is $10^{6,5}$ at pH=6.

In large quantity batches, the pH was not controlled and the supersaturation was probably not high enough everywhere in the reactor. The increase of the temperature should have been beneficial at constant supersaturation S: it is likely that the high temperature increase changes the equilibria in a way that the supersaturation decreases.

Using a Ce^{III} nitrate, the precipitation of a Ce^{III} oxalate is easier than with a Ce^{IV} nitrate. Following the conclusions drawn from our experience with cerium (rapid pouring, control of the reaction conditions…), it could be expected that similarly large quantity of Th^{IV} oxalate could be rather easily obtained from a Th^{IV} nitrate. However, the co-precipitation of an oxalate mixture of cation at the oxidation IV (Th) and III (Y) will require charge compensation (necessity of H_3O^+ integration). Thus, a high pH synthesis could be less favorable for the thorium oxalate nucleation, as the displacement of the reaction towards the oxalic acid decomposition at high pH could be balanced by the necessity to keep a proton concentration high enough to compensate the charge of the oxalate. The pH value selected in most of our syntheses (pH=6) allows us to obtain the target yield and microstructure, but it is likely that it also induces the co-precipitation of impurities.

It is clear that all the elaboration steps can lead to the incorporation of pollution in the pellets. However, it is assumed that the main polluting phase is the synthesis step, though it is known that the co precipitation itself is a purification step. It is expected that the next syntheses in carefully washed Teflon vessel, with high purity reactant, would yield better results. Also the manual milling process will be limited to a minimum.

The formation of a liquid phase during sintering by the impurities (e.g. $CaO\text{-}SiO_2$) or with Y (Al, Si) or Ce (Si) could lead to the formation of the observed large bulk spherical porosities, sometimes filled with impurities, sometimes only covered by them, sometimes empty. The liquid state might enhance their diffusion from the bulk towards the surface by capillarity. Then, they would crystallize as rods, dendrites, etc., during the cooling. The particular reactivity of the impurities during the sintering should be investigated.

CONCLUSION

The tests allow us to define the process conditions to obtain yttria doped ceria with the microstructure that was aimed at: dense with fine grains. The main parameters to be controlled are the mixing rate, the pH and the synthesis temperature.

The volume of effluents by synthesis could also be estimated, which is important information to properly manage the radioactive (thorium) wastes. As it is necessary to have an oxalic acid diluted enough in order to obtain an almost complete yield, the volume of liquid waste will be quite large but with few radioactive content.

The parameters concerning the calcination, pressing and sintering appeared to be less important. Calcination at 773K (500°C) produces an oxide powder with nanocrystallites ($\approx$ 10 nm) and a high specific area.

It is not clear whether carbon is incorporated in large amounts or not in the powder, and the influence of the calcination temperature on it should be investigated.

It was highlighted by these tests that many sources of impurities can pollute the pellets. Impurities at the grain boundaries might be responsible for local attack of the ceramic by sodium. Considering this, the choice of a co-precipitation process that suppresses the energetic milling step, which might induce large pollution, seems particularly relevant.

REFERENCES

[1.] L. Brissonneau, New considerations on the kinetics of mass transfer in sodium fast reactors: An attempt to consider irradiation effects and low temperature corrosion, *Journal of Nuclear Materials*, **423**, 67 (2012).

[2.] R. Thompson, R. G. Taylor, R. C. Asher, C. C. H. Wheatley and R. Dawson, in *LIMET Liquid Metal Technology*, Richland, USA (1980).

3. R. Ganesan, S. Vivekanandhan, T. Gnanasekaran, G. Periaswami and R. S. Srinivasa, Novel approach for the bulk synthesis of nanocrystalline yttria doped thoria powders via polymeric precursor routes, *Journal of Nuclear Materials*, **325**, 134 (2004).

4. J. Van Herle, T. Horita, T. Kawada, N. Sakai, H. Yokokawa and M. Dokiya, Oxalate coprecipitation of doped ceria powder for tape casting, *Ceramics International*, **24**, 229 (1998).

5. N. Hingant, N. Clavier, N. Dacheux, N. Barre, S. Hubert, S. Obbade, F. Taborda and F. Abraham, Preparation, sintering and leaching of optimized uranium thorium dioxides, *Journal of Nuclear Materials*, **385**, 400 (2009).

6. G. Bryan Balazs and R. S. Glass, ac impedance studies of rare earth oxide doped ceria, *Solid State Ionics*, **76**, 155 (1995).

HIGH TEMPERATURE CORROSION OF STRUCTURAL ALLOYS IN MOLTEN Li_2BeF_4 (FLiBe) SALT

Guiqiu Zheng*, David Carpenter*, Lin-Wen Hu*, and Kumar Sridharan**
**Nuclear Reactor Laboratory, Massachusetts Institute of Technology, Cambridge, MA 02139
**Department of Engineering Physics, University of Wisconsin-Madison, WI 53706
gqzheng@mit.edu

ABSTRACT

In support of structural material development for the fluoride salt-cooled high-temperature reactors (FHRs), the out-of-reactor and in-reactor corrosion tests of Hastelloy N® and 316L stainless steel were carried out in the potential primary coolant, molten Li_2BeF_4 (FLiBe) at 700°C for 1000 hours. With specially designed facilities for safety and maintenance of salt purity, the out-of-reactor and in-reactor corrosion tests were successfully conducted under identical conditions in both metal-lined graphite crucibles and bare graphite crucibles. The corrosion behavior of the alloys in molten salt and the effects of graphite and radiation on the alloys' corrosion behavior were analyzed through microstructural evaluations and post-irradiation examinations (PIE). Various corrosion test conditions resulted in different levels and types of corrosion attack including intergranular corrosion, carbides phase formation, and the development of porous surface structures. Overall comparisons suggest that both effects of graphite and neutron radiation accelerate the corrosion rate 2 to 10 times for the two tested alloys. Based on these evaluations a thermal diffusion controlled model has been developed to roughly predict the long-term corrosion of Cr-containing alloys in high-temperature molten FLiBe salt.

INTRODUCTION

The success of the Molten Salt Reactor Experiment (MSRE) at the Oak Ridge National Laboratory (ORNL) in the period 1950s-70s has rekindled the interested in molten salt cooled reactors. The most recent form of this type of reactor, the Fluoride salt-cooled High-temperature nuclear Reactors (FHRs) is emerging as a leading reactor concept for the next generation nuclear reactors[1–3]. Unlike the MSRE, the recent FHR design proposes to use non-fuel bearing Li_2BeF_4 (FLiBe) molten salt as primary coolant with TRistructural-ISOtropic (TRISO) fuel pebbles submerged in this coolant in order to provide a number of potential benefits such as low spent fuel, high thermal efficiency, and a high degree of passive safety[4–6]. Two alloys, nickel-based Hastelloy N® and 316 stainless steel, are being actively considered for structural components for FHR vessel in which the intended outlet temperature is expected to be 700°C[7–9]. In support of materials development for FHR, the corrosion behavior of these two alloys in FLiBe salt must be well understood for precisely evaluating alloys performance in such extreme environment and for extended duration.

Hastelloy N® (UNS N10003 in the unified numbering system) was originally developed during the MSRE program specifically for a combination of corrosion resistance in molten fluoride salts and oxidation resistance on hot air-side[10–13]. Most corrosion investigations for this alloy were conducted in fuel-bearing molten fluoride salts used in the MSRE program. A long-term corrosion experiment of Hastelloy N® in a flow loop showed a Cr-depleted attack depth of 100 µm in the hot section, and an oxide layer of approximately 50 µm in thickness in the outer hot air-side after 9.2 years continuous running in the temperature range of 560°C to 700°C[11]. This corrosion test and other investigations suggested that Hastelloy N® has good compatibility with molten fluoride salts[10]. However, Hastelloy N® is not code certified in the ASME Boiler Pressure Vessel Code under section I, III, B31.1, and B31.3 and therefore extensive testing is required for its application in a commercial nuclear reactor[14]. Meanwhile, 316 stainless steel is being considered because of its existing code certification at the designed FHR operation temperatures. Corrosion experiments of 316 stainless steel in FLiBe have been conducted in a thermal-convection loop at ORNL in 1970s for selecting first-wall and blanket materials for fusion reactors[15,16]. Results showed that 316 stainless steel had only 10 μm/year corrosion rate in as-received FLiBe salt at 650°C. This corrosion rate was significantly reduced by the addition of beryllium metal into molten salt. This suggests that 316 stainless steel has acceptable corrosion resistance in molten FLiBe, particularly when the salt redox potential is controlled. Nuclear graphite has been selected as a main in-core material for the FHR because of its high temperature strength and chemical resistance to high-temperature molten fluoride salts. Recent research has shown that the presence of graphite in molten fluoride salts (FLiBe and FLiNaK) affects the corrosion behavior of structural alloys[8,9,17–19].

In this study, the static high-temperature corrosion tests of 316 stainless steel and Hastelloy N® in molten FLiBe in graphite crucibles and metal-lined crucibles were accomplished at the University of Wisconsin-Madison. In-reactor corrosion tests of both alloys in identical crucibles and salt were conducted in the research nuclear reactor at Massachusetts Institute of Technology. The post-corrosion test examinations to evaluate alloys' corrosion performance in molten FLiBe at 700°C included weight change measurements and microstructural analysis of the near surface regions of the materials, and supported by a data based prediction model for long-term corrosion attack depth.

EXPERIMENTAL METHOD

Salt and Alloy Samples Preparation

The ^{7}Li-enriched Li_2BeF_4 (FLiBe) salt for this study was provided by ORNL from the MSRE program. This salt was re-melted for the present study and further purified by purging H_2-HF gas mixture. Details of the purification process have been described previously[8,9,20].

Commercial purity, 316L stainless steel and nickel-based Hastelloy N® are selected in this study for the corrosion tests in molten FLiBe. Both alloys were procured in sheet form from commercial vendors in the U. S. and their nominal chemical compositions are listed in Table 1. Prior to loading samples into the molten FLiBe crucibles, samples' surface was mechanically polished to a 1200-grit final finish with silicon-carbide paper, and cleaned in deionized water and acetone.

Table 1: Nominal chemical compositions 316L stainless steel and Hastelloy N® (wt.%)

Alloy	Vendor	Weight percent (wt.%)								
		Ni	Fe	Cr	Mn	Mo	Si	C	Cu	others
Hastelloy N®	HAYNES International	71**	5*	7	0.80*	16	1*	0.08*	0.35*	Co=0.20* W=0.50* Al+Ti=0.35*
316L stainless steel	North American Stainless	10.03	68.81**	16.83	1.53	2.01	0.31	0.02	0.38	N=0.05 P=0.03

* Maximum, ** As balance

Out-of-Reactor Corrosion Tests

Tests were conducted in graphite crucibles and metal-lined graphite crucibles. For testing 316L stainless steel the liner material was 316 stainless steel and for Hastelloy N® the liner material was nickel. To ensure that liners were free of leakage, 10psi pressurized gas was applied into the liners while immersing them in water. FLiBe salt was precisely filled into each crucible using a specially designed molten salt dropping system, following which 316L stainless steel and Hastelloy N® coupons were suspended in the molten salt using 316 stainless steel and pure nickel wires, respectively. The wires were attached to the top graphite lid. Then the cylindrical graphite crucibles were then transferred into a ceramic furnace in glove box to initiate out-of-reactor corrosion tests. During corrosion tests, the temperature in ceramic furnace was controlled at 700±2°C, and the moisture and oxygen in glove box were monitored online and maintained below 0.7ppm and 10ppm, respectively.

In-Reactor Corrosion Tests

In-reactor corrosion tests of the two alloys in molten FLiBe were carried out in the MITR, a 6 MW, light water-cooled, heavy water-reflected tank-type, nuclear research reactor at the MIT Nuclear Reactor Laboratory (MIT-NRL). To accommodate the free space available in-core (5cm in diameter, 56cm in height) for these tests, the use of specially designed facilities was needed, such as S-bend shape titanium thimble and an outer nickel capsule[21]. Corrosion temperature was achieved by utilizing the gamma heating of high-Z materials, which was mainly controlled by reactor power and capsule design. Temperature control (700±3°C) was achieved by adjusting the gas mixture in the gap between nickel capsule and thimble. More details about the in-reactor corrosion tests are described in references[21,22].

Post-Corrosion Test Examination

After 1000 hours static corrosion testing in furnace, the tested samples were cleaned in deionized water and dried in air, as shown in Figure 1 (a-d). After finishing the 1000-hour test in MIT reactor, the S-shape thimble was transferred into a hot cell, and then the corrosion assembly was disassembled in the hot cell as it cooled down. The cylindrical graphite crucible was transferred into a specially designed glove box for re-melting salt. Alloy samples were pulled out and cleaned using the same procedures as used for the out-of-reactor tests, as shown in Figure 1 (e-h). The photos of post-irradiation samples were captured by using a remote controlled optical microscope due to the high radioactivity from activation products, primarily ^{58}Co and ^{60}Co.

The dimensions and weights of all tested samples were measured by using a digital caliper and balance. The weight change was calculated using the following equation:

$$\Delta W = \frac{W_1 - W_0}{S_0} \qquad \text{Eq.(1)}$$

where W_1 and W_0 is the weight of post-corrosion and pre-corrosion respectively, and S_0 is the surface area of pre-corrosion sample.

The microstructure of the tested samples was characterized by using x-ray diffraction (XRD) and scanning electron microscope (SEM) in conjunction with energy dispersive spectrometry (EDS). Focused ion beam (FIB) technique was applied to prepare samples for transmission electron microscopy (TEM) analysis. The main analyses for out-of-reactor corrosion tested samples were completed, and the analysis is ongoing for the in-reactor corrosion tested samples.

Under the assumption of Cr diffusion controlled corrosion for Cr-containing alloys in molten fluoride salts, the effective diffusion coefficient (D_{eff}) of Cr through alloy matrix was obtained from the following formula[12]:

$$W_0 - W_1 = 2S_0C_{0,Cr}\sqrt{\frac{D_{eff}t}{\pi}} \qquad \text{Eq.(2)}$$

where $C_{0,Cr}$ is the initial Cr concentration in as-received alloys, and t is corrosion time.

Based on the calculated D_{eff} value the Cr depletion profiles in the near surface region of tested alloys were calculated by using a solution of Fick's second law[9]:

$$C_{Cr}(x,t) = C_{0,Cr}\,\mathrm{erf}\left(\frac{x}{2\sqrt{D_{eff}t}}\right) \quad \text{Eq.(3)}$$

where x is the distance from alloy surface.

RESULTS AND DISCUSSION

Visual Observation

All tested samples' surface was visually inspected after cleaning in deionized water. Figure 1 shows the tested 316L stainless steel and Hastelloy N® samples after both out-of-reactor and in-reactor corrosion testing in molten FLiBe in metal-lined crucibles and graphite crucibles. During cleaning in deionized water, some rust formed on the 316L stainless steel samples' surface, particularly on the samples tested in the graphite crucible. This qualitatively indicates that the oxidation resistant element Cr was significantly depleted from the near surface region of the 316L stainless steel samples. No significant change was observed on the Hastelloy N® samples' surface. For the samples tested in reactor, the surface is relatively rough compared to the out-of-reactor tested samples, especially for the 316L stainless steel samples tested in graphite crucible. This probably suggests that neutron radiation accelerates the corrosion attack to the alloys tested in molten FLiBe salt.

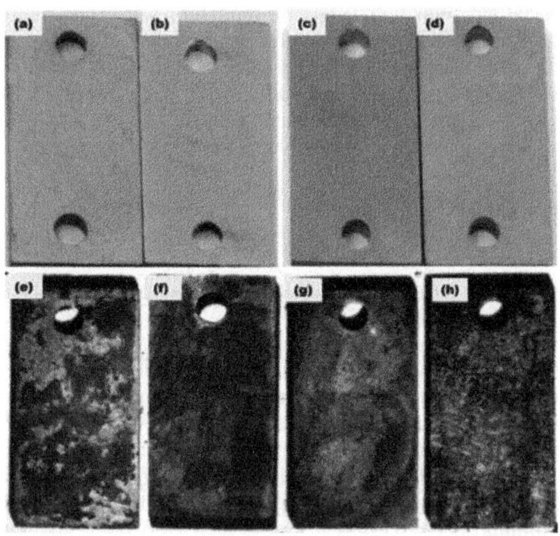

Figure 1: Out-of-reactor corrosion tested samples: (a) 316L stainless steel in graphite crucible, (b) 316L stainless steel in 316 stainless steel crucible, (c) Hastelloy N® in graphite crucible, and (d) Hastelloy N® in nickel crucible; and in-reactor corrosion tested samples: (e) 316L stainless steel in graphite crucible, (f) 316L stainless steel in 316 stainless steel crucible, (g) Hastelloy N® in graphite crucible, and (h) Hastelloy N® in nickel crucible, in molten FLiBe salt for 1000 hours.

Weight Change

The weight change per unit area of the out-of-reactor and in-reactor corrosion tested 316L stainless steel and Hastelloy N® in molten FLiBe at 700°C for 1000 hours was calculated using Equation (1), as shown in Figure 2. The samples tested in a graphite crucible lost more weight than ones tested in a metal-lined crucible, and the samples tested in-reactor in the same salt conditions lost more weight than out-of-reactor tested samples. The weight gain for the out-of-reactor tested Hastelloy N® in a graphite crucible results of the carbide phase formation due to the infusion of carbon from graphite into the alloy via the molten salt medium[9].

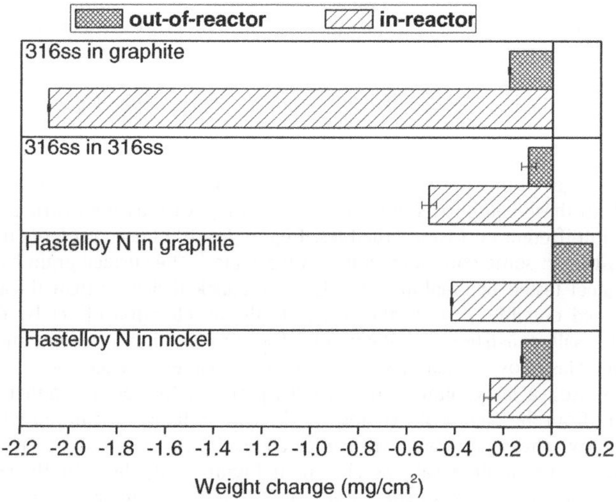

Figure 2: Weight change of the 316L stainless steel and Hastelloy N® coupons in out-of-reactor and in-reactor corrosion tests in molten FLiBe salt at 700°C for 1000 hours.

The mean weight changes of each type corrosion test samples are listed in Table 2. The neutron radiation results in approximately 10 times more weight loss for the 316L stainless steel samples in graphite crucible, and approximately 5 times more weight loss for the samples in 316 stainless steel lined crucible. Radiation also causes about 2 times more weight loss for the Hastelloy N® in nickel lined crucible.

Table 2: Mean weight change of 316L stainless steel and Hastelloy N after out-of-reactor and in-reactor corrosion tests in molten FLiBe for 1000 hours (unit: mg/cm^2).

Samples/crucible materials	Out-of-reactor	In-reactor
316ss in graphite	-0.18	-2.09
316ss in 316ss	-0.10	-0.51
Hastelloy N in graphite	0.17	-0.42
Hastelloy N in nickel	-0.13	-0.26

In addition to the fluorination of Cr, the chromium carbides formation on alloys' surface is another driving force to the Cr depletion from alloys, resulting in weight loss[8,9]. The infusion of carbon into Hastelloy N® and then eventual formation of carbides phase particles appears to the reason for weight gain in this alloys. This will be further detailed in the next section on microstructural observation. Since the Cr depletion preferentially along grain boundaries results in weight loss, it is reasonable to presume that the radiation (mainly neutron flux and gamma ray) stimulates Cr outward diffusion at 700°C, and also activates the carbon in molten salt to react with the outwardly diffused Cr at salt-alloy interface. Moreover, it is likely that the radiation induces Cr outward diffusion faster for high Cr alloy (17 wt.% in 316L stainless steel) than low Cr alloy (7 wt.% in Hastelloy N®) during corrosion in molten FLiBe. To fully understand the radiation effect on both alloys' corrosion behavior in molten FLiBe, other post-irradiation examinations such as microstructural analyses by using SEM, EBSD, and TEM are needed. These analyses are challenging because of the radioactivity of the tested alloys.

Microstructure of Corrosion Surface

The surface images of the out-of-reactor corrosion tested samples are shown in Figure 3. The corrosion attack to these samples is evident by presenting various microstructural features for different alloys in different corrosion crucibles. Figure 3(a, b) clearly presents the intergranular corrosion in addition to some corrosion attack on the grains. The surface grain boundaries of the 316L stainless steel tested in graphite crucible were attacked deeper than those tested in 316 stainless steel lined crucible. This further suggests the acceleration effect by the presence of graphite in molten salt, consistent with the trends of weight loss measurements shown in Figure 2. The surface of the Hastelloy N® samples also shows conclusive evidences of corrosion attack. As shown in Figure 3(c), a large number of carbides particles formed and adhered on the alloy surface, resulting from the chemical reactions of the carbon liberated from graphite with the Cr and Mo in the alloy. These particles were identified as Cr_3C_2, Cr_7C_3, Mo_2C by XRD analysis[9]. Without graphite in the molten salt, as shown in Figure 3(d), the initially polished surface exhibited an increase in surface roughness and developed a porous structure layer on surface. The dissolution of alloying elements, most likely Cr, from the alloy into the molten salt resulted in the porous surface. We will conduct the same microstructural observation under SEM for all in-reactor corrosion tested samples to study neutron radiation effects on the corrosion.

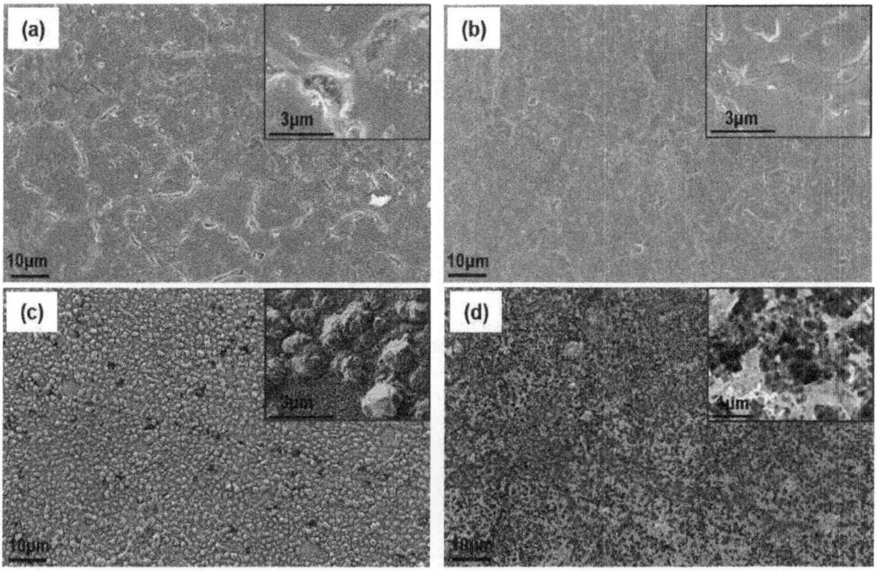

Figure 3: SEM images of the corrosion surface of (a) 316L stainless steel tested in graphite crucible, (b) 316L stainless steel tested in 316 stainless steel crucible, (c) Hastelloy N® tested in graphite crucible, and (d) Hastelloy N® tested in nickel crucible after out-of-reactor corrosion tests in molten FLiBe for 1000 hours. The inset in each SEM image shows a local microstructure with high magnification.

Prediction of Corrosion Attack Depth

Based on the understanding of out-of-reactor corrosion tested alloys, a thermal diffusion controlled corrosion model has been developed for the Cr-containing alloys in high-temperature molten FLiBe in metal lined crucibles because their weight loss mainly resulted from the Cr depletion from the alloys matrix[9,23]. Using Equation (2), the effective diffusion coefficient (D_{eff}) for out-of-reactor testing at 700°C was calculated to be $8.72x10^{-19}$ m^2/s for the Hastelloy N® and $1.22x10^{-19}$ m^2/s for the 316L stainless steel. For in-reactor testing the values are $3.83x10^{-18}$ m^2/s and $3.12x10^{-18}$ m^2/s for the Hastelloy N® and 316L stainless steel, respectively. These D_{eff} calculations suggest that neutron radiation stimulates Cr outward diffusion from alloys during corrosion in high-temperature molten FLiBe. Additionally, the D_{eff} of 316L stainless steel is lower than the Hastelloy N® for both out-of-reactor and in-reactor corrosion tests.

By substituting the calculated D_{eff} into Equation (3), the Cr concentration profiles $C_{Cr}(x, t)$ of the tested alloys in metal lined crucibles for 1000 hours are shown in Figure 4. These calculations neglect the preferential Cr depletion along grain boundaries, and assume homogeneous Cr distribution within the alloys. In our previous study, this prediction was validated by the experimental data for out-of-reactor corrosion tested alloys[9,23]. Therefore, based on the calculations, it is reasonable to expect that the corrosion attack depth is approximately 13.5μm for the 316L stainless steel and Hastelloy N® samples tested molten FLiBe for in-reactor testing for 1000 hours. From these profiles, the neutron radiation increases the corrosion attack depth by 10.5μm for 316L stainless steel, and by 5.5μm for Hastelloy N®. Other long-term corrosion of the Cr-containing alloys in high-temperature molten fluoride salts can also be predictive by using this model.

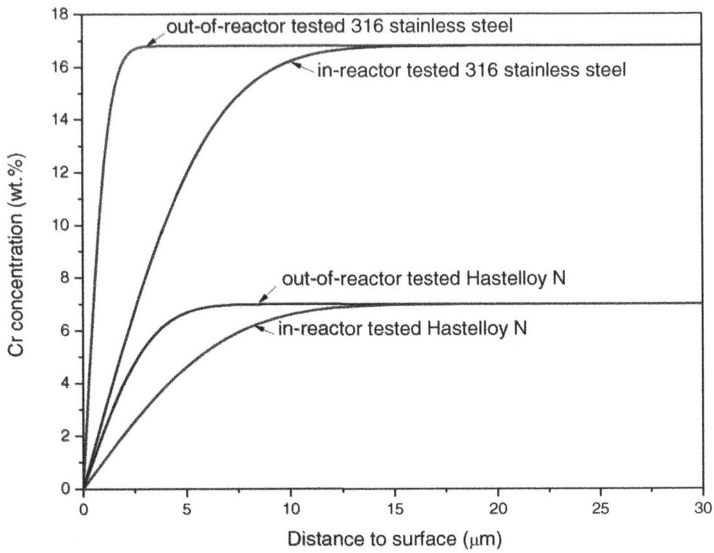

Figure 4: Predicted Cr concentration profiles of the alloys tested in molten FLiBe at 700°C in metal lined crucibles in out-of-reactor and in-reactor testing for 1000 hours.

CONCLUSIONS

Static corrosion tests of 316L stainless steel and Hastelloy N® were conducted in molten FLiBe salt at 700°C for 1000 hours in graphite and metal-lined crucibles both out-of-reactor and in a radiation field in the MIT reactor. The effects of graphite and neutron radiation on corrosion behavior were evaluated by microstructural analysis after the respective corrosion tests. Both graphite and the neutron radiation accelerated corrosion. The manifestations of the corrosion attack included intergranular corrosion, carbides phase deposition, and the formation of porous structure on tested alloys' surface. Additionally, based on the understanding of out-of-reactor corrosion tested alloys, a thermal diffusion controlled corrosion model has been developed and applied for predicting the corrosion attack depth of in-reactor tested alloys. From these predictions, the neutron radiation increases the corrosion attack depth by 10.5μm for the 316L stainless steel and by 5.5μm for the Hastelloy N® in molten FLiBe at 700°C in metal-lined crucibles.

REFERENCES

1. Robertson, R. C. MSRE Design and operations report : Part I - Description of reactor design. Oak Ridge National Laboratory Reports (1965).
2. Forsberg, C. W., Peterson, P. F. & Pickard, P. S. Molten-Salt-Cooled Advanced High-Temperature Reactor for Production of Hydrogen and Electricity. in Nuclear Technology, American Nuclear Society 1–25 (2003).

3. Allen, T. R. et al. Fluoride-Salt-Cooled, High-Temperature Reactor (FHR) Integrated Research Project Workshop 1. (2013).
4. Delpech, S., Cabet, C., Slim, C. & Picard, G. S. Molten fluorides for nuclear applications. Mater. Today 13, 34–41 (2010).
5. Williams, D. F. & Clarno, K. T. Evaluation of salt coolants for reactor applications. Nucl. Technol. 163, 330–343 (2008).
6. Khokhlov, V., Ignatiev, V. & Afonichkin, V. Evaluating physical properties of molten salt reactor fluoride mixtures. J. Fluor. Chem. 130, 30–37 (2009).
7. Flanagan, G. F., Holcomb, D. E. & Cetiner, S. M. FHR Generic Design Criteria, ORNL/TM-2012/226. (2012).
8. Zheng, G. et al. Corrosion of 316 stainless steel in high temperature molten Li_2BeF_4 (FLiBe) salt. J. Nucl. Mater. 461, 143–150 (2015).
9. Zheng, G. et al. High Temperature Corrosion of UNS N10003 in Molten Li_2BeF_4 (FLiBe) Salt. Corrosion 71, 1257–1266 (2015).
10. Ren, W. J., Wilson, D. F., Muralidharan, G. & Holcomb, D. E. Considerations of Alloy N for fluoride salt-cooled high-temperature reactor applications. in ASME 2011 Pressure Vessels and Piping Division Conference 1–12 (2011).
11. Koger, J. W. Evaluation of Hastelloy N alloys after nine years exposure to both a molten fluoride salt and air at temperatures from 700 to 560°C, ORNL-TM-4189. (1972).
12. DeVan, J. H. & Evans, R. B. Corrosion behavior of reactor materials in fluoride salt mixtures, ORNL-TM-0328. (1962).
13. McCoy, H. E. & McNabb, B. Intergranular cracking of INOR-8 in the MSRE, ORNL-4829. (1972).
14. White, H. J. Haynes International Capabilities: HASTELLOY N alloy, ORNL. (HAYNESS, International, 2010).
15. Keiser, J. R., DeVan, J. H. & Manning, D. L. The Corrosion Resistance of Type 316 Stainless Steel to Li_2BeF_4, ORNL/TM-5782. (1977).
16. Keiser, J. R., DeVan, J. H. & Lawrence, E. J. Compatibility of molten salts with type 316 stainless steel and lithium. J. Nucl. Mater. 86, 295–298 (1979).
17. Olson, L. C. Materials corrosion in molten LiF-NaF-KF eutectic salt. (PhD dissertation, University of Wisconsin-Madison, 2009).
18. Sellers, R. S. et al. Materials corrosion in molten LiF-NaF-KF eutectic salt under different reduction-oxidation conditions. in Proceedings of ICAPP (2012).
19. Cheng, W. J. et al. Zirconium effect on the corrosion behavior of 316L stainless steel alloy and Hastelloy-N superalloy in molten fluoride salt. Nucl. Technol. 183, 248–259 (2013).
20. Kelleher, B., Zheng, G., Anderson, M., Sridharan, K. & Cao, G. Purification of Non Uranium Bearing Fluoride Salts for Nuclear Applications. in Transactions of the American Nuclear Society 109, 1079–1081 (2013).
21. Carpenter, D. M., Ames, M., Kohse, G., Ostrovsky, Y. & Hu, L. Fluoride Salt High-Temperature Reactor Materials Irradiation Test at the MIT Research Reactor. Proc. ICAPP 14285 (2014).
22. Carpenter, D., Ames, M., Kohse, G., Ostrovsky, Y. & Hu, L. Findings of the Second Round of Fluoride Salt High Temperature Reactor Materials Irradiation Tests At the MIT Research Reactor. Proc. ICAPP 15105 (2015).
23. Zheng, G. Corrosion Behavior of Alloys in Molten Fluoride Salts. (PhD dissertation, University of Wisconsin-Madison, 2015).

CRACK INITIATION DUE TO LIQUID METAL EMBRITTLEMENT FOR THE STEEL T91 AND TWO ODS STEELS IN LIQUID LEAD

L.Rozumová [1], F. Di Gabriele [1], A. Hojná [1], and H. Hadraba [2]

[1]Centrum vyzkumu Rez (CVR), Hlavni 130, Husinec-Rez, 250 68, Czech Republic
[2]CEITEC IPM, Institute of Physics of Materials, AS CR, Brno, Czech Republic
Email: lucia.rozumova@cvrez.cz

ABSTRACT

Ferritic–martensitic steels and oxide dispersion strengthened steels are materials intensively studied because of their potential to be used for construction of Gen IV fast nuclear reactors. Steels were developed for fast breeder reactors (high-temperature strength and swelling resistance under irradiation).

In this study the effect of liquid lead on the crack growth rate of the steels was evaluated. Steels T91, MA957, PM2000 were used to evaluate their interaction with liquid lead under stress corresponding to maximum requested design condition. Liquid Metal Embrittlement (LME) initiation tests were conducted on pre-loaded specimens in flowing liquid metal loops. The experiments were carried out at 400°C and post-test evaluation was performed. Some specimens were pre-treated with a chemical flux, to induce wetting of Pb on the steel; some were polished or ground. Liquid Metal Embrittlement was not observed on the surface of any of the pre-loaded specimens after the exposure in liquid lead at 400°C for 500 hours. The results are discussed and compared with previous accelerated tests.

INTRODUCTION

The concepts for Generation IV nuclear reactors foresee operating temperatures beyond those realized in the current fleet of nuclear plants, 510–550°C for the supercritical water-cooled reactor (SCWR), the sodium-cooled (SFR) or lead-cooled fast reactor (LFR), or up to 1000°C for the very-high temperature gas-cooled system (VHTR)[1,2] and coolant media which have very different properties compared to water Therefore, the interaction of materials with aggressive media at high temperature is a strong concern in the research community.

Ferritic or martensitic steels with 9–14 wt. % chromium (Cr), that exhibit swelling-resistance under irradiation superior to austenitic steels. The T91 steel is among the candidate materials for fuel cladding, wrapper and near-core internals in Pb-cooled fast reactors.

This is especially true for the oxide dispersion-strengthened (ODS) steels currently developed, showing improved mechanical strength at temperatures beyond the application limit of the respective conventional steels[3,4].

Conventionally, ODS alloys have fallen into one of two groups – content of Cr, based on their intended application. Those designed for use in nuclear environments typically contain between 9 and 14 wt. % Cr and are characterised by the formation of high number densities of 1–5 nm diameter yttrium–titanium oxides. Alloys designed for application at high homologous temperatures generally have greater concentrations of Cr (16–20 wt. %) and incorporate aluminium in order that a protective alumina scale can be formed[5,6,7,8]. Y_2O_3 oxide particles introduced into the matrix increase not only the high-temperature mechanical strength by serving as obstacles for mobile dislocations, but, additionally may improve the resistance to neutron irradiation by acting as a sink of the point defects induced by irradiation atom displacement. The microstructure formed during the production of ODS steels can be influenced by small additions of elements like Ti, Nb, V and Zr, as was shown for ferritic 12Cr-ODS steel. The addition of titanium reduces the oxide particle size significantly in comparison to zirconium, improving the creep rupture strength of the steel considerably. Those particles, along with the fine-grained microstructure, additionally improve high-temperature oxidation resistance in comparison to conventional steels with similar chromium content[9,10,11].

The interaction of these materials with liquid Pb is a critical issue which is now widely studied. In fact, the environment can have a strong influence on creep-rupture strength and corrosion resistance. Liquid metal embrittlement (LME) may occur under stress owing to direct contact of a steel and a heavy liquid metal (HLM), leading to apparent degradation of mechanical properties of the steel and can give rise to fast, uncontrollable crack growth.[11].

In particular, liquid lead can be very aggressive towards the steel, because of the high solubility of some alloying elements into the liquid. However, when a sufficient amount of oxygen (about 10−6 wt. %, below 500 °C [1]) is present in the HLM, the T91 can grow a protective oxide scale.

On manifestation of LME of structural materials in contact with HLM is the observation of a transition from ductile to brittle fracture (described as a change from ductile to trans granular (TG) cleavage, cleavage-like, or inter granular (IG) brittle fracture). In addition, there is a decrease in ductility, namely lower values of mechanical properties such as elongation, reduction of area, fatigue life and fracture resistance.

Moreover, too low oxygen concentration leads to dissolution of the steel in HLM and penetration of HLM into the steel as well as too high oxygen concentration can lead to high/non-controlled formation of oxide scale that can crack and spall from the steel surface and, therefore, open new paths for HLM to fresh steel surface[11].

The overall objective of this research is to determine the susceptibility of two types of ODS steel to LME using environmental conditions determined to likely promote LME, but applying reasonable stress relevant to operation condition. It is hypothesis that oxidation and corrosion resistance of ODS steels will have a similar response mechanism as those seen in other FM steels although based on the mixed results of FM steels, it is not clear whether LME will or will not occur in ODS steels.

This paper reports some preliminary study on the simultaneous effect of liquid lead and stress for three steels: The 14Cr ODS represent the most perspective material, the 19Cr ODS is the optimal one with respect to corrosion resistance, but lower toughness, and the 9Cr FM, T91, as the reference material, the most studied for LME behaviour in past. The performance of T91 steel and ODS steels in oxygen-controlled Pb was investigated at 400°C, using post-test evaluation. The temperature was selected because the materials are more sensitive to LME at lower temperatures[12]. Although the operating temperatures for ODS materials are expected to be higher, the transient temperatures have to be taken into account when considering such fast phenomena like LME.

EXPERIMENTAL

Materials

The ferritic-martensitic steels T91 (Grade 91 Class 2/S50460) was produced by Industeel, ArcelorMittal group. The material was standardized at 1150°C for 15 min with following water cooling to the room temperature and finally annealed at 770°C for 45 min, slow cooling in air. The typical microstructure induced by this heat treatment consists of laths of martensite and original austenitic grains. The composition of the T91 steel is given in Table 1.

Table 1 Chemical composition of T91 [wt. %]

Element	Fe	C	N	P	Cr	Ni	Mo	Mn	Si	V	Cu	Nb	Al	W
T91	Bal.	0.1	0.04	0.02	8.9	0.1	0.9	0.4	0.2	0.2	0.06	0.06	0.01	0.01

MA957 and PM2000 are oxide dispersion strengthen ODS steels, the composition is given in Table 2. The MA957 and PM2000 were manufactured by IPM ASCR, v.v.i. from commercially available atomized iron powder, ferrotitanium powder, ferrochrome powder, yttrium powder and molybdenum powder. The ODS steels were produced by mechanical alloying process. Those powders were mixed in exact proportions and processed in high energy ball mills for 24 hours in air. Afterwards the degassed powder was cold pressed to compact pellet and a rod of diameter 30 mm was hot-extruded at 1150°C from the pellet[13].

Table 2 Chemical composition of MA957 and PM2000 [wt. %]

Element	Cr	Ti	Mo	Y2O3	Fe	Al
MA957	13.6	0.85	0.29	0.25	Bal.	0.06
PM2000	19	0.5	0.3	0.5	Bal.	5.5

Table 3 Mechanical properties of the test materials

(bolds indicate the tensile test results of present ODS materials)

Temperature [°C] / Yield strength [MPa]	20	300	400	500
T91	580	503	456	464
MA957 ODS	**1023**	850	**851**	700
PM2000 ODS	**853**	800	**713**	650

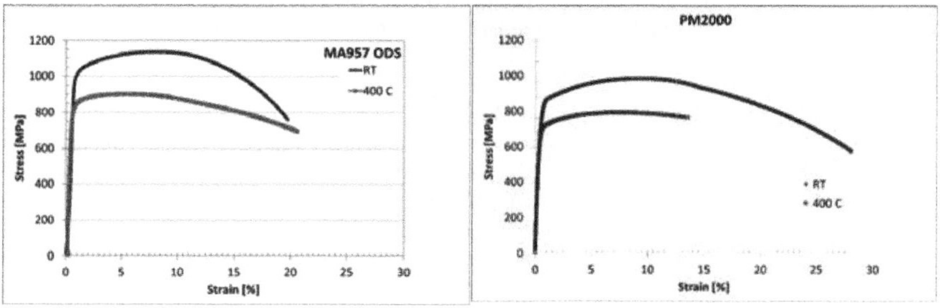

Fig. 1 Tensile properties of the two ODS steels.

Specimens

Specimens were machined from the base material. All specimens were manufactured as T–L cuts. Flat smooth specimens of dimensions 14.9x3x1 mm (Fig. 2 a) were designed to fit the size of the holders in the COLONRI-II loop. The sample surface was cleaned with acetone in an ultrasonic bath. Inside each holder six specimens (Fig. 2 b) were pre-loaded to 1.3 times the value of the yield stress with the holder screw movement at room temperature, respective elastic deflection was calculated according the standard ISO7539-2: 1989[14]. Then the three holders were inserted into the loop.

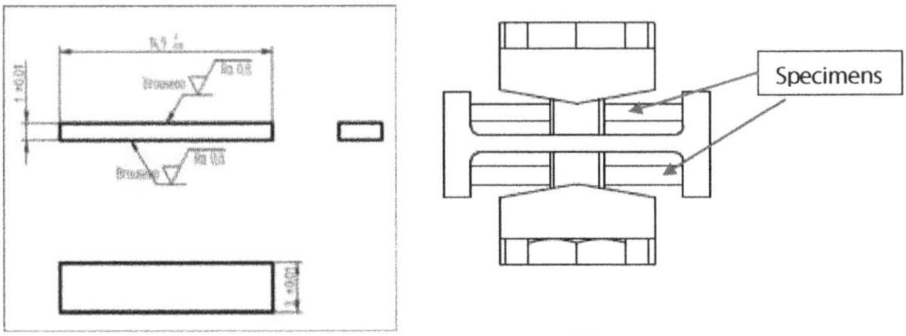

(a) (b)

Fig. 2 (a) Test specimen with the original surface after fabrication with EDM and grinding; (b) schematic of the specimen holder for the six specimens

COLONRI II loop

The natural convection loop COLONRI II is used for corrosion tests of structural materials in flowing Pb. It is a vertical, square structure manufactured from the austenitic stainless steel AISI 321 with 5 calibrated sensors. The loop has a high temperature and a low temperature section, heating and cooling legs. Each section is about 1m long with constant temperature along the test sections. This type of loop allows dosing of the gas mixture, in the upper equalising tank. The liquid metal flows in the equalising tank and then flows back into the main circulation loop. The two experimental sections can work at different temperatures. The oxygen content is monitored by oxygen sensors immersed in the liquid metal, in the point where the liquid metal went out of the experimental section. The oxygen control is performed by adding a mix of gases (O_2, H_2 and Ar) and monitoring by use of sensors, based on the Bi/Bi_2O_3 reference electrode[15,16].

Exposure in liquid lead

The exposure of pre-loaded specimens of MA957, PM2000 and T91 steels was performed in the COLONRI-II loop, a natural convectional loop with flowing liquid lead. The concentration of dissolved oxygen in the liquid Pb, C_{O2}, was controlled by means of mass exchange with an oxygen-containing gas and was continuously monitored using oxygen sensors; the corresponding oxygen concentration was oscillating between 10^{-7} and 10^{-5} wt. %. A constant temperature (400°C) was maintained for 500h. The oxygen content in liquid lead was measured with oxygen sensors.

After exposure to liquid lead, the samples were cut and cross section samples prepared. The remaining specimens were immersed in a solution of H_2O_2, CH_3COOH and CH_3CH_2OH, in the ratio 1:1:1, in order to remove the residual Pb[18].

Characterization of exposed materials

Standard metallographic cross-sections were prepared from the steel specimens after removal of the residual Pb from the specimen surface. Light microscopy was used to observe the specimen surface for any crack occurrence. A scanning electron microscope (SEM) equipped with energy-dispersive X-ray spectrometer (EDX) was used to investigate the corrosion scales. Qualitative EDX elemental analysis was mainly used for recording line scans and elemental maps of the corrosion scales that formed on the exposed steel surfaces. Post-test analyses were performed on the surface and cross-sections prepared from the gauge section close to or in some distance from the site of rupture.

Samples exposed to liquid Pb were additionally cleaned chemically in CH_3COOH-H_2O_2-C_2H_5OH mixture (1:1:1) at room temperature to remove adherent Pb from the surface[18].

RESULT AND DISCUSSIONS

Liquid metal corrosion can proceed by various processes dissolution formation of intermetallic compounds, and penetration of liquid metal along grain boundaries. All these processes depend on experimental factors such as, temperature, thermal gradients, solid and liquid compositions and liquid metal velocity. In the case of liquid lead, one of the fundamental parameters on which the corrosion process depends is the oxygen content dissolved in the liquid metal.

The interaction mode of steels with liquid lead at 550 – 650°C, as a function of the oxygen concentration in the liquid metal, changes from oxidation in the oxygen-saturated melt ($O \sim 10^{-3}$ wt. %), to a thin film formation (passivation) in the oxygen-added melt ($O \sim 10^{-6}$ wt. %) and then to the severe dissolution at lower oxygen contents.

T91

The original surface was ground to 600 grid, only in the vertical direction. Homogeneous grinding lines were observed (Fig. 3a). After exposure the ground surface disappeared for a bumpy and irregular oxidised surface (Fig. 3b). Even after chemical removal of the residual Pb, the EDX detector measured large amount of Pb on the oxidised surface. This was possibly due to residual Pb particles trapped on the rough oxidised surface. From the surface observation, no transversal cracks were observed, also in the areas which underwent the highest load, the oxide was homogenously developed, with no visible defects.

In the cross section, the presence of oxides was confirmed (Fig.4a). There appeared to be:

-an inner layer (about 1.5μm thick), enriched in oxygen (inwards diffusion), slightly depleted in Fe and enriched in Cr (Fig. 4b),

-an outer oxide scale well adherent, compact and homogeneous (about 2μm thick), enriched in Fe.

EDX qualitative analyses (Fig. 4b) highlighted that the outer oxide was enriched only in Fe, while the inner oxides were a mix of Fe and Cr, indicating the possibility of a spinel oxide in the inner layer.

However, it was observed and measured by EDX that the outer oxide had particles of Pb incorporated. In particular, where the detector measured the presence of Pb, there was also presence of Ni. This is a strong indication of the presence of impurities in the liquid Pb (possibly coming from the steel of loop).

In general, the oxides developed for 500h at 400°C were sufficient to protect the steel from the direct contact with the flowing Pb. The oxygen content in the loop was ranging between 10^{-7} and 10^{-5} wt. %. This amount of oxygen was sufficient to develop oxides preventing since the early stage the contact with the liquid metal. The level of load applied was not sufficient to damage the surface of the steel nor the oxide formed during exposure.

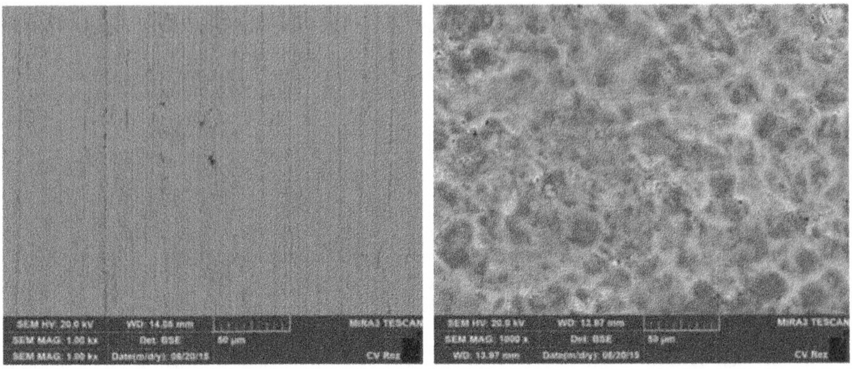

(a) (b)

Fig. 3 T91 (a) Original surface – post-test surface observation (b)

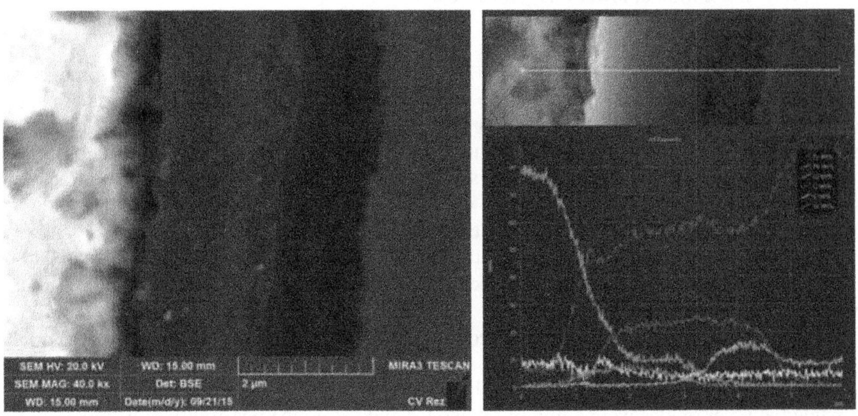

(a) (b)

Fig. 4 Cross sections of T91 steel after exposure in liquid lead (a) and the line/scan across the oxide (b)

ODS steels

Original surface was ground to 600 grid, only in the vertical direction. Homogeneous scratches were observed (Fig. 5a and Fig. 6a). Observation of the surface after exposure to environment and removal of the residual Pb was carried out. In both ODS steels, the original grinding marks were still clearly visible (Fig. 5b and Fig. 6b). In the case of the MA957 isolated oxidised areas (large fluffy, no scratch) were observed (Fig. 5a). No transversal cracks were observed, also in the areas which underwent the highest load.

The cross section of the ODS steels confirmed that the oxidation occurred only in localised areas, producing nodules of outer oxides and inner oxidation zones (Fig. 7). The visible oxides of Fe and Cr nucleated in localised areas. The rest might be covered by an extremely thin oxide, which was not detected with the metallographic techniques available.

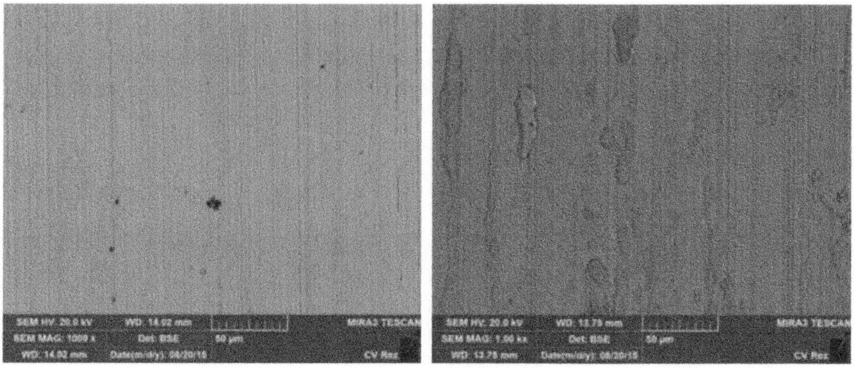

(a) (b)

Fig. 5 ODS steel MA957 – original surface (a), post-test surface observation (b)

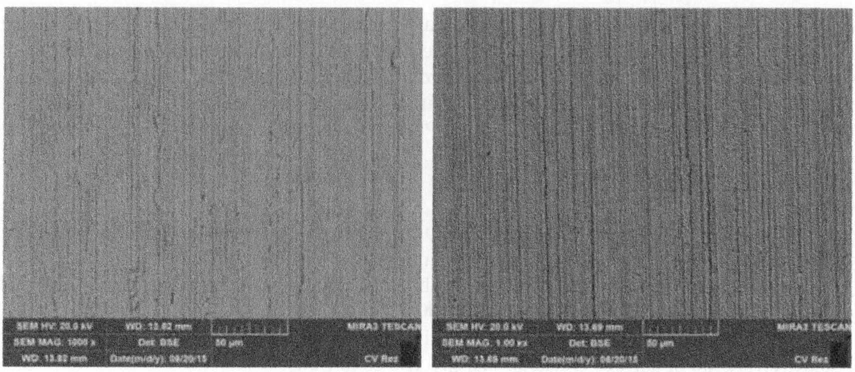

(a) (b)

Fig. 6 ODS steel PM2000 – original surface (a), post-test surface observation (b)

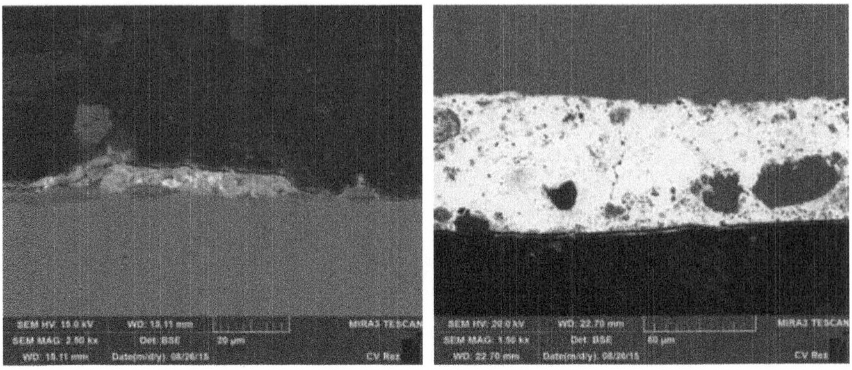

(a) (b)

Fig. 7 Cross sections of ODS steels after exposure in liquid lead – (a) MA957, (b) PM2000

The applied load was meant to represent a conservative value (which designers would have to respect in phase of calculations), limiting the load to the elastic domain. This value was not sufficient to develop cracks in the specimens at room temperature and, due to the stress relaxation, neither during exposure. Moreover, oxide formation, especially in the case of T91, prevented the direct contact between the steels and the liquid Pb, preventing the formation of the pre-conditions for LME.

CONCLUSIONS

-The interaction of three steels (T91, ODS – MA957, PM2000) with liquid metal and simultaneously applied load was studied using coupon specimens.

-Liquid Metal Embrittlement was not observed on the surface of any of the pre-loaded specimens after the exposure in liquid lead at 400°C for 500 hours.

-The applied load on level of 1.3 times the yield strength was conservative enough to not stimulate formation of surface cracks in this specific conditions.

-The amount of oxygen (between 10^{-7} and 10^{-5} wt. %) in the liquid Pb was sufficient to develop a thin but homogeneous oxide scale on the T91 steel, thus preventing the direct contact with the liquid metal.

-The amount of oxygen in the Pb was sufficient to develop visible localised oxides in the ODS steels.

ACKNOWLEDGEMENT

This work has been supported by the SUSEN Project CZ.1.05/2.1.00/03.0108 realized in the framework of the European Regional Development Fund (ERDF) and by the GACR project No. 14-25246S.

REFERENCES

[1] R.L. Klueh, Elevated temperature ferritic and martensitic steels and their application to future nuclear reactors, *Int. Mater. Rev.*, **50**, 287-310, (2005).

[2] E. Arzt, Creep of dispersion strengthened materials: a critical assessment, *Res. Mech.*, **31**, 399-453, (1991).

[3]B. Raj, and M. Vijayalakshmi, Ferritic Steels and Advanced Ferritic–Martensitic Steels, *Compr. Nucl. Mat.,* **4**, 97-121, (2012).
[4]M. Yurechko, C. Schroer, O. Wedemeyer, A. Skrypnik, and J. Konys, Creep-rupture tests on chromium-containing conventional and ODS steels in oxygen-controlled Pb and air at 650 °C, *Nucl. Eng. Des.,* **280**, 686-696, (2014).
[5]K. Dawson, S.J. Haigh, G.J. Tatlock, and A.R. Jones, Nano-particle precipitation in mechanically alloyed and annealed precursor powders of legacy PM2000 ODS alloy, *J. Nucl. Mat.,* **464**, 200-209, (2015).
[6]B. Dubiel, M. Wróbel, P.J. Ennis, and A. Czyrska-Filemonowicz, Microsturture of INCOLOY MA956 after low and high temperature deformation, *Scr. Mater.,* **37**, 1215-1220, (1997).
[7]S. Ukai, M. Harada, H. Okada, M. Inoue, S. Nomura, S. Shikakura, K. Asabe, T. Nishida, and M. Fujiwara, Alloying Design of Oxide Dispersion Strengthened Ferritic Steel for Long Life FBRs Core Materials, *J. Nucl. Mater.,* **204,** 65-73, (1993).
[8]G. R. Odette, Recent Progress in Developing and Qualifying Nanostructured Ferritic Alloys for Advanced Fission and Fusion Applications, *JOM,* **66,** 2427-2441, (2014).
[9]M. Yurechko, C. Schroer, A. Skrypnik, O. Wedemeyer, and J. Konys, Creep-to-rupture of 12Cr- and 14Cr-ODS steels in oxygen-controlled lead and air at 650 °C, *J. Nucl. Mat.,* **450**, 88-98, (2014).
[10]T. Okuda, S. Nomura, S. Shikakura, K. Asabe, S. Tanoue, and M. Fujiwara, Proc. Symp. Sponsored by the TMS Powder Metallurgy Committee, Indiana, 195, (1989).
[11]T. Kaito, T. Narita, S. Ukai, and Y. Matsuda, High temperature oxidation behavior of ODS steels, *J. Nucl. Mat.,* **329-333,** 1388-1392, (2004).
[12]A. Hojná, F. Di Gabriele, J. Klecka, J. Burda, "Behaviour of the steel T91 under uniaxial and multiaxial slow loading in contact with liquid lead", Journal of Nuclear Materials 466 (2015) 292-301
[13]L. Rozumova, A. Hojna, J. Kalivodova, P. Bublikova, and H. Hadraba, The effect of high temperature helium gas on the fracture behaviour of ODS MA957, 24th International Conference on Metallurgy and Materials, METAL 2015, Brno, ISBN 978-80-87294-58-1.
[14]Corrosion of Metals and Alloys–Stress Corrosion Testing – Part 2: Preparation and use of Bent-Beam Specimens. International Standard. ISO7539: 1989(E).
[15]F. Di Gabriele, S. Amore, C. Scaiola, E. Arato, D. Giuranno, R. Novakovic, and E. Ricci, Corrosion behaviour of 12Cr-ODS steel in molten lead, *Nucl. Eng. Des.*, **280**, 69-75, (2014).
[16]F. Di Gabriele, and L. Kosek, Oxygen Monitoring in the Natural Convection Loop Colonri I, 21st International Conference on Nuclear Engineering, ICONE 2013, Chengdu, ISBN 978-0-7918-5578-2.
[17]H.J. Grabke, E.M. Muller-Lorenz, S. Strauss, E. Pippel, and J. Woltersdorf, Effects of Grain Size, Cold Working, and Surface Finish on the Metal-Dusting Resistance of Steels, *Oxid. Metals,* **50,** 241, (1998).
[18]M. Del Giacco, Investigation of Fretting Wear of Cladding Materials in Liquid Lead. KIT Scientific Publishing, (2013), ISBN 978-3866449602.

Nanotechnology for Energy, Environment, Electronics, and Industry

STABILIZATION OF NANO-SCALE NICKEL ELECTRO-CATALYSTS AT HIGH TEMPERATURE

David R. Driscoll
Mechanical & Industrial Engineering - Montana State University
Bozeman, MT, USA

Stephen W. Sofie PhD.
Mechanical & Industrial Engineering - Montana State University
Bozeman, MT, USA

ABSTRACT

Coarsening of nano-metal particulate at elevated temperatures is a significant impediment to the efficiency and longevity of numerous catalyst and energy conversion/storage systems. Utilizing a model high temperature fuel cell system of nickel metal-stabilized zirconia doped with Al_2TiO_5, we have identified a novel approach to increase the thermal stability of nickel microstructures by fostering interfacial chemical reactions or 'anchors.' The mechanisms and specific species that confer enhanced catalyst stability through localized reactions are not well understood, even though the performance impact is simply demonstrated. X-ray diffraction has demonstrated the formation of interfacial $Zr_5Ti_7O_{24-x}$, a potential oxygen ion conductor known to form in the presence of yttria. Anchoring phase stability in reducing conditions at high temperatures is also examined.

INTRODUCTION

Nickel metal supported on yttria-stabilized zirconia (YSZ) substrates has been selected as a model system for the development of chemical anchors to inhibit thermal coarsening of high temperature metal catalysts. In solid oxide fuel cells (SOFCs), instability of the nickel microstructure within the cell's anode is a primary source of degradation; particularly for high performance infiltrated catalyst cells.[1–6] Beyond SOFCs, high temperature stability of metal catalysts is problematic in numerous applications including naphtha reformation systems and automotive catalytic converters.[7–12]

The function of SOFC anodes requires mixed electronic and ionic conductivity as well as catalytic activity. This is traditionally achieved through combination of nickel and YSZ. While YSZ is ionically conductive, metallic nickel provides electronic conductivity and good catalytic function at operational temperatures. The requirement of two phases does limit the functional portion of the cell to a triple phase boundary where Ni, YSZ and the gaseous reactant stream are in simultaneous contact.[13] It is from this perspective that infiltrated catalyst techniques are very attractive.

Porous cermets of nickel and YSZ have become the standard for SOFC anodes, but the infiltrated catalyst alternative is based on porous YSZ scaffolds that are then coated with catalyst through solution infiltration techniques.[4, 14–16] In this way, the catalyst is placed effectively, and small particle sizes of 100 nm and less are easily realized. These small particles sizes provide greatly increased triple phase boundary length for chemical reaction and surface area for H_2 adsorption. Unfortunately, fine metallic catalysts are also prone to very rapid coarsening at high temperatures.[12, 17] When coarsening occurs, not only are triple phase boundary length and surface area lost, but the percolated nickel network is broken and electronic resistivity of the anode rises. In this way, the exciting merits of using this fabrication technique are largely negated.

The use of aluminum titanate (Al_2TiO_5, ALT) to facilitate enhanced electro-catalyst binding with metallic nickel on YSZ has been examined in prior studies w the outcomes predicted non-active species that would deteriorate fuel cell performance.[18, 19] Metastable ALT is known to easily decompose to Al_2O_3 and TiO_2. Upon thermal treatment of ALT doped anodes, further reaction occurs between Al_2O_3/Ni and TiO_2/YSZ. In this report, the importance of thermal processing conditions is examined and functional properties of the established anchoring phases are considered.

METHODS

Electrochemical Cells

Electrochemical cells were fabricated and after ALT doping, heat treated to different temperatures for formation of anchoring phases to test for performance differences. Cells were supported with commercially available 300μm electrolytes which were fabricated from 8 mol% yttria YSZ. Anodes and cathodes were applied via aerosol spraying with a target thickness of 50μm. The anode spray was an equivolume mixture of NiO, YSZ and cornstarch in a 60 vol% xylene 40 vol% ethanol solvent with appropriate dispersant and binder. All anodes were sprayed from the same batch of spray mixture with a portion of the spray mixture having been separated for ALT doping. The extent of doping was 5% of the NiO + YSZ mass. The cathode mixture was similar with an equivolume mixture of YSZ, LSM, and cornstarch. Fabrication proceeded such that that the anode was applied and sintered, then the cathode applied and sintered. One ALT anode was sintered at 1400°C for 1 hour while the other ALT anode and un-doped anode were sintered at 1200°C for 1 hour. The high temperature run is referred to the ‘activation’ process in which the new secondary phases form to improve the thermal stability of the nickel catalyst.

Testing of fabricated cells was performed in a platen-type test fixture with Inconel platens enclosed in a clamshell furnace. The cells were interfaced with the platens using silver mesh at the cathode and nickel foam at the anode with silver paste used for the connections. All cells were run at 850°C and a constant 0.7V.

X-Ray Diffraction

X-Ray Diffraction was performed to correlate two SOFC reaction temperatures with phase evolution and to investigate anchoring phase stability under temperature and pO_2 conditions resembling an SOFC anode. Powders were prepared as equivolume mixtures of YSZ, NiO, and ALT by ball milling, flash freezing, and freeze drying. These powders were compacted into pellets using a 1.9 cm (3/4”) diameter die at a pressure of 250 MPa. These pellets were then sintered at 1200°C or 1400°C in air or a 5% H_2 + 95% N_2 mixture.

Following sintering, pellets were ground in a mortar and pestle for XRD examination to eliminate preferred orientation effects on the diffraction patterns. Analysis was performed on a Scintag XRD using Cu-kα radiation with scans that ranged 20-70° 2θ.

RESULTS & DISCUSSION

Electrochemical Cells

Three electrochemical runs which include a baseline cell, a 1200°C activated ALT and a 1400°C activated ALT cell are shown in Figure 1 below. Cell specific power output was recorded for 45 hours. Initial power output was observed to vary from roughly 200mW*cm^{-2} to 300mW*cm^{-2} which was considered reasonable given the relatively thick 8YSZ electrolyte supports. While the 1200°C and non-doped cell showed the highest initial power output, the lower power output of the 1400°C cell is associated with decreased anode porosity and coarsening of the nickel oxide prior to reduction. Nonetheless, the stability of the 1400°C activated cell shows significant improvement in degradation rates.

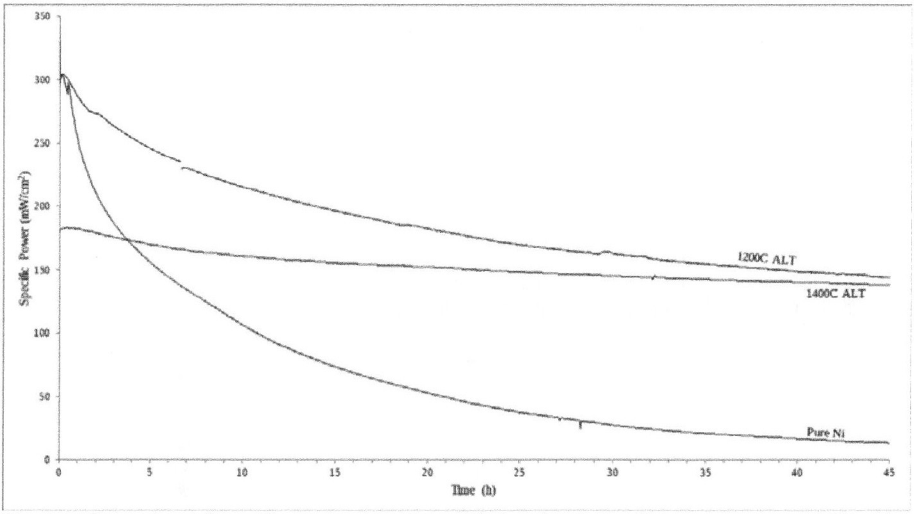

Figure 1: Comparison of specific power output loss with time for three cells run at a constant 0.7V and 850°C. Cells are varied by anode composition and include a baseline Ni/YSZ and ALT doped Ni/YSZ with 1200°C and 1400°C thermal treatments.

The advantage of ALT doping is readily apparent upon initial inspection with the non-doped cell having degraded to nearly 20mW*cm^{-2} in 45 hours of operation. Despite impaired initial power output, the ALT cell treated at 1400°C appears to show significantly increased stability over the 1200°C treatment. After 45 hours of operation, specific power outputs had degraded to 5.5% of initial for the pure Ni cell, 47.4% of initial for the 1200°C activation cell, and 75.4% of initial output for the 1400°C activation cell as summarized in Table I. The discrepancy in stability between thermal treatments for ALT doped cells is further discussed in conjunction with XRD data below.

Table I: Power output of ALT-doped and un-doped cells after 45 hours of operation reported as a percentage of initial power output.

Un-doped	1200°C - ALT	1400°C - ALT
5.5%	47.4%	75.4%

X-Ray Diffraction

The processing temperatures of 1200°C and 1400°C for electrochemical cells were based on a combination of a multitude of previous electrochemical tests and previously reported XRD.[18] The prior and current XRD analyses have demonstrated formation of anchoring phases including nickel aluminates and zirconium titanates at temperatures as low as 1200°C. Given the propensity for the coarsening of NiO at high temperature and potential for deleterious reactions in fabricating complete cells (formation of $La_2Zr_2O_7$ at the cathode) it is sensible to form the anchoring phases at the lowest possible temperatures. However, electrochemical tests have demonstrated improved stability after 1400°C activation relative to lower activation temperatures. This XRD analysis investigates this seeming inconsistency between the lower activation temperature suggested by previous XRD and the 1400°C suggested by electrochemical tests.

Figure 2 contains a portion of three XRD patterns extending from 27-34° 2θ. The narrow range is utilized to highlight representative peaks of the key phases to aid in identifying trends in thermal treatment. The lower two both represent powder mixtures treated at 1200°C. The first was heated to 1200°C for one hour in a manner consistent with the electrochemical cell processing and the second was treated for 24 hours to examine the influence of diffusion limitations in phase formation related to slower kinetics at 1200°C compared to 1400°C. The third mixture was heated to 1400°C for 1 hour in the same manner as the corresponding electrochemical cell.

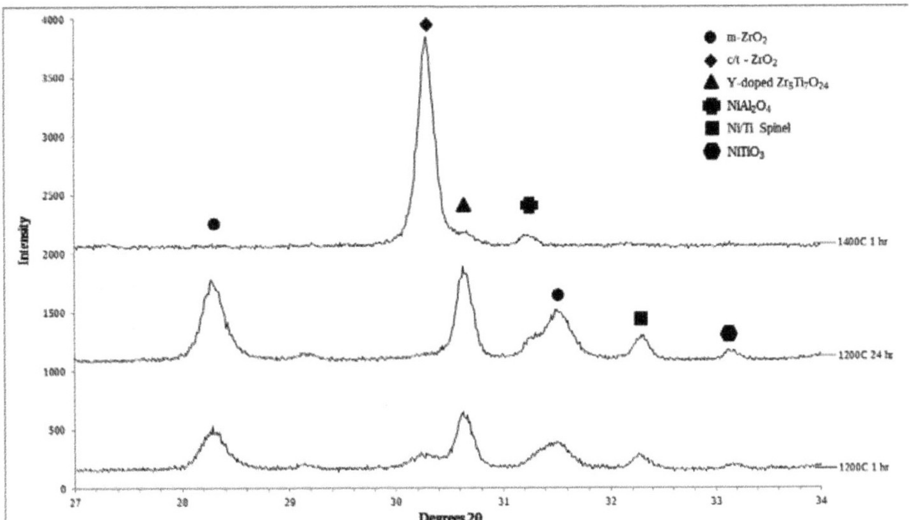

Figure 2: XRD patterns of equivolume NiO/YSZ/ALT mixtures heated to 1200°C for 1 & 24 hours and 1400°C for 1 hour. 27-34° 2θ, Cu-Kα radiation.

Examination of the patterns reveals dramatic discrepancy between thermal activation temperatures for this system. It is apparent in Figure 2 that the cubic polymorph of zirconia has been destabilized to the monoclinic phase at 1200°C after 24 hours with some retained cubic phase at 1200°C for 1 hour, recognizing that the typical detection limits of XRD are approximately 3-5%. The destabilization of the cubic phase is notable here given the starting mixture contained YSZ at 8 mol% Y_2O_3. This destabilization corresponds to strong formation a new Zr/Ti complex with the stoichiometry, $Zr_5Ti_7O_{24}$. Formation of this phase stands in contrast to the models of previous work where the formation of $ZrTiO_4$ had been assumed. [18] Literature identifies that $Zr_5Ti_7O_{24}$ is only preferred over $ZrTiO_4$ in the presence of Y_2O_3 such that $Y_x(Zr_5Ti_7)_{1-x}O_{24-\delta}$, an oxygen deficient compound, is a more accurate representation of the reaction product.[20-22]

In this way, the harvesting of Y to form a new phase provides an explanation for the destabilization of the cubic phase in YSZ. It is appropriate to re-iterate that the powder mixtures for XRD contained dramatically more ALT than actual anode compositions to ensure detection of secondary phases despite the relatively coarse detection limits of XRD. Given the lower volume of ALT in the cell fabrication, it is unlikely that a large fraction of the YSZ in the actual electrochemical cells was destabilized. However, local phase dependent variations in oxygen ion conductivity are still a concern in the triple phase boundary region, hence the motivation for electrochemical testing in concert with XRD analysis.

Furthermore, the presence of Y doping in a Zr/Ti/O compound yields exciting potential for functional properties in what was previously thought to be an otherwise non-functional anchoring phase. Y^{3+} substitution on $(Zr,Ti)^{4+}$ could very easily be compensated by oxygen vacancies thus inducing permeability to O^{2-} ions in the same manner as YSZ. Beyond ionic conduction the very low pO_2 of an anode environment should be considered for its potential to reduce Ti within the defected structure and potentially induce electronic conductivity. This is in addition to the potential electronic conductivity derived from the nature of O^{2-} as an intrinsic electron donor during the formation of vacancies.[23] Observation of $Y_x(Zr_5Ti_7)_{1-x}O_{24-\delta}$ is a profound result in its potential to explain how introducing anchoring phases near the triple phase boundary does not dramatically impede cell performance as in Figure 1.

Activation at 1200°C also reveals important interactions between nickel and titanium. These phases were not anticipated and are not significant in concentration, but potentially influential given their inherent proximity to the triple phase boundary. However, it should be questioned whether these interactions occur at all in an anode. In that environment, the ratio of ALT/Y_2O_3 is much lower than in the XRD studies. As a result, Ti is likely to be the limiting reactant in $Zr_5Ti_7O_{24}$ formation which would not allow for an excess of Ti to react with Ni. It is noted that $NiAl_2O_4$ is strongly present at both temperatures. The peak for $NiAl_2O_4$ which appears in this window is simply not a primary one, however a primary peak does occur at 37° 2θ. A key outcome of this study is that the primary phases forming from ALT additions are $NiAl_2O_4$ and $Zr_5Ti_7O_{24}$.

Figure 3 demonstrates that the formation of the "anchoring" phases appears partially reversible upon introduction to a low pO_2. In the lower pattern, the mixture was treated at 1400°C in air and then reduced at 850°C for 50 hours in an attempt to understand the anode chemistry after some time in real fuel cell operational conditions. The upper pattern represents a powder mixture with no initial heat treatment (no activation) in air.

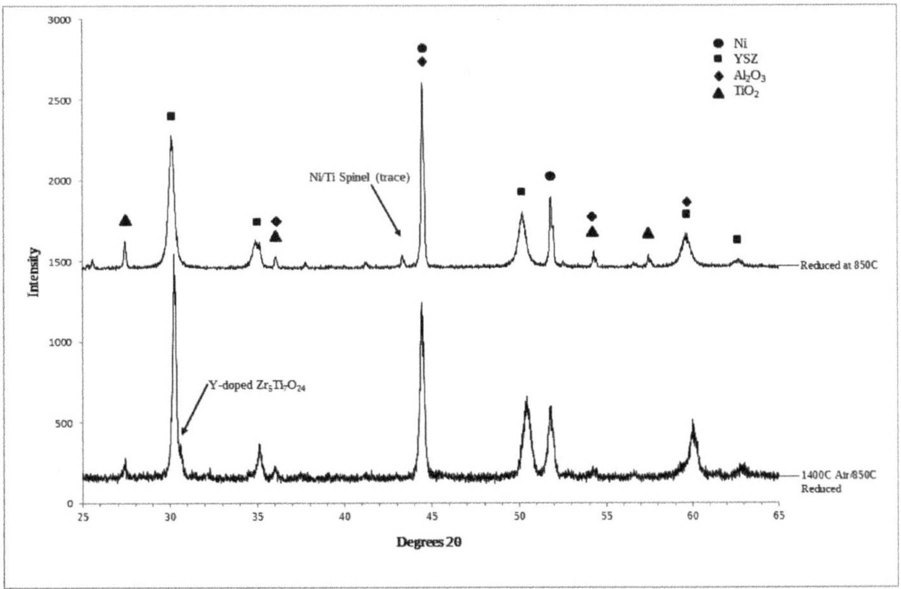

Figure 3: XRD patterns of equivolume mixtures of NiO/YSZ/ALT that were reduced at 850°C for 50 hours following a 1400°C initial treatment in air (lower pattern), and no initial treatment (upper pattern).

The formation of $NiAl_2O_4$, which was considered a primary anchoring phase, appears to be completely reversed upon reduction. Nickel metal is clearly observed and while peaks corresponding to Al_2O_3 are convoluted with other peaks, no other Al-containing compounds are observed. The identifying peak for $NiAl_2O_4$ is seen at 37.00° 2θ when the compound is present. The formation of $Y_x(Zr_5Ti_7)_{1-x}O_{24-\delta}$ is partially, but not completely, reversed. An identifying peak for TiO_2 is observed just above 27° 2θ and the most intense peak for $Y_x(Zr_5Ti_7)_{1-x}O_{24-\delta}$ is observed as a barely perceptible shoulder on the YSZ peak only in the mixture which had an initial 1400°C treatment. It is confirmed that a high temperature treatment is required if any $Y_x(Zr_5Ti_7)_{1-x}O_{24-\delta}$ is to be observed.

The loss of $NiAl_2O_4$ and most of $Y_x(Zr_5Ti_7)_{1-x}O_{24-\delta}$ does not imply that the heat treatment is of limited/no value. In order for $Y_x(Zr_5Ti_7)_{1-x}O_{24-\delta}$ or $NiAl_2O_4$ to have formed, homogeneous mixing of Ni/Al and Zr/Y/Ti had to have occurred over the extent of the anchoring phases. Upon reduction to Ni(m)/Al_2O_3 and YSZ/TiO_2, the extent of phase segregation is unsure. This of course, is not to discount that some $Y_x(Zr_5Ti_7)_{1-x}O_{24-\delta}$ is likely still present. The remnants of a known $Y_x(Zr_5Ti_7)_{1-x}O_{24-\delta}$ or $NiAl_2O_4$ grain have yet to be observed. These results do indicate that kinetic arguments such as the extent of inter-diffusion, or anchoring phase grain size should be considered in defining the disparity of stability between 1200°C and 1400°C activated anodes.

CONCLUSIONS

In this study, aluminum titanate was used to stabilize nickel metal in SOFC anodes after varying thermal treatments. It was observed that ALT yielded substantial performance retention over an un-doped baseline anode, but was observed that the doped cell treated at 1400°C appeared to be better stabilized than the cell treated at 1200°C. XRD identified the potential anchoring phases of $NiAl_2O_4$ and $Y_x(Zr_5Ti_7)_{1-x}O_{24-\delta}$ after both levels of heat treatment, but also observed the destabilization of the cubic phase of zirconia in the 1200°C treatment where $Y_x(Zr_5Ti_7)_{1-x}O_{24-\delta}$ concentration was maximized.

Further work needs to consider the impact of the thermal treatment and anchoring phase formation on a functioning anode given that formation of the anchoring phases has been showing to reverse in an environment that resembles anode conditions. It is promising that the remaining portion of the anchoring phases, $Y_x(Zr_5Ti_7)_{1-x}O_{24-\delta}$, potentially exhibits conductive properties. This material needs to be more fully characterized in additional work. Ultimately, success in understanding the mechanisms by which increased catalyst stability is imparted via ALT will allow extension of the technology to numerous systems of importance.

1 A. Buyukaksoy, V. Petrovsky, and F. Dogan, "Stability and Performance of Solid Oxide Fuel Cells with Nanocomposite Electrodes," *J. Electrochem. Soc.*, **159** [6] B666 (2012).

2 A. Buyukaksoy, V. Petrovsky, and F. Dogan, "Redox Stable Solid Oxide Fuel Cells with Ni-YSZ Cermet Anodes Prepared by Polymeric Precursor Infiltration," *J. Electrochem. Soc.*, **159** [2] B232 (2012).

3 T. Klemensø, K. Thydén, M. Chen, and H.J. Wang, "Stability of Ni-yttria stabilized zirconia anodes based on Ni-impregnation," *J. Power Sources*, **195** [21] 7295–7301 (2010).

4 H. Yokokawa, H. Tu, B. Iwanschitz, and A. Mai, "Fundamental mechanisms limiting solid oxide fuel cell durability," *J. Power Sources*, **182** [2] 400–412 (2008).

5 Z. Liu, B. Liu, D. Ding, M. Liu, F. Chen, and C. Xia, "Review Fabrication and modification of solid oxide fuel cell anodes via wet impregnation/infiltration technique," *J. Power Sources*, **237** 243–259 (2013).

6 S.P. Jiang, "Nanoscale and nano-structured electrodes of solid oxide fuel cells by infiltration: Advances and challenges," *Int. J. Hydrogen Energy*, **37** [1] 449–470 (2012).

7 R. Mathur, V. Balaram, M. Satyanarayanan, S.S. Sawant, and S.L. Ramesh, "Anthropogenic platinum, palladium and rhodium concentrations in road dusts from Hyderabad city, India," *Environ. Earth Sci.*, **62** [5] 1085–1098 (2011).

8 J.C. Ely, C.R. Neal, C.F. Kulpa, M. a. Schneegurt, J. a. Seidler, and J.C. Jain, "Implications of platinum-group element accumulation along U.S. roads from catalytic-converter attrition," *Environ. Sci. Technol.*, **35** [19] 3816–3822 (2001).

9 S. Chellam, *Platinum Metals in the Environment.* 2015.

10 K. Ravindra, L. Bencs, and R. Van Grieken, "Platinum group elements in the environment and their health risk.," *Sci. Total Environ.*, **318** [1-3] 1–43 (2004).

11 M.R. Rahimpour, M. Jafari, and D. Iranshahi, "Progress in catalytic naphtha reforming process: A review," *Appl. Energy*, **109** 79–93 (2013).

12 F. Behafarid and B. Roldan Cuenya, "Coarsening phenomena of metal nanoparticles and the influence of the support pre-treatment: Pt/TiO 2(110)," *Surf. Sci.*, **606** [11-12] 908–918 (2012).

13 I. EG&G Technical Services, "Fuel Cell Handbook," *Fuel Cell*, **7 Edition** [November] 1–352 (2004).

14 B. Shri Prakash, S. Senthil Kumar, and S.T. Aruna, "Properties and development of Ni/YSZ as an anode material in solid oxide fuel cell: A review," *Renew. Sustain. Energy Rev.*, **36** 149–179 (2014).

[15] S.. Aruna, M. Muthuraman, and K.. Patil, "Synthesis and properties of Ni-YSZ cermet: anode material for solid oxide fuel cells," *Solid State Ionics*, **111** [1-2] 45–51 (1998).

[16] W.Z. Zhu and S.C. Deevi, "A review on the status of anode materials for solid oxide fuel cells," *Mater. Sci. Eng. A*, **362** [1-2] 228–239 (2003).

[17] H. Kishimoto, A. Suzuki, T. Shimonosono, M.E. Brito, K. Yamaji, T. Horita, F. Munakata, and H. Yokokawa, "Agglomeration behavior of nickel particles on YSZ electrolyte," *Solid State Ionics*, **225** 65–68 (2012).

[18] C.H. Law and S.W. Sofie, "Anchoring of Infiltrated Nickel Electro-Catalyst by Addition of Aluminum Titanate," *J. Electrochem. Soc.*, **158** [9] B1137 (2011).

[19] D. Driscoll, C. Hunt, J. Muretta, and S. Sofie, "*Corresponding Author," **53** [9] 63–72 (2013).

[20] E. López-López, M.L. Sanjuán, R. Moreno, and C. Baudín, "Phase evolution in reaction sintered zirconium titanate based materials," *J. Eur. Ceram. Soc.*, **30** [4] 981–991 (2010).

[21] E. López-López, C. Baudín, and R. Moreno, "Synthesis of zirconium titanate-based materials by colloidal filtration and reaction sintering," *Int. J. Appl. Ceram. Technol.*, **5** [4] 394–400 (2008).

[22] F. Azough, R. Freer, and J. Petzelt, "A Raman spectral characterization of ceramics in the system ZrO2-TiO2," *J. Mater. Sci.*, **28** [9] 2273–2276 (1993).

[23] A. J. Weisentein, N. Childs, R. Amendola, D. Driscoll, S.W. Sofie, P. Gannon, and R. Smith, "Processing and characterization of Sr2-xVMoO6-?? double perovskites," *Mater. Chem. Phys.*, **139** [2-3] 706–718 (2013).

NANOTECHNOLOGY ADVANCEMENTS AND APPLICATIONS

Stephen Miranda
Sales Director, Grinding and Dispersing, North America
NETZSCH Premier Technologies, LLC, Exton, PA USA

ABSTRACT

This presentation will discuss methods for Comminution and dispersion of agglomerates and particles into the nanometer (Nano) size range using dry and wet milling techniques. Comminution and dispersion of solids in slurries and suspensions has been more commonly used due to energy efficiency compared to dry grinding processes. However, when a dry powder is the final product, a dry method may be preferable and economical. Both are possible with the newest comminution and dispersion technology available.

WHY NANOTECHNOLOGY?

The list of products that benefit from nano size particles is large; and growing each day. Here are a few examples in three industries that affect our daily life – food, healthcare and energy.

Crop protection companies are now manufacturing herbicides and other products in the range of 10 nm to 100 nm. A nanoparticle size herbicide contains trillions of particles of active ingredient per liter. The increased surface as a result of the product fineness area improves the potency of the material, accelerates the absorption by the plant, and can significantly reduce the amount of active ingredient needed for the same effectiveness, thereby reducing potential toxicity and negative environmental effects.

Active pharmaceutical ingredients also benefit from nano size particles. Many potentially effective drug discoveries are abandoned because they are insoluble. By reducing the particles to nano size, we can increase their surface area and solubility dramatically. Other known benefits are improved dissolution rate, increased bioavailability and higher activity leading to lower dosages and a lower risk of side effects for the patient.

Another example is the growing field of reusable energy products, especially lithium ion based batteries. Battery manufacturers are always looking to increase overall efficiency by improving conductivity, lengthening charge life and shortening recharge time. All three are needed to make a commercially viable battery solution for advanced technology like cell phones, laptops and transportation. Second generation lithium batteries require a precise and repeatable manufacturing approach that relies on advances in nanotechnology. Anode and cathode coatings with nano sized particles help manufacturers produce a better performing, more efficient battery. As a worldwide manufacturer of bead mills used in the production of dispersions in the ultrafine size range, and dry fluidized bed jet mills for ultrafine grinding of industrial minerals and ceramic materials, NETZSCH has witnessed the trend toward nano particle size solid material in dispersions, those with median particle sizes below 100 nm (nm), and for many applications much finer; and dry powders with median particle sizes in the range of 130 nm with 100% of the distribution finer than 400 nm.

A primary goal of size reduction to the nano range is to improve product performance. This is usually due to the increased surface area that will be created by grinding finer. However, the impetus to develop new products is also huge. Innovation has driven many ceramics industry researchers to look to nanoparticles – materials ground to 200 nm and finer in size – to enhance product performance or unlock new applications for ceramic materials.

Production of nanoparticles using media milling was achieved many years ago, and the benefits are well documented. Nanoparticles are used to improve reactivity, provide increased surface area, and reduce material consumption (particularly with rare and high-value ingredients). A description of this process follows.

WET GRINDING - AGITATOR BEAD MILL TECHNOLOGY

For grinding down to the nano size range or for the dispersion of nano sized particles using wet methods, the use of very small grinding media is necessary. For some products it is also necessary to avoid changes in the crystal structure of the product, therefore mild dispersion is applied. The difference between real grinding – reducing particles below their primary particle size - and mild dispersion – disaggregation or de-agglomeration – is a matter of energy input. This energy input is largely a function of the peripheral speed of the agitator shaft which drives the grinding media. The diameter and density of grinding media is also a factor. If a finer product is required, a smaller diameter grinding media is needed. As grinding media becomes smaller, the process of separation becomes more critical.

Before discussing bead separation systems, it is a good idea to describe what an agitator bead mill is and how it functions. Agitator bead mills evolved from ball mills used since the 19th century. In these horizontally rotating machines, stones or steel balls cascaded due to the rotation of the ball mill drum and milled solids in a liquid suspension, usually water (or without liquid in the case of dry ball mills) in a batch operation. These mills used grinding media measured in inches. Separation of the grinding media from the suspension was simply a matter of emptying the mill through a steel grate at the discharge port.

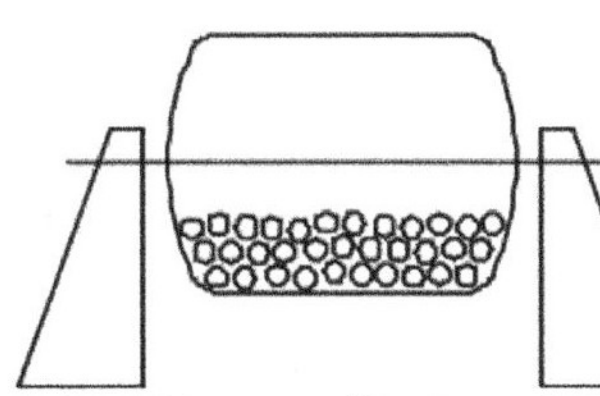

Figure 1 – Diagram of horizontal ball mill

Later, in the middle of the 20th century, attritors developed by Dr. Andrew Szegvari began to replace ball mills in many applications. These were vertical stationary drums with a rotating shaft driving the grinding media. Smaller media could be used in attritors, typically 2mm – 10mm, enabling attritors to grind finer or reach the desired fineness in a shorter time. Attritors can be used in batch or continuous processes. Retention of grinding media in an attritor was similar to a ball mill – passing the suspension through a steel grate at the bottom discharge of the mill.

Ball mills and attritors are still widely used today, along with the use of newer agitator bead mills. In principle an agitator bead mill consists of a grinding chamber, an agitator shaft equipped with agitator elements, a pump and a drive motor. The agitator elements are typically discs or pins. The grinding chamber is filled with grinding media up to 95 % of the mill volume. Grinding media with diameters between 9 mm and 30 µm are used. The grinding media is made from materials such as stainless steel and glass, but also high technology ceramic materials such as yttrium stabilized zirconium oxide and cerium stabilized zirconium oxide are used.

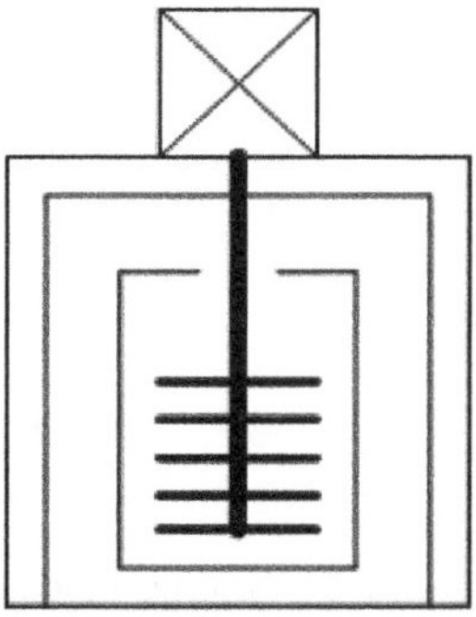

Figure 2 - Diagram of vertical attritor

The agitator rotates with typical tip speeds between 4 and 20 m/s. The grinding media is separated from the product suspension with a separation system at the outlet of the mill to keep the media inside the grinding chamber. The grinding media separation system can be a rotating separation gap, a screen cartridge or a system utilizing centrifugal force (with or without a screen cartridge).

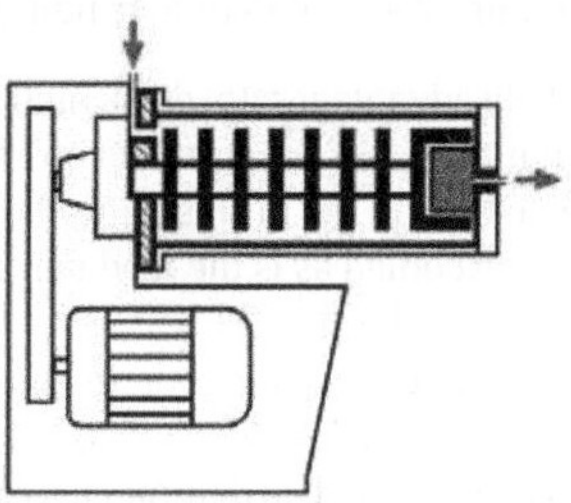

Figure 3 - Diagram of an agitator bead mill

The product suspension is pumped through the mill from a feed tank. There are several methods of passage through the mill. In some cases the suspension is passed through the mill once and collected in a product tank (single pass). Another method is multiple pass through the same mill from tank to tank (pendular) or through a series of mills (cascade). Finally, the suspension can be pumped through the mill multiple times and continuously with short residence time from a single tank until the desired particle size is reached (circulation).

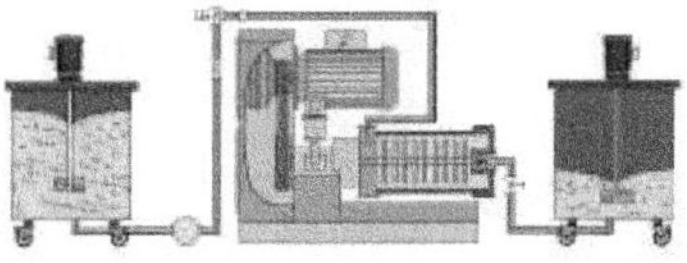

Figure 4 – Single pass grinding

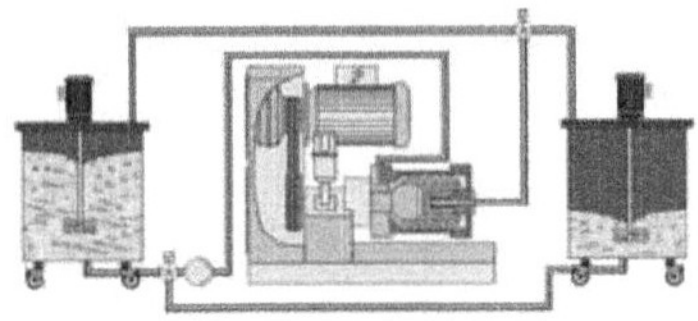

Figure 5 – Pendular grinding

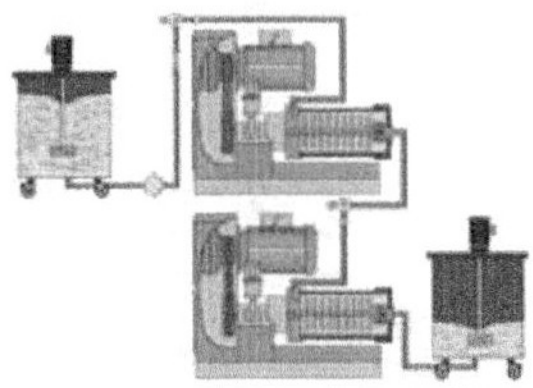

Figure 6 – Cascade grinding

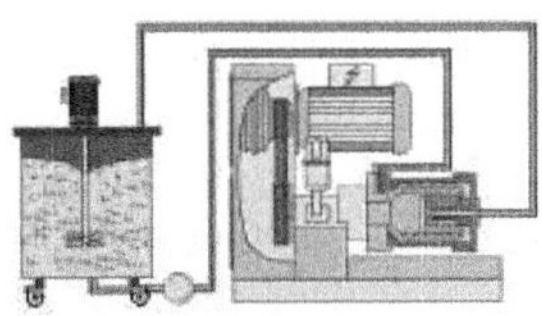

Figure 7 – Circulation grinding

There are advantages and disadvantages to each method. Among the most significant advantages of circulation with a high flow rate are product cooling and residence time distribution. In circulation grinding, the residence time in the mill is quite short during each passage of product. The mill housing and also the rotor are cooled as is the feed tank. This ensures that the temperature of the product can be accurately controlled.

As the product suspension is circulated many times through the mill, this assures that all particles pass through the high energy area of the mill, statistically leading to a more even residence time. This results in a very steep particle size distribution curve as shown in the graph below.

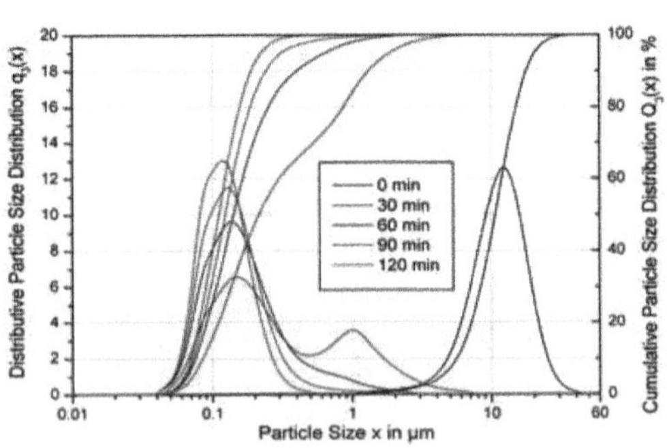

Figure 8 - Development of the PSD during circulation grinding

Separation systems

The simplest and earliest system is a rotating separator gap. This design uses an annular gap between the rotating inner ring and stationary outer ring. The gap is smaller than the diameter of the grinding beads and the parts are made from hard ceramic materials. Product flows through the gap and the beads are retained in the mill.

Figure 9 - Gap separation system

Another type of separation system uses a combination of centrifugal force and a slotted screen. The disc nearest the outlet side of the mill is designed to pump the media back toward the inlet side, while the classifying rotor "rejects" the grinding beads away from the outlet. Some who may be familiar dry powder processing, will recognize the classifying rotor is similar to an air classifier. The function of an air classifier will be discussed later in this paper.

Figure 10 - Dynamic classifying cartridge

The slotted screen located at the material outlet, positioned inside the classifying rotor, prevents beads from exiting the mill during startup before the rotor reaches operating speed and on shutdown while the rotor decelerates. Also broken grinding beads, which may be light enough to be carried by the suspension during normal operation, are prevented from exiting the mill. The following photo demonstrates the effectiveness of this system. It clearly shows the concentration of beads in the proximity of the discs, while there are nearly no beads in the area of the screen and mill outlet.

Figure 11 - Dynamic classifying cartridge in operation

As a further development, a change in the aspect ratio of the mill chamber was made to take advantage of high flow rates and circulation grinding. By increasing the diameter / length dimension of the grinding chamber, the drag force, or the tendency of the suspension to carry the grinding beads toward the outlet side of the mill is reduced. This combined with a larger classifying cartridge and higher screen surface area allows the use of the smallest grinding beads. So far all the agitator bead mills described use a stationary screen or gap separator. The size of the screen is dependent on the size of the grinding beads – normally the screen opening is one half to one third the diameter of the beads. Therefore, if you change the bead size, the screen must also be changed.

A mill with a rotating screen (or no screen at all) for the use of the smallest grinding media

As a general rule of thumb, the final median particle size (d50) of the product will be approximately 1/1000 the diameter of the grinding media. Therefore, to reach a median particle size of 100 nm, a grinding bead diameter of 100 microns is selected. Grinding media as small as 30 μm is sometimes used to reach median particle sizes less than 30 nm.

When using grinding media smaller than 200 μm, coupled with an increasing viscosity of the product suspension when producing nanometer dispersions; it is possible that the media will be transported to the separator screen by the flow forces of the suspension though the mill causing blockage of the screen. The standard in agitator bead mills is the rotating media separation system with a stationary screen; the dynamic classifying cartridge, as described above. This system ensures separation of grinding media from the suspension by centrifugal forces generated only by the classifying rotor.

In one of the newest mill designs, these centrifugal forces are increased further by rotating the separator screen with agitator shaft. This design guarantees grinding media separation even if the viscosity rises during dispersion process. It is especially useful when using very fine grinding media and low agitator tip speeds as are typical in mild dispersion processes.

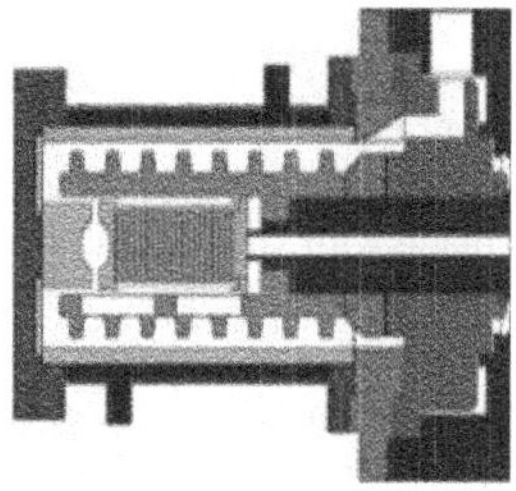

Figure 12 - Agitator bead mill with rotating screen

For low viscosity suspensions, where there is less a danger of beads being transported with the suspension, a screen-less design can be employed, known as an open dynamic classifier. This has several obvious advantages.

- Variable grinding media sizes can be used without having to change the screen size
- The decrease of the grinding media size by wear during long operation is no problem
- Screen plugging is impossible
- Significantly lower pressure increase in the mill leads to higher throughput rates
- The open dynamic classifier system is easily disassembled and cleaned

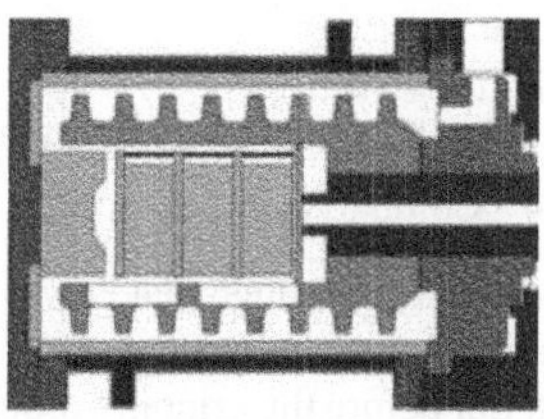

Figure 13 - Agitator bead mill with open dynamic classifier

Mild dispersion and real comminution – practical examples

In real comminution, the primary particles are ground within a liquid phase by high shearing, pressure and impact forces. Typically the agitator tip speed is in the range of 13 – 16 m/sec. Grinding media size and density is selected based on optimal efficiency. The rotor tip speed is maximized in consideration of motor power and product temperature requirements. Particles are de-agglomerated but also crushed and the structure of the particles may be destroyed.

In de-agglomeration or mild dispersion, agglomerates and aggregates are dispersed by lower hearing, pressure and impact forces. The surface air is also removed and the surface of the particles is wetted. Typical agitator tip speeds are in the range of 4 – 6 m/sec. The smallest grinding media is selected based on separation techniques, and the lowest tip speed is used that will provide effective bead separation. The tip speed can be increased as long as the specific energy (kW/kg) does not increase. Particles strings and agglomerates are merely separated into discrete particles. Naturally in both processes, appropriate additives are used to stabilize the suspension and prevent re-agglomeration.

Dispersion of titanium oxide

A comparison of dispersion techniques was made with titanium dioxide. The objective was to de-agglomerate the TiO_2 without degrading the surface as measured by X-ray diffraction. Since the primary particle size of the raw material is already 6 nm, further size reduction was not a consideration. The comparison was made with yttrium stabilized zirconium oxide grinding beads, 0.1 mm diameter with tip speeds of 4, 6, 10 and 13 m/sec. The measured size of the agglomerated feed material was 0.6 microns.

The graphic below shows the median particle size change compared to specific energy input and time. In both comparisons, the lower tip speed proved to be more effective providing a finer dispersion at lower specific energy input.

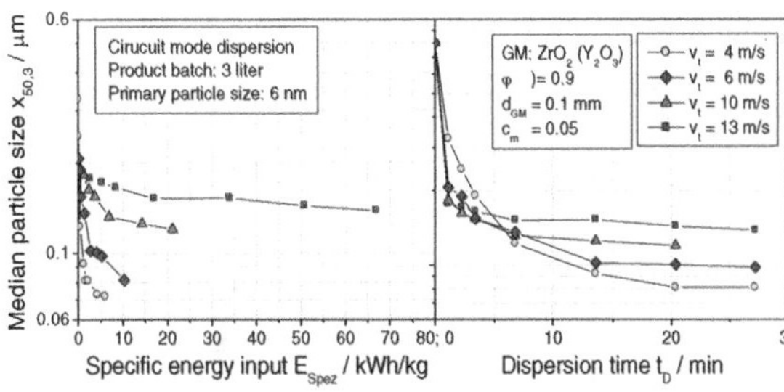

Figure 14 - Comparison of tip speeds for TiO_2

More important however is the advantage the lower tip speed gives in the material properties. From the X-ray diffraction study below, the TiO_2 after mild dispersion maintained quality much closer to the raw material before the grinding process, while the sample processed at high tip speed was significantly different.

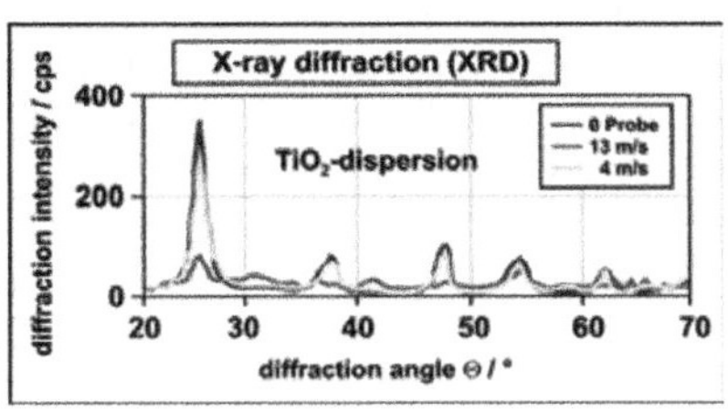

Figure 15 - X-ray diffraction results by tip speed

Real comminution of titanium dioxide

For the production of functional coatings titanium dioxide with the particle size distribution d50 ~ 200 nm and d99 ~ 375 nm was to be ground as fine as possible. It was a water based suspension with a solids content of 48.5 wt%, which was stabilized by an appropriate additive. Two grinding tests were carried out in a 4 liter agitator bead mill.

During test A, yttrium-stabilized zirconium oxide grinding media with a diameter of 0.2 to 0.3 mm were used. The grinding media separation system employed a rotating screen as described above (Figure 12). Test B was made with yttrium-stabilized zirconium oxide grinding media with a diameter of 0.1 mm. To avoid a pressure increase in the mill, the open dynamic separating system (Figure 13) was used. All other operating parameters were constant. The test results are shown in Figure 16 and Figure 17.

During the comminution test with larger grinding media (Test A) particle sizes d50 ~ 60 nm, d90 ~ 92 nm and d99 ~ 125 nm were obtained after a grinding time of 6 hours and a specific energy input of 1.83 kWh/kg. By using smaller grinding media (Test B) significantly better results were obtained with one third of the specific energy input (0.61 kWh/hg) and after a grinding time of only 2.5 hours. The particle size results were d50 ~ 45 nm, d90 ~ 76 nm and d 99 ~ 110 nm.

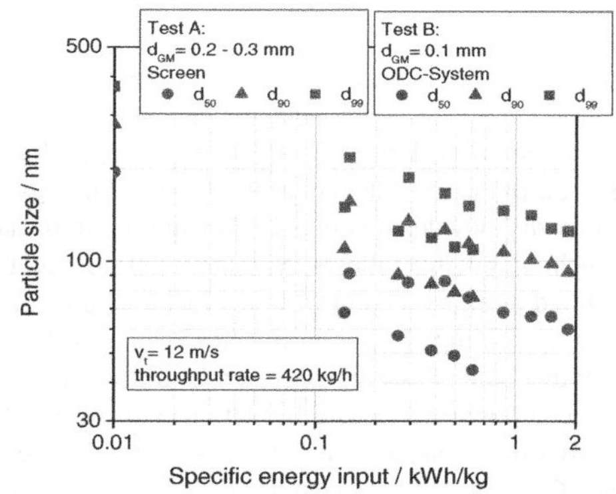

Figure 16 - Comparison of specific energy input by grinding media diameter

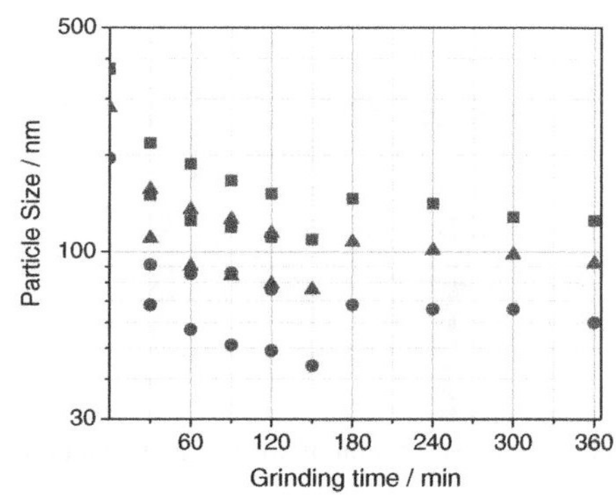

Figure 17 – Comparison of grinding time by grinding media diameter

Media milling technology – the use of grinding beads (media) to break down larger particles into smaller ones — still plays a significant role in many areas of ceramics processing. Media milling processes typically entail dispersing the materials into a suspension; using media mills for grinding or dispersing; then drying the suspension to recover the nanoparticles. (In many cases, the final product is used as a suspension or slurry, and the drying step can be avoided.)

Each successive process is energy intensive and can result in product contamination and wear on the equipment. These challenges have kept many potential products from being produced in efficient commercial-scale manufacturing processes when a dry powder is ultimately needed. However, since dry processing was limited in its capacity to produce particles in the nano size range, wet processing was the only possibility. The demand for finer dry powder products in the submicron or nano scale has led to the development of a new dry milling technology that allows real comminution in this range.

Dry Grinding – Fluidized Bed Jet Mills Using Superheated Steam

The term MICRONIZATION has long been used to describe the miniaturization of particles by jet milling with compressed air or other gas to a range of 2-200 microns (μm). For most dry grinding processes, 2 μm is the practical limit of the technology.

The demand for finer dry powder products in the submicron or nano scale has led to the development of a new technology that allows real comminution in this range. Contrary to conventional dry fluidized bed jet mills, this new system uses superheated steam as the grinding gas. Superheated steam as the grinding gas in jet mills has been used for many decades in simple spiral or loop jet mills without an integrated air classifier. Until now it was not possible to produce a material with a well-defined upper particle size limitation. This new technology provides a mill with an integrated air classifier for the separation of the exact particle sizes, and hence the fineness of the product to be milled. Creation of nano sized dry solids in a dry process with a fluidized bed jet mill, with integral classification is now possible, with the added benefit of being energy efficient and economical.

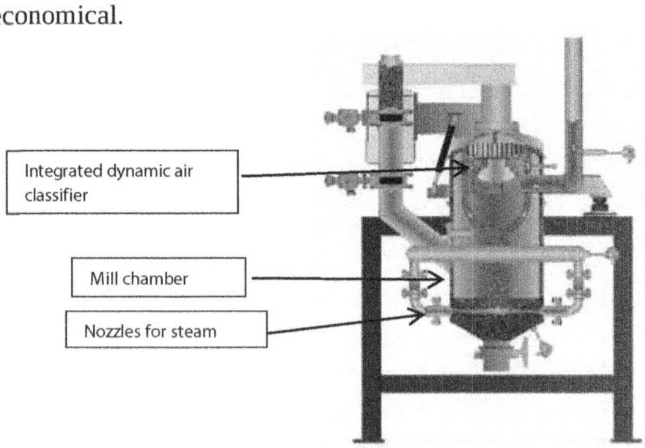

Figure 18. Cross section of a fluidized bed jet mill

Properties of Steam That Enhance the Jet Milling Process

The factors make steam fluidized bed jet milling successful and commercially viable:
Steam can be provided to a jet mill at very high pressures compared to air. At higher grinding pressures, higher jet speeds can be attained. At a pressure of 100 BAR absolute, the jet speed

exceeds 1200 m/sec. Compared to a near maximum of 600 m/sec when using air, the kinetic energy at the nozzles is approximately four times higher. This increased grinding energy results in much finer size reduction in the milling process, therefore finer particles are presented to the integrated classifier (separation system).

While steam jet milling provides the ability to grind solids to higher fineness because of higher kinetic energy, the precise control of the milling process itself is only possible when a grinding mill has an internal classifier. The classifier controls the particle size by not allowing material to leave the grinding system unless the particle size reaches the correct fineness. The performance of a conventional classifier with air is limited to a d_{97} of approximately 2 μm. This of course varies with the material properties, such as specific gravity. However, we are grinding to a much higher fineness than this in steam. How then can we control the classification process to produce a finer product?

As the medium for dynamic classification of particles, steam allows a finer cut size than air. The properties of steam, which has a lower dynamic viscosity and lower density than air, as well as a higher speed of sound than air, makes possible a higher flow velocity within a classifier wheel. Thus the acceleration forces acting on the material being separated is higher and the possible cut size is finer. In a jet mill, this means the particle size distribution of the product is finer.

There are two forces acting on particles in a classifier field (see Figure 19). These are mass force and drag force. Mass force, which in the case of a dynamic classifier is a result of centrifugal force, tends to move particles to the coarse fraction – away from the classifier field and back into the grinding process. Drag force, which is a result of the fluid medium transporting the particles, tends to move particles through the classifier field to the fines fraction where they are discharged with the fluid medium. The gas properties of steam allow for finer particle sizes to be produced by influencing these forces. As mentioned above, steam has a much lower dynamic viscosity and density than air and a higher speed of sound than air.

The higher speed of sound in steam results in a significantly higher peripheral velocity of the gas flow within the classifier (at the same classifier speed). This increases the acceleration of the particles in the classifier field. For example, if the peripheral velocity increases from 200 to 300 m/s, the equivalent effect of the mass force on the particles is a relationship of the square of the difference. Therefore in this example the mass force on the particles is increased by a factor of 2.25, moving the cut size to a finer range by that same factor.

The reduced dynamic viscosity reduces the drag force on the particle. Drag force tends to move the particle to the fines fraction. Reducing the drag force lessens the tendency to drag coarser particles to the fine fraction, also moving the cut size to the finer range. These two factors combined reduce the median particle size of the product by a factor of 0.3 times finer or more. This means – if an air jet mill is capable of making a particular product to a fineness of 1.0 μm, it is now possible to make that same product to a fineness of 0.3 μm or even finer using a steam fluidized bed jet mill. The forces acting on the particles are really quite simple – but the result is dramatic. The use of dry fluidized bed jet mill technology with steam opens new opportunities for nano dry powders.

Table I. Process Conditions

Process Conditions	Air Jet Mill	Steam Jet Mill
Grinding gas pressure (BAR abs)	4 BAR, 225°C	40 BAR, 350°C
Pressure relieved (internal mill) pressure (BAR abs)	1.1 BAR, 150°C	1.1 BAR, 150°C
Jet speed [m/s]	542	1190
Density (relived) [kg/m^3]	0.87	0.53
Sound velocity (relieved) [m/s]	410	504
Dynamic viscosity (relieved) [kg/ms]	$2.41 * 10^{-5}$	$1.42 * 10^{-5}$

The graphic below represents the forces at work in the classification process.

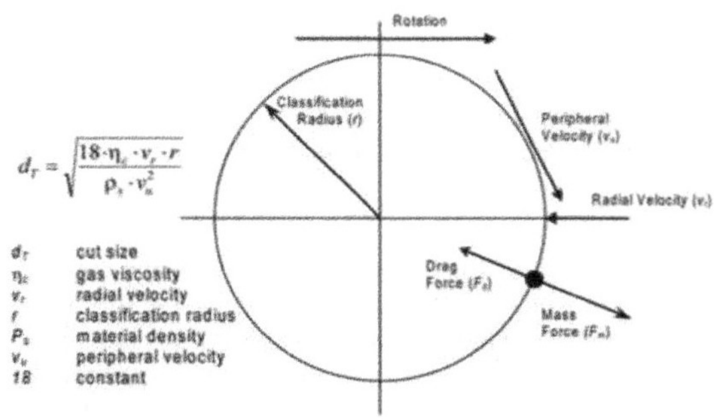

Figure 19. Effective forces of classification

Higher jet energy, as mentioned above, can also result in significant increases in capacity and improved energy efficiency. Jet speeds of 1,200 m/s and the increase of the kinetic impact energy can also lead to a significant increase in mill throughput compared to air operation with comparable finesse. The increase in throughput is proportional to the increase in kinetic energy. For materials of the same particle size and with equivalent gas flow rates, throughput rates are increased by two to three times depending on the friability of the materials.

Test Material	Feed Size	Final Fineness		Throughput
	d_{99} µm	d_{50} µm	d_{99} µm	kg/hour
Ceramic oxide (steam)	70	0.60	2.30	52
Ceramic oxide (air)	70	0.70	2.45	10

Figure 20. Capacity advantage using steam jet mill at the same fineness

Ceramic Materials

A full range of ceramic materials are now being processed. These include aluminum oxide, zirconium oxide, silicon carbide, and calcium carbide to name a few. As an example, aluminum oxide with conventional dry milling can result in a fineness of about 2.3 µm (d_{99}). We are now able to reach d_{50} = 130-140 nm with a d_{99} of .340 to .350 nm.

A Wide Range of Other Applications

Among the other materials that have been tested and processed are ceramic pigments, iron oxide, silica, various battery materials, titanium dioxide, glass, and activated carbon. Prior to the integration of stream in a fluidized bed jet mill, particle size distributions such as these could only be achieved by wet grinding.

Table II. Example of products

Product examples	D_{50}	D_{99}
Alumina	0.13	0.35
Amorphous Graphite	0.16	0.58
Barium	0.13	0.34
Iron oxide	0.07	0.37
Glass frit	0.57	1.89
Limestone	0.54	1.87
Silicon carbide	0.24	1.04
Wollastonite	0.30	2.60
Zirconium oxide	0.44	2.59

The results show that with this technology, it is now possible to achieve particle sizes in the nano range with dry processing. Previously this only could be achieved by wet grinding. However, the disadvantage of wet grinding is the added equipment and energy required for drying when a dry product is required.

Steam Jet Milling Is Greener

Steam is the oldest utilization of thermal energy known to humankind, and is the driving force of almost all energy production worldwide. In 2012, about 87% of the electrical energy in the US was generated by large power plants (coal, gas and nuclear). Large power plants operate on average with a degree of primary energy efficiency of around 40%. Transformation and line losses cause an additional loss of about 10%. Therefore when the electricity arrives at the plant, it has a degree of efficiency (compared to the primary energy) of about 36%. When you factor in compressor efficiency, which is about 45%, the overall energy is only about 16% from primary energy to kinetic (grinding) energy in the mill! By using steam directly, the process becomes two or three times more energy efficient. Grinding with steam is greener!

Table III - US electrical energy generation by source

2012 US Electricity Generation by Source	
Source	Percent of US Electricity
Coal	37
Natural Gas	30
Nuclear	19
Conventional hydroelectric	7
Other renewables	5
Oil	1.0
Other gases	<1

CONCLUSIONS

As the trend toward finer products manufactured in slurries or suspensions continues, the use of smaller grinding media, and techniques suited to their application – whether real comminution or mild dispersion, are further advanced. As the examples above demonstrate, it is not always a matter of using the highest energy input, largest media or highest tip speeds. Indeed, using more energy than necessary results in unnecessary wear on the mill and grinding media, increasing costs and downtime, and decreasing product quality by adding contamination.

The needs of companies developing dry materials in the nanometer particle size range can be met with the use of steam jet milling technology. Steam jet milling technology can apply to ceramics, alternative energy, optical glass, pigments, and industrial minerals markets. New materials may be developed based on this new technology.

Mill designs of both types will continue to evolve to meet the demands of the industry. The engineers designing a product should consult with the engineers designing the milling equipment early in the process. This often involves conducting product trials at the milling company to learn about the capabilities and limitations of the equipment. NETZSCH conducts hundreds of customer trials each year in our customer service laboratories. Development of a product and appropriate processing technology allows the process equipment to be matched to the application for high purity, process reliability and energy efficiency.

THE SENSING PROPERTIES OF FUZZY CARBON NANOTUBE BASED SILICA FIBERS

M. Radetić[1], P. Cortes[1, 2], G. Kubas[2], Jim Cook[3], Ravi Chandra Reddy Gade[4], and T. Oder[4]

[1]Department of Civil/Envtl. & Chemical Engineering. Youngstown State University. Youngstown, OH. 44555
[2]Materias Science and Engineering Program. Youngstown State University. Youngstown, OH. 44555
[3]Department of Electrical and Computer Engineering. Youngstown State University. Youngstown, OH. 44555
[4]Department of Physics and Astronomy. Youngstown State University. Youngstown, OH. 44555

ABSTRACT

The present research program investigates the mechanical sensing properties of fuzzy silica fibers either coated with carbon nanotubes (CNTs), or with a CNT-Zinc Oxide (ZnO) layered arrangement. Here, the CNTs have been grown directly on the silica fiber using a Chemical Vapor Deposition (CVD) process. In the case of the CNT-ZnO based Silica fibers, the semiconducting coating was performed using a sputter deposition system. The sensing properties of the manufactured fuzzy fibers have been investigated through a curing process on a thermoset resin as well as through a series of compressive tests. It has been observed, that the fuzzy fibers are capable of detecting the curing time of the polymeric resin as well as measuring the mechanical deformation of the cured resin after being subjected to a series of compressive tests.

INTRODUCTION

Carbon nanotubes have become a material of great interest in a large number of scientific and engineering fields, due to their exceptional thermal, electrical, and mechanical properties[1]. One area of research that particularly benefits from their unique electrical properties is the area of sensors. Indeed, employing carbon nanotube based electrical sensors onto structural components may allow for selective response of the materials as a result of the nanosensor detecting certain degrees of stimuli. It is well known that when an assembly of carbon nanotubes experience some form of deformation, neighboring nanotubes are formed/broken inducing a change in the electrical resistance of the system[2-4]. In the case of broken nanotubes, the "tunneling current" process maintains the sensing performance of the nanotubes in small-induced strains[5,6].

Nowadays, piezoelectric sensors embedded into composite materials are quite popular systems capable of detecting damages on the structural properties of a material[7]. However, the use of piezoelectric devices can downgrade the fatigue life of a composite[7]. On the other hand, the use of an integrated part of a composite (such as the reinforcing fibers) as the sensing system, is an attractive structural feature since the fiber could act as the sensing phase, in addition to providing its reinforcing function. Indeed, a relatively large number of initiatives on the development of mechanical sensors have been investigated. For instance, Zhao et al. fabricated a sensitive sensor based on covalently bonded nitrogen-doped carbon-nanotube-supported Ag hybrid sponges and described applications in flexible conductors and strain gauges[8]. MacQueen et al. also investigated the strain sensing capabilities of carbon nanotubes for elastomeric membrane deformation, showing reliable sensing responses[9]. It has also been shown that the addition of different coatings and materials on the CNTs may also increase the electrical properties of the nanotubes[10]. In fact, the incorporation of zinc oxide on the CNTs, seems to enhance the mechanical sensing properties of the nanotubes[11]. Additionally, it has been shown that the piezoelectric character of zinc oxide carbon nanotubes complex does not fade even when exposed to continual stress cycles[11]. Certainly, carbon nanotube sensors are responsive to different forms of mechanical sensing deformation such as bending, twisting, or compression[12,13]. Mechanical deformations on the CNTs such as bending, twisting, or flattening will influence the electronic properties of the material[11]. It seems, that the exceptional electrical properties of carbon nanotubes allow for easy sensing customization and will undoubtedly improve the platform of mechanical sensors.

The present paper investigates the sensing properties of fuzzy inorganic fibers based on carbon nanotubes for determining the feasibility of detecting the curing point of thermoset materials as well as their compressive profile.

EXPERIMENTAL

Synthesis of fuzzy fibers and ZnO coating

Silica fibers pre-treated with a ferrocene solution were placed into the chemical vapor deposition (CVD) chamber for the superficial radial growth of CNTs. Figure 1 shows the CNT coated fiber following the deposition process.

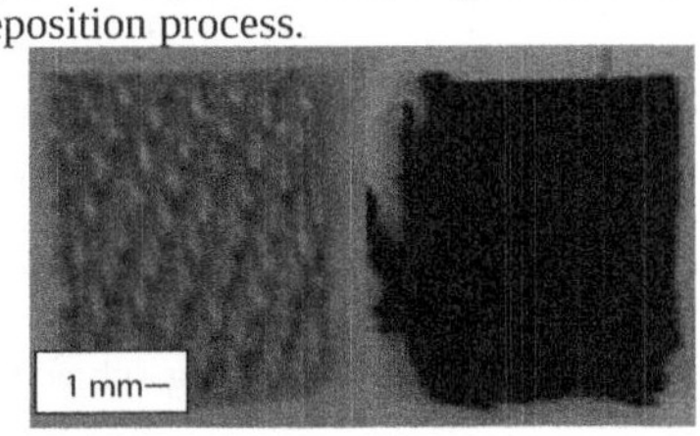

Figure 1. Fuzzy fiber following the growth process of CNTs on the radial surface of silica fibers (Left-plain fiber, Right-CNT coated fiber)

Briefly, the CVD process consisted of introducing the coated sample into a deposition system, filling the chamber with Argon gas to remove any presence of oxygen, and setting the CNT synthesis temperature to 750°C. Once a continuous flow of Argon (@ 40 ml/min) as the carrier gas was maintained, a mixture of hydrogen and acetylene (in a 1:3.5 flow ratio respectively) was introduced into the system for approximately 20 minutes. Following the synthesis step, the hydrogen and acetylene flow streams were stopped with argon continuing to flow through the system for another five minutes. The furnace was turned off and the samples were removed from the chamber once the system cooled naturally to room temperature.

Some of the Silica-CNT coated fibers underwent an additional coating of zinc oxide using a sputter deposition chamber. Here, the pressure in the sputter deposition chamber was brought to a high vacuum and the furnace was set to a temperature of 700°C. An argon/oxygen gas mixture continuously flowed into the system once the deposition began. In order to add a zinc oxide coating of 40 nm, the deposition process lasted for 12 minutes on each side of the fiber.

Preparation of Epoxy/CNT composite material

The CNT "fuzzy" fibers were placed into a mold with the dimensions of 2 cm (length), 1.7 cm (width), and 1 cm (thickness). A 5 minute curing Loctite epoxy resin was used as the matrix material for the fiber/polymer composite. An equal volume of hardener and resin were combined in a separate beaker and mixed. An amount of 3.4 ml (corresponding to the volume of the mold) of the combined hardener and resin solution was added to the mold. Figure 2 illustrates the mold and the final polymer/fiber composite material investigated. Prior to the addition of the polymer solution into the mold, the ends of the CNT-silica fibers were connected to an electrical circuit and a data acquisition (DAQ) device, which continuously recorded (for two hours) the voltage drop across the fibers. Here, the DAQ provided information about the variation of resistance across the fuzzy fibers during the curing process of the sample.

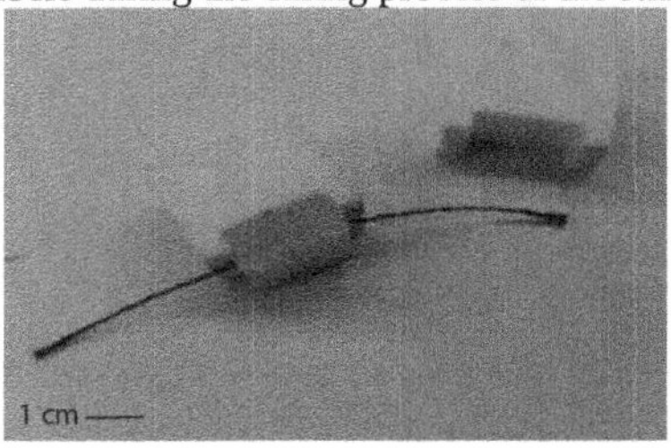

Figure 2. The mold design and final sensor composite

Compression Test

The samples from the curing process were allowed to harden for one day before being subjected to compression tests. The mechanical tests were performed on an Instron universal testing machine with three or more trials repeated for each sample (including control samples). The compression tests were performed at a loading rate of 1 mm/min at room temperature. During the compression tests, the fuzzy fiber was connected to a DAQ device in order to constantly monitor the voltage profile across the sample. This voltage information was subsequently converted into resistance data using Ohm's law. During the mechanical testing, the Instron and the DAQ software were operated simultaneously, in order to match stress-strain profile with the resistance outcome.

DATA AND DISCUSSION

Curing Process

It was of interest to continuously monitor the electrical output of the fuzzy fiber during the curing process of the epoxy system in order to investigate the mechanical detection properties of the CNTs as potential sensing materials. Figure 3 presents a normalized resistance versus time profile, for the sample coated only with carbon nanotubes following the "manufacturing" curing process of the epoxy. The normalized resistance may be determined by taking the difference between the present resistance and the original resistance (prior to testing) and dividing this value by the original resistance. From the figure, it is observed that the resistance increased immediately with the addition of the epoxy up to a normalized value of nearly 0.08 (50.23 kohms). It is here theorized that the increase of the resistance during the curing process is associated to the incorporation of the epoxy resin between the nanotubes, which can increase the electrical percolation of the nanotubes. Figure 4 presents the curing results for the CNT-ZnO trial. A similar profile was observed where the normalized resistance increased to 0.03 (278 Ohms) at the onset of the trial. A normalized steady-state value was also here observed. The slight deformation that resulted by pouring the epoxy on the mold describes this initial increase in resistance.

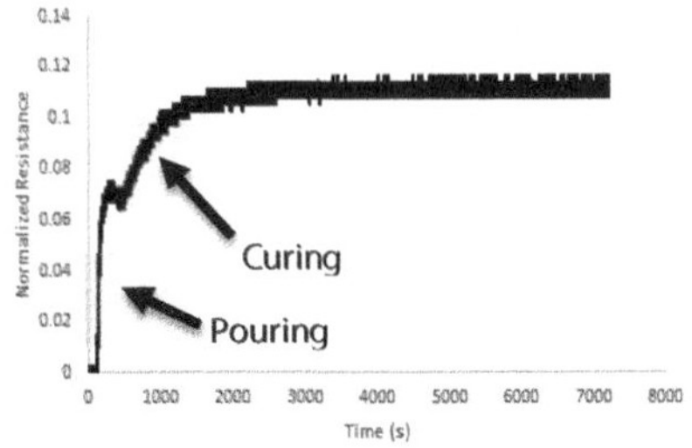

Figure 3. A plot of the normalized resistance for a CNT coated silica fiber during the curing process

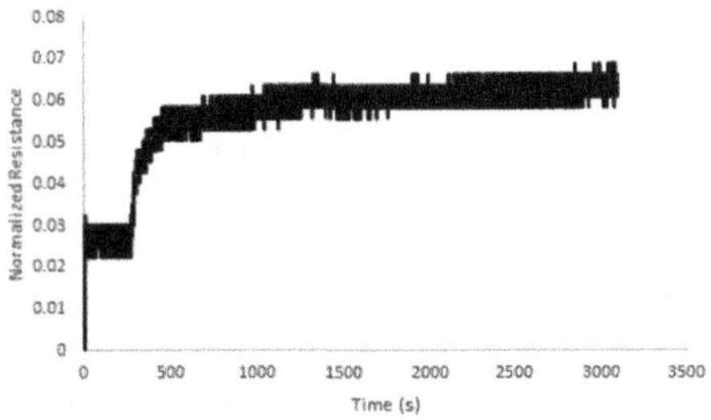

Figure 4: A plot of the normalized resistance for a CNT-ZnO coated silica fiber during the curing process

Following the pouring process, the resistance continued to slowly climb eventually reaching a steady-state value. This increase in resistance appears to be associated to the hardening process of the resin. Here, the observed increase in resistance occurred for both the CNT coated fiber and the CNT-ZnO coated fibers. These results suggest that the fuzzy fibers are able to detect the curing process of a thermoset system.

Compression tests

In order to characterize the mechanical profile of the investigated epoxy material, it was tested under compression at quasi-static loading rates. The stress- strain curve of the plain Loctite material is shown in figure 5. From the figure, a compressive strength of 36 MPa, and a brittle fracture at a strain of 0.913 were observed.

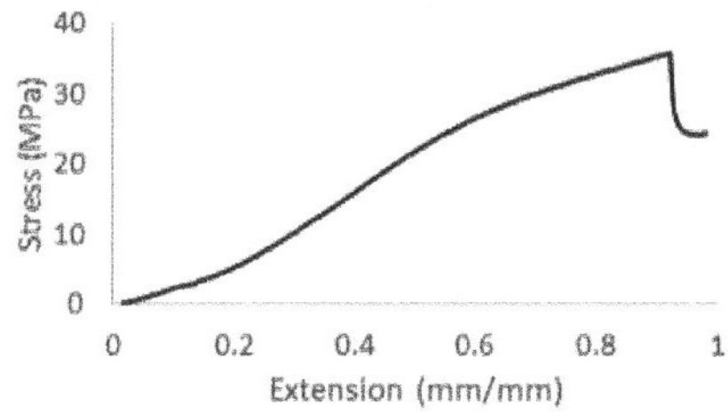

Figure 5. Stress vs. Strain curve of the plain epoxy material, following a quasi-static compression test

The epoxy material containing the CNT-fuzzy fiber was also tested. Figure 6 presents a resistance vs. stress plot following a compressive test for the CNT coated fiber. From the figure, it appears that a stress of 4.05 MPa was required in order to excite the CNT system and detect an initial change in resistance on the nanotubes. In some samples, it was observed that the resistance yielded a plateau value, which could be associated to a permanent damage of the CNTs. The stress on the tested samples was removed, and additional loading trials in a cyclic action were performed, in order to provide information regarding the durability of the sensor. It was observed that the CNT fuzzy fibers appeared to become more sensitive with every successive trial applied to the system. Here, it was seen that the change in resistance increased more rapidly at lower stresses compared to previous trials. The increase of electrical resistance during the compression tests, can be associated to local nanotubes deformations, which could result in electron scattering deformation and consequently on the lattice distortion of the CNTs wall. These deformations appear to increase the intrinsic resistance of carbon nanotubes[14].

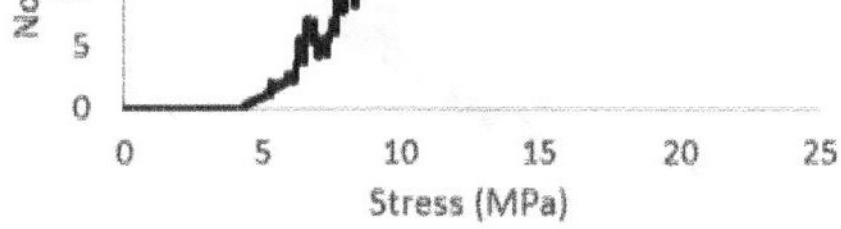

Figure 6. The resistance response to the sensor subjected to a compression test

The ZnO-CNT fuzzy fibers were also mechanically tested, and it was observed that they also produced a change in resistance during the compression process on the epoxy. In contrast to the CNT-fuzzy fibers, these (ZnO-CNT) fibers seemed to be less responsive to the applied stresses, although this system yielded a more consistent pattern of resistance change during the testing as well as displayed a measurable electrical resistance above the 12 MPa (pressure at which the fuzzy system without Zn, yielded a plateau value). Figure 7 shows the resistance change as a function of stress on a CNT-ZnO fuzzy fiber-epoxy system following a quasi-static compression testing. The figure shows that the fiber displays a very responsive electrical profile, as the compression stress increases. These initial results suggest that a reinforcing fiber could also be used as a mechanical sensing platform if subjected to a coating process prior to being assembled on a polymeric matrix. The compression tests were also repeated several times in a cyclic manner to ensure that the sensor continually identified a change in resistance on the system (see figure 8). It was observed that the system experienced a slight enhancement of sensitivity to lower stresses, since a vertical shift was observed when increasing the number of compressive cycles (trials). It was also observed that the shape of the plots for each trial was very similar. It seems that the additional coating of ZnO on the CNT yielded a more stable resistance reading during mechanical deformations. The increase in resistance following a compression test on the fuzzy fibers is a feature that needs further studied.

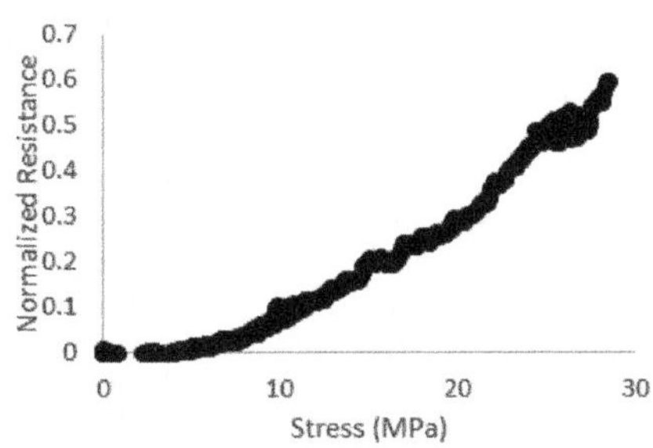

Figure 7. Resistance vs. Stress curve CNT-ZnO compression test (trial 2)

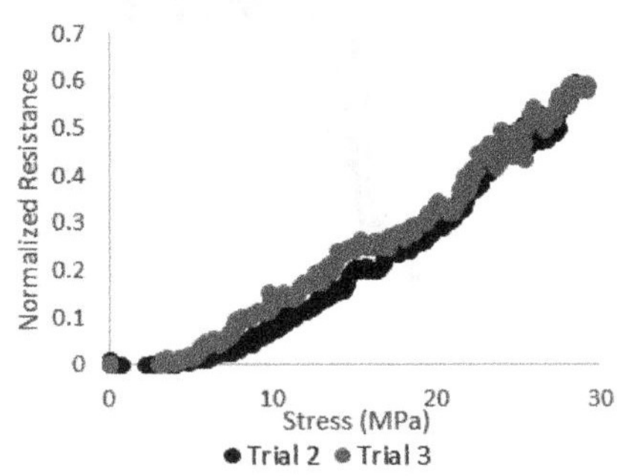

Figure 8. Resistance vs. Stress curve of the CNT-ZnO fuzzy fiber-epoxy system after being subjected to several compression cycles.

CONCLUSIONS

It has been observed that fuzzy fibers based on either CNT or ZnO-CNT are capable of detecting the curing process of a 5 minute Loctite epoxy material. Additionally, the resistance of the fuzzy-fibers increases during the curing process reaching a steady-state resistance value after the curing. It has also been observed that the fuzzy fibers were capable of detecting compressive deformation on an epoxy material. Furthermore, the ZnO-CNT fuzzy fibers yielded a sensitive sensing compression performance which was repetitive after several compression cycles. The increase of cycling trials seemed to result in a fuzzy-fiber slightly more responsive to lower compression stresses. The reason behind the increase in resistance during the curing and compression process requires further investigation. However, these preliminary results indicate that the reinforcing fiber can be used as an intrinsic sensing mechanical platform on composite material following previous CNT or ZnO-CNT coating process.

REFERENCES

[1]A. Naghashpour and S.V. Hoa, In situ monitoring of through-thickness strain in glass fiber/epoxycomposite laminates using carbon nanotube sensors, *Composites Science and Technology.* **78,** 41 (2013)

[2]G. Ehlert and H. Sodano, Fiber strain sensors from carbon nanotubes self-assembled on aramid fibers, *Journal of Intelligent Material Systems and Structures.* **25,** 2117 (2014)

[3]AH. Barber, Q. Zhao et al. Characteriza-tion of E-glass–polypropylene interfaces using carbon nanotubes as strain sensors. *Composites Science and Technology.* **64,** 915 (2004).

[4]I. Kang MJ. Schulz et al. A carbon nano-tube strain sensor for structural health monitoring. Smart Materials and Structures 15, 737 (2006).

[5]Hu N, Karube Y, Arai M, et al. Investigation on sensi-tivity of a polymer/carbon nanotube composite strain sensor. *Carbon* **48,** 680 (2010).

[6]Alamusi, N. Hu, et al. Piezoresistive Strain Sensors Made from Carbon Nanotubes Based Polymer Nanocomposites. *Sensors.* **11,** 10691 (2011).

[7]ND Alwxopoulos, C. Bartholome, et al., Damage detection of glass fiber reinforced composites using embedded PVA-carbon nanotube (CNT) fibers, *Composites Science and Technology.* **70,** 1733-1734 (2010)

[8]S. Zhao, Y. Gao, et al., Covalently bonded nitrogen-doped carbon-nanotube-supported Ag hybrid sponges: Synthesis, structure manipulation, and its application for flexible conductors and strain-gauge sensors, *Carbon.* **86,** 223-234

[9]L. MacQueen, C. Oleg, et al., Miniaturized platform with on-chip strain sensors for compression testing of arrayed materials, *Lab on a Chip.* **12,** 4178-4184 (2012)

[10]L. Zuccaro, K. Kern, et al., Identifying Chemical Functionalization on Individual Carbon Nanotubes and Graphene by Local Fingerprinting, *ACS Nano.* **9,** 3314 (2015)

[11]JH. Chia, HL Yu, et al., ZnO-Coated Carbon Nanotubes: Flexible Piezoelectric Generators, *Advanced Materials.* **23**, 2941 (2011)

[12]P. Dharap, Z. Li, S. Nagarajaiah, et al., Nanotube film based on single-wall carbon nanotubes for strain sensing, *Nanotechnology.* **15**, 379 (2004)

[13]M. Maschmann, G. Ehlert, et al., Bioinspired Carbon Nanotube Fuzzy Fiber Hair Sensor for Air-Flow Detection, *Advanced Materials,* **26**, 3230-3234 (2014)

[14]S. Gong, ZH. Zhu, et al. Modeling electrical conductivity of nanocomposites by considering carbon nanotube deformation at nanotube junctions. Journal of Applied Physics. 114 074303 (2013).

NANOMODIFIED LOW-COST BIOLOGICAL MATERIAL FOR THE REMOVAL OF HEAVY METAL IONS

Rozumová L.[1,2], Seidlerova J.[2], Safarik I.[3]

[1]Research Centre Rez, Hlavní 130, Husinec-Řež, 250 68, Czech Republic

[2]VSB – Technical University of Ostrava, Nanotechnology Centre, 17. listopadu 15/2172, Ostrava, 708 33, Czech Republic

[3]Institute of Nanobiology and Structural Biology of GCRC AS CR, Department of Nanobiotechnology, Na Sadkach 7, Ceske Budejovice, 370 05, Czech Republic

ABSTRACT

This work deals with sorption properties of magnetically modified biological materials using low-cost biological material peanut husks, a waste from agriculture industry. Peanut husks represent a possible replacement for the current expensive adsorbent for removal of heavy metals from wastewater.

Magnetic iron oxide nanoparticles (ferrofluid), ranging from 10 to 20 nm in size according to electron microscopy measurements, were used to prepare magnetically modified peanut husks. Chemical compositions, phase analysis, scanning electron microscopy were used for characterization of prepared material.

Adsorption studies were carried out using the batch technique at a laboratory temperature to obtain the sorption and desorption equilibrium data. AAS-FA was used for determination of the heavy metals concentration before and after both experimental studies. The experimental data were assessed by several types of adsorption isotherms. The changes of Fe content in magnetically modified materials were observed as well.

KEYWORDS: Lead; Zinc; Peanut husks; Sorption

INTRODUCTION

Clean water is an essential and limited natural resource for the development of a series of living organisms in aquatic environments as well as for humans, all of which require its preservation. The quality of such resources is negatively affected by industrial development. Common sources of metal polluted wastes include electroplating plants, metal finishing operations, as well as many mining, nuclear and electronics industries. Therefore, any type of aquatic pollution is always an environmental concern with a high risk for animal and human health. Conventional methods (physical and chemical) for removal of toxic metals are available, but they have drawbacks such as incomplete metal removal, reagent requirements and excessive generation of toxic sludge[1,2].

Therefore, it is necessary to develop new environmentally friendly ways to clean up contaminants using low-cost methods. Among low-cost techniques, adsorption is an efficient and economic method, based on simple and flexible operating conceptions and use of regenerative adsorbents, for the removal of organic or inorganic pollutants with a high efficiency in many cases[3,4].

Biological sorbents provide an alternative to existing technologies because they are more cost effective, eco-friendly and can be applied to removal of metal ions and other contaminants from wastewater[3].

Sorption of metal ions by various biological materials have been used extensively in the last decade due to its potential for a wastewater treatment. Several studies showed that non-living microorganisms are effective in the removal of heavy metal ions[3]. As a result, a great variety of alternative low cost adsorbents obtained from different types of biomass have been proposed, such as nut shells[5], wool, olive cake, pine needles, almond shells, cactus leaves, papaya wood[6], maize leaf[7], leaf powder[8], peanut hull pellets[9], tree fern[10], rice husk ash[11], neem bark[11], grape stalk wastes[12], sago waste[13], peanut hulls[14], hazelnut shell[15], saw dust[16], chitin beads[17], thermally treated rice husk ash[18], waste banana[19], orange peels[19], cocoa shells[20], coffee residue[21], palm kernel fibre[22], olive stone waste[23], orange peel[24], grape stalk[25], coir[26] ,bagasse fly ash[27], cane bagasse[28], marine algae[29] and dead aquatic plants[30, 31] have been reported for the removal of heavy metals from aqueous solutions.

The peanut is an oil plant, which is extensively cultured in India and China. Thus, peanut husk is an abundant and inexpensive agricultural by-product. Exploitation and utilization of this biomaterial should bring obvious economic and social benefits. Peanut husk has been used as an adsorbent for the removal of heavy metal ions from solutions[32].

Peanut husks were magnetically modified for adsorption. Magnetic derivatization of adsorbents is a very important modification which was proven to improve the manipulation and increase the adsorption capacity[33,34,35]. Biological materials are mainly diamagnetic, which means that they do not interact significantly with an external magnetic field. However, their magnetic modification significantly improves the whole adsorption and separation process[36].

In this paper, the sorption capacity of a new magnetically modified biosorbent obtained from peanut husks was investigated for removing Pb(II), Zn(II) ions and Pb(II)-Zn(II) binary metallic system from test metallic solutions. Batch sorption experiments were carried out, assessing shaking velocity on metal sorption performance. The objective of the present study was to investigate the efficiency of adsorption for the removal of heavy metal ions from different concentration of solutions.

MATERIALS AND METHODS

Preparation of biosorbent

The peanut husks waste from locally available roasted peanuts was ground in a coffee mill. Fractions smaller than ca 0.5 mm were collected and used for magnetic modification. The peanut waste husks was stabilized by perchloric acid. Water-based ionic magnetic fluid was prepared using a standard procedure[37]. For the magnetic modification were used ferrofluid, magnetic iron oxide nanoparticles. Nanoparticles of ferrofluid had diameters ranging between 10 and 20 nm. The relative magnetic fluid concentration (25.2 mg/mL) was given as the iron (II, III) oxide content[38].

The peanut husk powder was suspended in methanol and then ferrofluid was added. The suspension was mixed for one hour. Then a sample was rinsed by methanol.

Reagents

Chemical reagents used in these experiments were of analytical grade. In all experiments fresh demineralized water was used for the preparation and dilution of aqueous solutions. Using water-soluble Pb and Zn salts ($Pb(NO_3)_2$ and $Zn(NO_3)_2 \cdot 6H_2O$ compounds), a set of stock solutions of Pb(II) and Zn(II) was prepared at concentrations ranging from 1 to 700 $mg.L^{-1}$. Stock solutions of Pb(II) and Zn(II) were prepared by dissolving lead nitrate and zinc nitrate in demineralized water and then conserved using concentrated nitric acid, analytical grade. Working solutions were obtained by diluting the stock solutions.

Analytical techniques

Chemical compositions of the samples were determined using X-ray fluorescence spectrometry (SPECTRO XEPOS). Iron oxides in the samples were detected by a power X-ray difractometer (BRUCKER D8 ADVANCE). The surfaces of the samples were characterized by scanning electron microscopy (SEM) combined with energy dispersive X-ray (QUANTA 450 FEG). The specific surface area was measured with the Sorptomatic 1990 using nitrogen and calculated by the Advance Data Processing software according to the BET isotherm. Particle size was measured by Laser Scattering Particle Size Distribution Analyzer (Horiba LA-950).

The Pb, Zn and Fe concentration in aquatic solution was determined by atomic absorption spectroscopy with flame atomization AAS – FA (UNICAM 969).

Sorption experiments

The sorption experiment was conducted in a stock solution (adsorbate). The stock solution containing 1g.l^{-1} of Zn and Pb was prepared from $Zn(NO_3)_2.6H_2O$ and $Pb(NO_3)_2$. The stock solution concentrations in a range of 1 to 700 mg.l^{-1} were prepared by a serial dilution in deionized water followed by a conservation by concentrated nitric acid of analytical grade (5 ml of 65 wt. % acid added per one liter of solution). Single system comprise only stock solution ions of Pb(II) and Zn(II) of known concentration and binary metallic system comprise together Pb(II)-Zn(II).

0.2 g of magnetically modified peanut husks were poured into 100 ml plastic bottle containing 50 ml of metal solution with defined initial ion concentration. The samples were stirred for 60 minutes at a laboratory temperature using overturn shaker (Heidolph Reax 20). The adsorbent was separated using a membrane filter (0.45 μm).

Adsorption studies were carried out using the batch technique to obtain the equilibrium data. Adsorption isotherms were obtained from data from AAS-FA of experiment Zn^{2+} and Pb^{2+} ions separately as well as with the mixture of the two ions.

Peanut husks adsorption capacities (mg.g^{-1}) were calculated on the base of the following Equation (1):

$$\text{Sorption capacity} = (x_0 - x_s) / x_s \quad (1)$$

where x_0 is the total mass of sorbent after adsorption and x_s (g) is the mass of the sorbent before adsorption. All experiments were conducted at room temperature (25 °C ± 2 °C).

Isotherm study and data evaluation

Batch adsorption isotherm studies were performed using magnetically modified peanut husks. The obtained experimental data were analysed using Langmuir (type 1-4), Freundlich, Temkin and Dubinin-Radushkevich adsorption isotherm models.

The Langmuir model is obtained under the ideal assumption of totally homogenous adsorption surface. There are four different ways how to linearize the Langmuir isotherms[39]. The values of Langmuir constants can be calculated from the plot between c_e/q_e versus c_e, $1/q_e$ versus $1/c_e$, q_e versus q_e/c_e, and q_e/c_e versus a_e for linearized equation types 1-4, of Langmuir isotherms, respectively. Linear regression analysis was applied to determine the best corresponding adsorption isotherm. Langmuir Equation (2) is given by:

$$q_e = \frac{q_m K_L c_e}{1 + K_L c_e} \quad (2)$$

where q_e (expressed in mg.g^{-1} or mg.mL^{-1}) is the amount of the adsorbed metal ion per sorbent unit and c_e (expressed in mg.L^{-1}) is the unadsorbed metal concentration in solution at equilibrium. q_m is the maximum amount of the metal ion per mass unit to form a complete monolayer on the surface and K_L is a constant related to the affinity of the binding sites (expressed in L.mg^{-1}).

The Freundlich isotherm is an empirical equation employed to describe highly heterogeneous systems. The Freundlich equation (3) is expressed:

$$q_e = K_F c_e^{1/\beta} \quad (3)$$

where q_e is the amount of adsorbed metal ions per unit of adsorbent weight, K_F and β are the Freundlich constants related to the adsorption capacity and adsorption intensity, respectively, and c_e is the concentration of unadsorbed metal ion in solution at equilibrium[40].

The Temkin isotherm model[41] contains a factor that explicitly takes into account adsorbing species – adsorbate interactions. In this model it is assumed that the heat of adsorption of all the molecules in the layer decreases linearly with coverage due to adsorbent – adsorbate interactions, and that the adsorption is characterized by a uniform distribution of binding energies, up to a maximum binding energy. The Temkin equation (4) has been applied in the following form:

$$q_e = \frac{RT}{b_T}\ln(A_T c_e) = B\ln(A_T c_e) \qquad (4)$$

where b_T is Temkin isotherm constant, A_T ($L.g^{-1}$) is Temkin isotherm equilibrium binding constant, R is the ideal gas constant, T (K) is the temperature of adsorption and B ($J.mol^{-1}$) is a constant related to heat of the sorption.

The Dubinin–Radushkevich isotherm is applied to find out the adsorption mechanism based on the potential theory assuming a heterogeneous surface. Dubinin–Radushkevich Equation (5) is expressed as follows:

$$q_e = q_{mD} e^{-\beta\varepsilon^2} \qquad (5)$$

where q_{mD} ($mg.g^{-1}$) is the Dubinin–Radushkevich monolayer capacity, β is a constant related to the sorption energy, and ε is the Polanyi potential which is related to the equilibrium concentration[42].

RESULT AND DISCUSSIONS

Material characterizations

The chemical composition of the peanut husks sample is shown in Table 1. The chemical composition is expressed via oxides in original sample of peanut husks (OPH) and magnetically modified peanut husks (MMPH). It should be noticed that the process of magnetic modification lead to pronouncedly enhanced Fe content. Iron in the magnetically modified peanut husks took forms of magnetite ($FeO.Fe_2O_3$), wustite (FeO) and non-stechiometric iron oxide γ-$Fe_{21.34}O_{32}$ (Fig. 1). The ignition loss of the magnetic peanut husks was 92.3 %.

Table 1. Chemical composition of samples of original peanut husks (OPH) and magnetically modified peanut husks (MMPH). Content of elements is expressed through the oxides (wt. %)

Sample	Na_2O	MgO	Al_2O_3	SiO_2	P_2O_5	SO_3	K_2O	CaO	TiO_2	MnO	Fe_2O_3
OPH	0.73	0.19	0.09	0.08	0.11	0.25	0.48	0.31	0.01	0.01	<0.01
MMPH	0.67	0.17	0.08	0.07	0.10	0.23	0.44	0.28	0.01	0.01	6.22

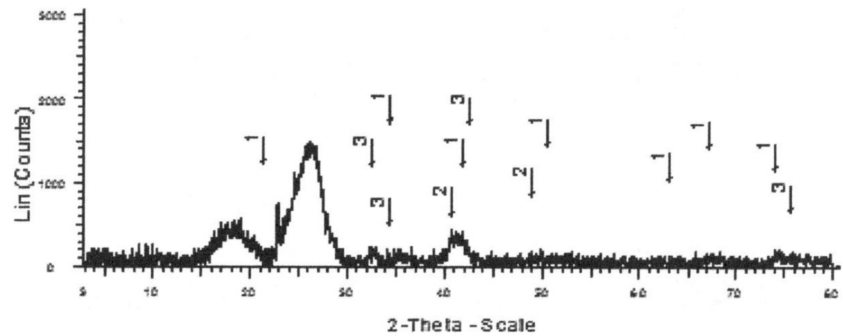

Fig. 1. X-ray diffraction pattern of magnetically modified peanut husks showing presence of the specific oxides (1 – magnetite; 2 – wustite; 3 – γ iron oxide).

Morphological appearance and maps of selected elements are shown in Fig. 2 and 3. On the view of the sample before adsorption is shown an area of clusters. By mapping it was proven that adsorbed particles of iron oxides were present on the surface of peanut husks. The SEM image reveals the porous structure of peanut husks surface. Comparing the SEM images of the peanut husks before and after adsorption demonstrates that the metal ions are adsorbed, as confirmed the maps of selected elements. After the adsorption of the metal ions (Fig. 4), the pores on peanut husks are covered. The maps of elements of the adsorbed metal ions demonstrate the adsorption of metal ions on the surface of the adsorbent (Fig. 5).

Specific surface area magnetically modified peanut husks was 7.99 $m^2.g^{-1}$ and the average particle size was 136 μm.

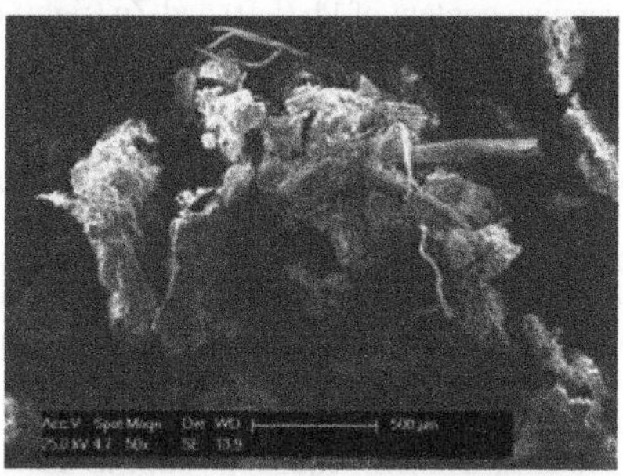

Fig. 2 SEM images of magnetically modified peanut husks before adsorption of metal ions

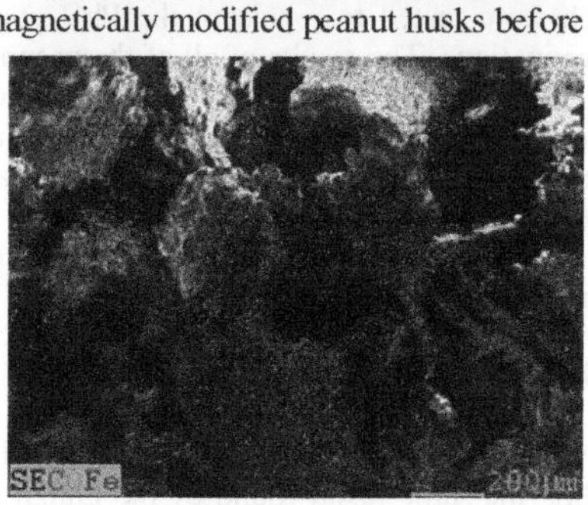

Fig. 3 SEM images of mapping content of C (red dots), O (green dots), Fe (blue dots) of magnetically modified peanut husks before adsorption of metal ions

Fig. 4 SEM images of magnetically modified peanut husks after adsorption of Pb (left) and Zn (right)

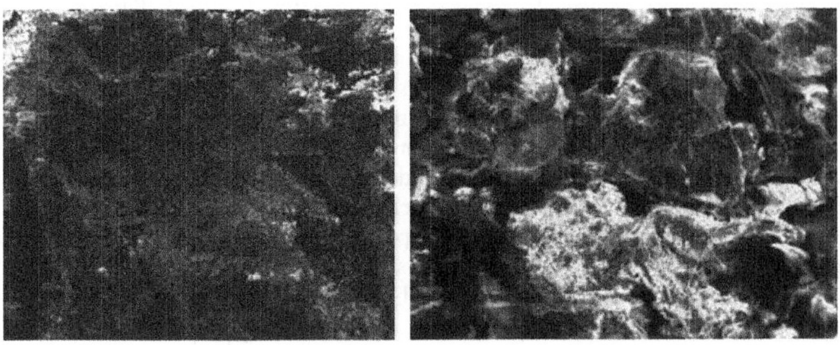

Fig. 5 SEM images of mapping content of Pb (left) and Zn (right) of magnetically modified peanut husks after adsorption.

Adsorption isotherms

Sorption experiments for single and binary systems using magnetically modified peanut husks as a biosorbent were performed under constant shaking for 60 min to reach the equilibrium at room temperature. After shaking, the samples were filtered through 0.45 μm membranes. Metal concentrations in the liquid phase were determined by AAS – FA. The experimental adsorption data were described by the linear form of individual adsorption isotherms and the linear coefficient of determination (R^2). R^2 represents a proportion of the dependent variable expressed by the regression line was examined.

For single metal systems, isotherm equations such as Langmuir, Freundlich, Temkin and Dubinin-Radushkevich were tested to represent the Pb^{2+} and Zn^{2+} sorption data with modelling parameter determinations. The results of all of the proposed models with regard to their adjustable modelling parameters are summarized in Table 2, corresponding to Pb^{2+} and Zn^{2+} sorption data, respectively, using magnetically modified peanut husks as biosorbent. Table 2 summarizes the adsorption constants derived from linearization and their respective coefficients of determination.

The results show that magnetically modified peanut husks remove effectively the metal ions from water solution in the single system.

According to the indicators of goodness of the fit, namely R^2, the Langmuir model fits the single metal equilibrium data better than the other models, followed by Temkin and Dubinin-Radushkevich.

It must be noted that for the single systems, the adjustable adsorption parameter values, were equal to one, providing the same result found by the Langmuir model, suggesting that Pb^{2+} and Zn^{2+} uptake by magnetically modified peanut husks could be interpreted as a monolayer adsorption process. In addition, it can be noticed that an improvement in metal sorption capacities was achieved by magnetic modification in comparison to the unmodified peanut husks.

Fig. 6 shows the amount of Pb and Zn ions sorbed per unit mass of material (q_e) plotted against the concentration of Pb and Zn single system of metal ion solution (c_e) after reaching the equilibrium. The maximal adsorption capacity of Pb^{2+} ions on the adsorbent is 34.3 mg.g^{-1} and Zn^{2+} is 17.8 mg.g^{-1}, suggesting that magnetically modified peanut husks have the following order of affinities: $Pb^{2+} > Zn^{2+}$. A larger number of available sorption sites are apparently available for lead compared to zinc in binary system. In binary system "Pb and Zn ions", ions of Pb fill the pores and Zn ions do not have available adsorption sites on the sorbent surface.

Table 2. Isotherms parameters obtained by using linear regression

Adsorption isotherms	Isotherm constants	Adsorbate Solution	
		Pb	Zn
Langmuir-type 1	q_m	34.4	25.5
	K_L	0.087	0.008
	R^2	0.997	0.755
Langmuir-type 2	q_m	28.3	9.60
	K_L	0.434	0.070
	R^2	0.999	0.999
Langmuir-type 3	q_m	30.6	19.5
	K_L	0.367	0.031
	R^2	0.900	0.685
Langmuir-type 4	q_m	31.3	23.2
	K_L	0.331	0.031
	R^2	0.900	0.685
Freundlich	K_F	4.69	0.463
	n	2.784	1.642
	R^2	0.877	0.985
Temkin	b_T	9767	9198
	A_T	$6.1.10^{-5}$	5.237
	R^2	0.985	0.758
Dubinin - Radushkevich	q_D	26.3	9.415
	K_D	$8.5.10^{-8}$	$2.35.10^{-7}$
	R^2	0.917	0.738

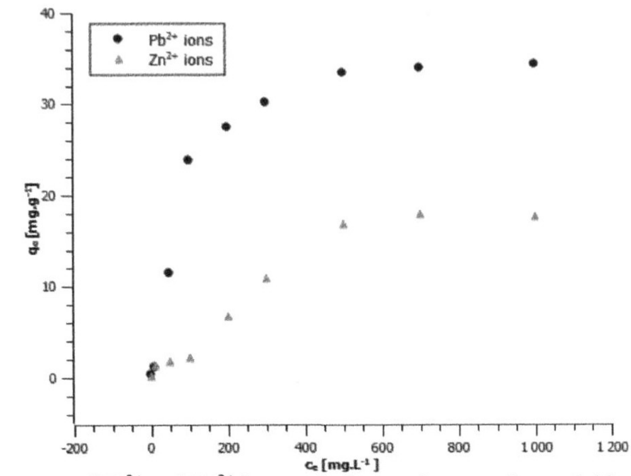

Fig. 6. The amount of Pb^{2+} and Zn^{2+} ions sorbed per unit mass of material in one-metal-ion solution

The equilibrium parameter (R), also referred to as separation factor. It is a constant (factor), which can be used to determine the suitability of the adsorbate. The values of equilibrium parameter were calculated according to the Equation (6):

$$R = 1/(1+Kc_0) \quad (6)$$

where c_0 [mg.L^{-1}] is the initial amount of adsorbate, K constant of adsorption. This factor R is a direct function of K which is related to the slope $1/q_m$ of the linearized Langmuir expression. R is known as the "constant separation factor" or "equilibrium parameter". The slope of the linearized Langmuir isotherm can be used to interpret the type of adsorption by using the values of R as follows: $R > 1.0$, not suitable; $R = 1.0$, linear; $0 < R < 1.0$, suitable and $R = 0$, irreversible[43]. The result of calculation using $K = K_L$ (calculated from linearized Langmuir isotherm - type II) is shown in Fig. 7. The values of equilibrium indicate that sorption is more favourable for higher initial concentrations of Pb^{2+} and Zn^{2+} ions than for their lower concentrations.

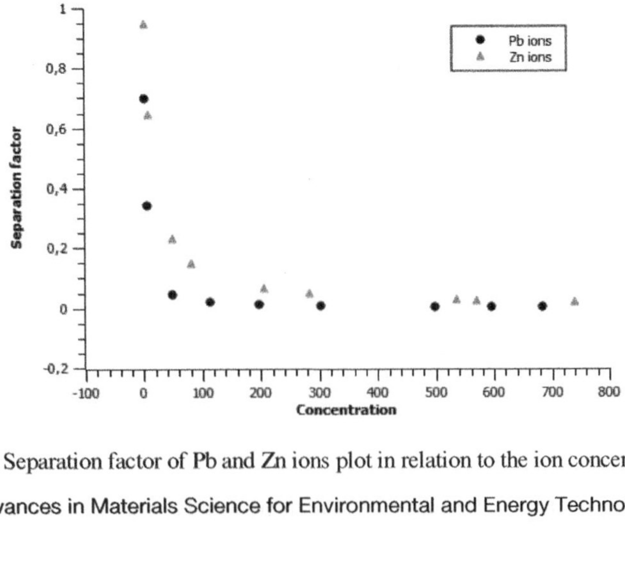

Fig. 7. Separation factor of Pb and Zn ions plot in relation to the ion concentration

The results indicate that adsorbate forms a monolayer on the outer surface of adsorbent while no further adsorption takes place. Maximum monolayer coverage capacities calculated from linearized Langmuir isotherm - type II of Pb^{2+} and Zn^{2+} ions are 28.3 $mg.g^{-1}$ and 9.60 $mg.g^{-1}$, respectively. The Langmuir constant (or affinity constant – K_L) expresses how strongly are the adsorbed ions attracted to a surface. Pb^{2+} (K_L =0.434) ions form stronger bonds on surface than Zn^{2+} ions (K_L= 0.070).

The curves of the Langmuir isotherms compared to the results of sorption from solution containing both Pb and Zn ions are plotted in Fig. 8. In the case of adsorption from solution containing Pb^{2+} and Zn^{2+} ions, the maximum monolayer coverage capacity of Pb^{2+} ions was reduced by approximately 76 % (6.72 $mg.g^{-1}$) and the maximum monolayer coverage capacity of Zn^{2+} ions was reduced by 92 % (0.72 $mg.g^{-1}$).

Such results could be caused by ionic mobility, diffusion coefficient, ionic radii and hydration energy. Although the ionic mobility and diffusion coefficients of Pb^{2+} ions are higher than those of Zn^{2+} ions, the hydration energy of Pb^{2+} ions is smaller than that of Zn^{2+} ions. Consequently, Pb^{2+} ions have a greater accessibility to the surface than Zn^{2+} ions.

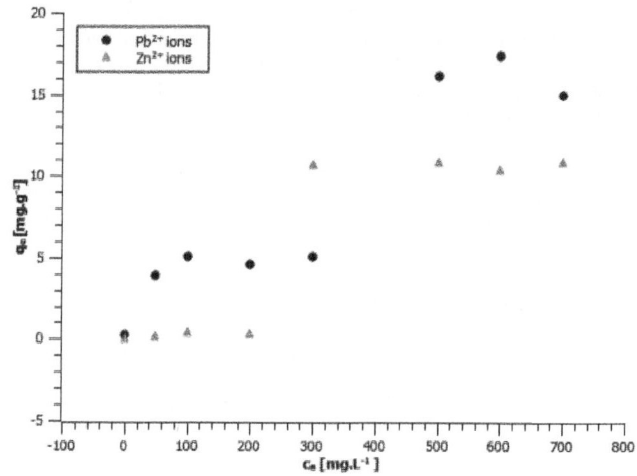

Fig. 8. The amount of Pb^{2+} and Zn^{2+} ions sorbed per unit mass of material in two-metal-ion solution

Negligible leaching of iron was observed during sorption experiments. The concentration of iron in solution was determined after sorption and desorption of metal ions. The concentration of iron in solution after sorption and desorption was low and the portion of dissolved iron did not exceed 0.20 % of iron content in magnetically modified peanut husks. The adsorbed amount of metal ions was not affected by the loss of iron from magnetic material. This means that the iron oxides are very strongly fixed onto the surface and that the material could be used as Pb^{2+} and Zn^{2+} ion sorbent.

CONCLUSIONS

The magnetically modified peanut husks were used to prepare a biological sorbent, the adsorption properties of which were studied. Results of experiments show that they represent a suitable material for sorption of Pb^{2+} and Zn^{2+} ions from water solution. Adsorption data of the biosorption of metal ions on magnetically modified peanut husks were described by the two-parameter Langmuir isotherm model. The results indicate that the adsorbate formed a monolayer on the outer surface of adsorbent, and that no further adsorption took place. During the sorption from two-ion solution, the maximum monolayer coverage capacity of Pb^{2+} and Zn^{2+} decreased. It was also observed that Pb^{2+} ions have the same diffusion profiles in both single and binary metal systems, but the diffusion of Zn^{2+} ions undergoes an inhibition due to the presence of Pb^{2+} ions, limiting its maximum sorption capacity in a binary system. The affinity of Pb^{2+} ions to the surface were better than that of Zn^{2+} ions as a consequence of ionic mobility, diffusion coefficient, ionic radii and hydration energy.

Acknowledgement

This work has been supported by the SUSEN Project CZ.1.05/2.1.00/03.0108 realized in the framework of the European Regional Development Fund (ERDF).

REFERENCES

[1]A. Sari, and M. Tuzen, Kinetic and equilibrium studies of biosorption of Pb(II) and Cd(II) from aqueous solution by macrofungus (Amanita rubescens) biomass, *J. Hazard. Mater.*, **164**, 1004-1011, (2009).

[2]A.N. Módenes, F.R. Espinoza-Quiñones, A. Colombo, C.L. Geraldi, and D.E.G. Trigueros, Inhibitory effect on the uptake and diffusion of Cd^{2+} onto soybean hull sorbent in Cd–Pb binary sorption systems, *J. Environ. Manag.*, **154**, 22-32, (2015).

[3]R. Kumara, D. Bhatiab, R. Singha, S. Rania, and N.R. Bishnoi, Sorption of heavy metals from electroplating effluent using immobilized biomass Trichoderma viride in a continuous packed-bed column, *Internat. Biodeter. Biodegrad.*, **65**, 1133-1139, (2011).

[4]L. Rozumová, J. Seidlerová, I. Šafařík, M. Šafaříková, and R. Gabor, Sorption of Pb on magnetically modified biological materials, In Nanocon 2012: 4th International Conference. First edition. Ostrava: AMOS repro Ltd., 623-628, (2012), ISBN 978-80-87294-35-2.

[5]Y. Orhan, and H. Buyukgungor, The removal of heavy metals by using agricultural wastes, *Water Sci. Technology,* **28**, 247–255, (1993).

[6]A. Saeed, A.M. Waheed, and M. Iqbal, Removal and recovery of heavy metals from aqueous solution using papaya wood as a new biosorbents, *Sep. Purif. Technology*, **45**, 25-31, (2005).

[7]N.A.A. Babarinde, J.O. Babalola, and R.A. Sanni, Biosorption of lead ions from aqueous solution by maize leaf, *Int. J. Phys. Science*, **1**, 23–26, (2006).

[8]M.A.K. Hanafiah, W.S.W. Ngah, H. Zakaria, and S.C. Ibrahim, Batch study of liquid phase adsorption of lead ions using Lalang (Imperata cylindrica) leaf powder, *J. Biol. Science,* **7**, 222–230, (2007).

[9]P.D. Johnson, M.A. Watson, J. Brown, and I.A. Jefcoat, Peanut hull pellets as a single use sorbent for the capture of Cu(II) from wastewater, *Waste Manage.*, **22**, 471-480, (2002).

[10]Y.S. Ho, W.T. Chiub, C.S. Hsub, and C.T. Huanga, Sorption of lead ions from aqueous solution using tree fern as a sorbent, *Hydrometallurgy,* **73**, 55-61, (2004).

[11]A.K. Bhattacharya, S.N. Mandal, and S.K. Das, Adsorption of Zn(II) from aqueous solution by using different adsorbents, *Chem. Eng. Journal,* **123**, 43-51, (2006).

[12]I. Villaescusa, N. Fiol, M. Martínez, N. Miralles, J. Pocj, and J. Serarols, Removal of copper and nickel ions from aqueous solutions by grape stalks wastes, *Water Researcher,* **38**, 992-1002, (2004).

[13]S.Y. Quek, D.A.J. Wase, and C.F. Forster, The use of sago waste for the sorption of lead and copper, *Water SA*, **24**, 251-256, (1998).

[14]P. Brown, I.A. Jefcoat, D. Parrish, S. Gil, and E. Graham, Evaluation of the adsorptive capacity of peanut hull pellets for heavy metals in solution, *Adv. Environ. Research,* **4**, 19-29, (2000).

[15]G. Cimino, A. Passerini, and G. Toscano, Removal of toxic cations and Cr(VI) from aqueous solution by hazelnut shell, *Water Research*, **34**, 2955-2962, (2000).

[16]A. Sukla, Y.-H. Zhang, P. Dubey, J.L. Margrave, and S.S. Sukla, The role of saw dust in the removal of unwanted materials from water, *J. Hazard. Mat.*, **95**, 137-152, (2002).

[17]D. Zhou, I. Zhang, J. Zhou, and S. Guo, Cellulose/chitin beads for adsorption of heavy metals in aqueous solution, *Water Research*, **38,** 2643-2650, (2004).
[18]G. Doner, and S. Akman, The removal of copper, cadmium and lead by rice husk ash, *Fresen. Environ. Bulletin,* **12,** 736–739, (2003).
[19]G. Annadurai, R.S. Juang, and D.J. Lee, Adsorption of heavy metals from water using banana and orange peels, *Water Science Technology*, **47,** 185–190, (2003).
[20]N. Meunier, Lead removal fromacidic solutions by the sorption on cocoa shells: effect of some parameters, *J. Environ. Eng. ASCE,* **129,** 693–698, (2003).
[21]V. Boonamnuayvitaya, C. Chaiya, W. Tanthapanichakoon, and S. Jarudilokkul, Removal of heavy metals by adsorbent prepared from pyrolysed coffee residues and clay, *Sep. Purif. Technology,* **35,** 11-22, (2004).
[22]Y.S. Ho, and A.E. Ofomaja, Kinetic and thermodynamics of lead ion sorption on palm kernel fibre from aqueous solution, *Proc. Biochemistry*, **40,** 3455-3461, (2005).
[23]N. Fiol, I. Villaesscusa, M. Martinez, N. Miralles, J. Poch, and J. Serarols, Sorption of Pb(II), Ni(II), Cu(II) and Cd(II) from aqueous solution by olive stone waste, *Sep. Purif. Technology,* **50,** 132–140, (2006).
[24]Z. Xuan, Y. Tang, X. Li, Y. Liu, and F. Luo, Study on the equilibrium, kinetics and isotherm of biosorption of lead ions onto pretreated chemically modified orange peel, *Bio. Eng. Journal,* **31,** 160-164, (2006).
[25]M. Martinez, N. Miralles, S. Hidalgo, N. Fiol, I. Villaeseasa, and J. Poch, Removal of lead (II) and cadmium (II) from aqueous solutions using grape stalk waste, *J. Hazard. Mat.,* **133,** 203-211, (2006).
[26]S.Y. Quek, B. Al Duri, D.A.J. Wase, and C.F. Forster, Coir as a biosorbent of copper and lead, *Proc. Safe. Environ. Protect.,* **76,** 50-54, (1998).
[27]V.K. Gupta, and I. Ali, Removal of lead and chromium from wastewater using bagasse fly ash-a sugar industry waste, *J. Colloid Interf. Science,* **271,** 321-328, (2004).
[28]O. Karnitz Jr., L.V.A. Gurgel, J.C.P. Melo, V.R. Botaro, T.M.S. Melo, R.P. Gil, and L.F. Gil, Adsorption of heavy metal ion from aqueous single metal solution by chemically modified sugarcane bagasse, *Bioresour. Technol.,* **98,** 291-1297, (2007).
[29]P.X. Sheng, Y. Ting, I.P. Chen, and L. Hong, Sorption of lead, copper, cadmium, zinc and nickel by marine algal biomass: characterization of biosorption capacity and investigation of mechanisms, *J. Colloid Interf. Sci.,* **275,** 131-141, (2004).
[30]A.N. Módenes, J.M.T.A. Pietrobelli, F.R. Espinoza-Quiñones, P.Y.R. Suzaki, V.L. Alflen, and M.R. Fagundes-Klen, Potencial de biossorção do zinco pela macrófita Egeria densa, *Eng. Sanit. Ambient.,* **14,** 465–470, (2009).
[31]J.M.T.A. Pietrobelli, A.N. Módenes, M.R. Fagundes-Klen, and F.R. Espinoza-Quiñones, Cadmium, copper and zinc biosorption study by non-living Egeria densa biomass, *Water Air Soil Poll.,* **202,** 385–392, (2009).
[32]B. Zhao, W. Xiao, Y. Shang, H. Zhu, and R. Han, Adsorption of light green anionic dye using cationic surfactant-modified peanut husk in batch mode, *Arabian J. Chem.,* In Press, (2014).
[33]A.B. Bourlinos, R. Zboril, and D. Petridis, A. simple route towards magnetically modified zeolites, *Micropor. Mesopor. Mater.,* **58,** 155–162, (2003).
[34]A. Mockovciakova, Z. Orolova, and J. Škvarla, Enhacement of the bentonite sorption properties, *J. Hazard. Mat.,* **180,** 274–281, (2010).
[35]L.C.A. Oliviera, R.V.R.A. Rios, J.D. Fabris, K. Sapag, V.K. Garg, and R.M. Lago, Clay-iron oxide magnetic composites for the adsorption of contaminants in water, *Appl. Clay. Sci.,* **22,** 169–177, (2003).
[36]L. Uzun, N. Saglam, M. Safarikova, I. Safarik, and A. Denizli, Copper biosorption on magnetically modified yeast cells under magnetic field, *Sep. Sci. Technol.,* **46,** 1045–1051, (2011).
[37]R.A. Massart, Preparation of aqueous magnetic liquids in alkaline and acidic media, *IEEE Trans. Magn.,* **17,** 1247–1248, (1981).

[38]I. Safarik, M. Safarikova, Magnetic fluid modified peanut husks as an adsorbent for organic dyes removal, *Phys. Proc.,* **9**, 274-278, (2010).
[39]A. Kumar, N.N. Rao, and S.N. Kaul, Alkali-treated straw and insoluble straw xanthate as low cost adsorbent for heavy metal removal. Preparation, characterization and application, *Biores. Techn.*, **1**, 133–142, (2007).
[40]U. Forstner, and G.T.W. Wittmann, Metal Pollution in the Aquatic Environment, 2nd Ed., *Berlin heidelberg: Springer Verlag*, (1983).
[41]M.J. Temkin, and V. Pyzhev, Recent modifications to Langmuir isotherms, *Acta Physicochim. URSS*, **12**, 217–225, (1940).
[42]R. Sudha, K. Srinivasan, and P. Premkumar, Removal of nickel(II) from aqueous solution using Citrus Limettioides peel and seed carbon, *Ecotoxicol. Environ. Saf.*, **117**, 115-123, (2015).
[43]Y. Bayrak, Application of Langmuir isotherm to saturated fatty acid adsorption, *Micropor. Mesopor. Mat.*, **87**, 203-206, (2006).

Materials Issues in Nuclear Waste Management in the 21st Century

EFFECTS OF Al_2O_3, B_2O_3, Li_2O, Na_2O, AND SiO_2 ON NEPHELINE CRYSTALLIZATION IN HANFORD HIGH LEVEL WASTE GLASSES

Jared O. Kroll, John D. Vienna, Michael J. Schweiger
Pacific Northwest National Laboratory
Richland, Washington, United States

ABSTRACT

Nepheline (nominally $NaAlSiO_4$) formation during slow cooling of high-alumina (25.4 - 34.5 mass% Al_2O_3) Hanford high level waste glasses may significantly reduce product durability. To investigate the effects of composition on nepheline crystallization, 29 compositions were formulated by adjusting Al_2O_3, B_2O_3, Li_2O, Na_2O, and SiO_2 around a baseline glass that precipitated 12 mass% nepheline during canister centerline cooling. Thirteen of these compositions were generated by adjusting one component at a time, while two or three components were adjusted to produce the other 16 (with all remaining components staying in the same relative proportions by mass). Quantitative X-ray diffraction was used to determine the nepheline concentration in each sample. Twenty two glasses precipitated nepheline, two of which also precipitated eucryptite (nominally $LiAlSiO_4$), and one glass formed only eucryptite upon slow cooling. Increasing Na_2O and Li_2O had the strongest effect in promoting nepheline formation. Increasing B_2O_3 inhibited nepheline formation. SiO_2 and Al_2O_3 showed non-linear behavior related to nepheline formation. The composition effects on nepheline formation in these glasses are reported.

INTRODUCTION

The United States Department of Energy will immobilize the roughly 200,000 m^3 of radioactive waste stored in underground tanks on the Hanford site in borosilicate glass. The Hanford Tank Waste Treatment and Immobilization Plant (WTP) is being constructed to accomplish this task. The WTP will segregate high-level waste (HLW) and low-activity waste, blend each with glass forming materials, heat the mixtures to ~1150 °C, and pour the molten glasses into stainless steel canisters to cool and solidify.[1]

According to the 2008 HLW feed vector reported by Vienna *et al.*, the composition of the HLW contains high concentrations of Al_2O_3 (up to 70 mass% Al_2O_3 on a calcined basis after caustic leaching).[2] Currently, a major effort is being made to increase the efficiency of WTP, which includes achieving maximum waste loading for HLW.[3]

The formation of nepheline (nominally $NaAlSiO_4$) upon slow cooling of highly–loaded glasses with high concentrations of alumina is a major concern. If nepheline precipitates from the melt during canister cooling, it will likely reduce the chemical durability of the resulting glass by removing 3 moles of glass formers (1 mole of Al_2O_3 and 2 moles of SiO_2) from the amorphous network for every mole of Na_2O.[4] It will also make it difficult to predict the product consistency test (PCT) response,[5] which under current HLW disposal criteria must be controlled and reported.[6] It has also been documented that eucryptite ($LiAlSiO_4$) has a similar effect on glass durability.[7] Figure 1 shows the impact of nepheline and eucryptite formation on previously reported simulated Hanford HLW glasses.[8]

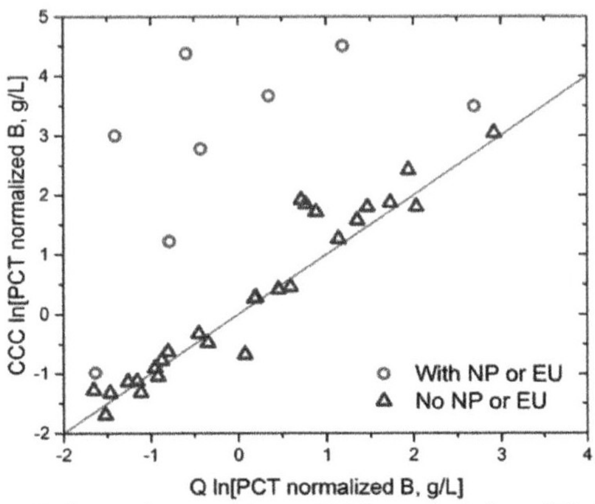

Figure 1 - Impact of nepheline and eucryptite on PCT response from [8], NP = nepheline, EU = eucryptite

In Figure 1, the glasses that precipitated either nepheline or eucryptite during simulated canister centerline cooling (CCC) heat treatment show higher releases than quenched (Q) samples of the same glass. To meet disposal requirements, nepheline precipitation must be avoided, or the amount of nepheline formed and its impact on PCT must be predicted.

A nepheline discriminator (ND) was developed to reduce the risk of nepheline formation upon slow cooling.[9] The ND is based on limiting the normalized SiO_2 concentration (N_{Si}) to the value just above the nepheline primary phase field in the SiO_2-Na_2O-Al_2O_3 ternary phase diagram at 62 mass%, as seen in Figure 2.

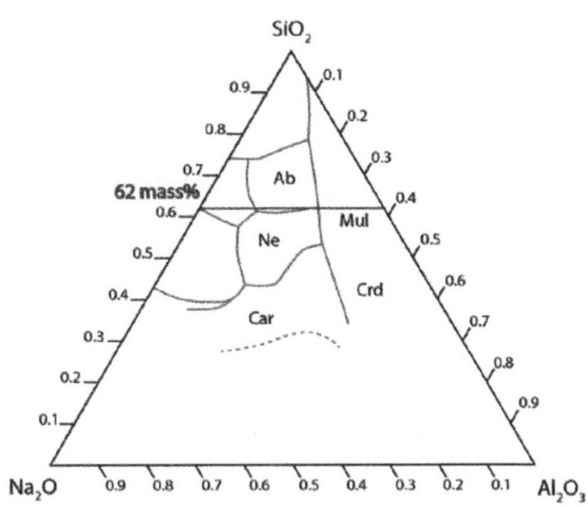

Figure 2 - SiO_2-Na_2O-Al_2O_3 ternary phase diagram with the N_{Si} limit in blue above the nepheline primary phase field (Ne)

The N_{Si} can be calculated from glass compositions using Equation (1):

$$N_{Si} = \frac{g_{SiO_2}}{g_{SiO_2}+g_{Al_2O_3}+g_{Na_2O}} < 0.62 \quad (1)$$

where g_i is the ith oxide mass fraction in glass. The ND states that glasses with $N_{Si} < 0.62$ are prone to nepheline precipitation. However, the ND constraint is overly conservative and drastically limits waste loading. As can be seen in Figure 3, several previously reported glasses with $N_{Si} < 0.62$ did not precipitate nepheline upon slow cooling.[10]

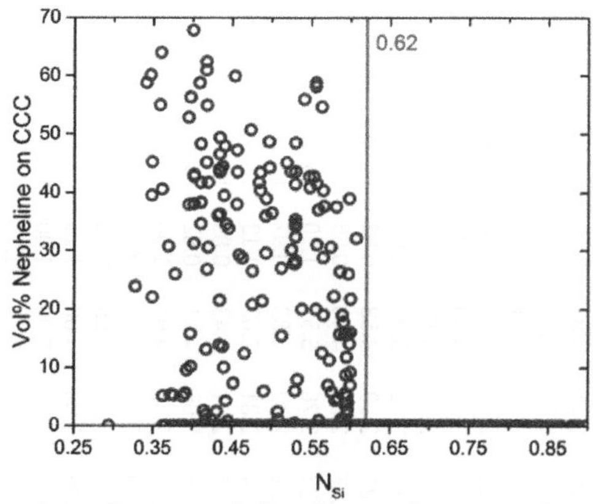

Figure 3 - Comparison of N_{Si} to nepheline fraction for previously reported glasses[10]

A new approach to limiting nepheline precipitation upon slow cooling of HLW canisters is needed to optimize waste loading. Part of the conservatism of the ND is likely due to that fact that it does not take into account the effects of other components in the melt. It has been demonstrated that components, such as B_2O_3, CaO, K_2O, and Li_2O effect nepheline formation.[9] Steps have been taken to develop a model that takes advantage of the success of the ND, but includes the effects of some of these components.[10] In order to generate data relevant to the Hanford HLW composition region for model development and provide some insight on component effects on nepheline crystallization two glass matrices were developed. This paper is focused on these two glass matrices and the findings associated with melt composition and nepheline crystallization.

EXPERIMENTAL APPROACH

Two glass matrices were developed by varying Al_2O_3, B_2O_3, Li_2O, Na_2O, and SiO_2 around a baseline glass formulated to immobilize a typical Hanford high-alumina HLW while meeting all processing and product quality constraints. The baseline glass was adjusted so that roughly 12 mass% nepheline would precipitate after CCC heat treatment. The first matrix consisted of 14 glasses that were formulated by adjusting Al_2O_3, B_2O_3, Li_2O, Na_2O, and SiO_2 one-component-at-a-time (OCAT). The second matrix consisted of 16 glasses formulated around the same baseline glass, but here the compositions were varied two- and three-components-at-a-time (2,3CAT). The targeted compositions of the baseline glass, OCAT glasses, and 2,3CAT glasses are presented in Table 1.

Table 1 - Glass compositions (mass%)

Glass ID	Al_2O_3	B_2O_3	Bi_2O_3	CaO	Cr_2O_3	F	Fe_2O_3	Li_2O	MnO	Na_2O	P_2O_5	RuO_2	SiO_2	SO_3	ZrO_2
Baseline	28.50	17.20	0.65	0.65	1.10	0.30	2.50	5.00	1.00	12.50	0.70	0.05	29.35	0.25	0.25
NP2-High Al	31.50	16.48	0.62	0.62	1.05	0.29	2.40	4.79	0.96	11.98	0.67	0.05	28.12	0.24	0.24
NP2-High B	26.85	22.00	0.61	0.61	1.04	0.28	2.36	4.71	0.94	11.78	0.66	0.05	27.65	0.24	0.24
NP2-High Li	28.20	17.02	0.64	0.64	1.09	0.30	2.47	6.00	0.99	12.37	0.69	0.05	29.04	0.25	0.25
NP2-High Na	27.52	16.61	0.63	0.63	1.06	0.29	2.41	4.83	0.97	15.50	0.68	0.05	28.34	0.24	0.24
NP2-High Si	27.03	16.31	0.62	0.62	1.04	0.28	2.37	4.74	0.95	11.85	0.66	0.05	33.00	0.24	0.24
NP2-Low Al	25.50	17.92	0.68	0.68	1.15	0.31	2.60	5.21	1.04	13.02	0.73	0.05	30.58	0.26	0.26
NP2-Low B	29.60	14.00	0.68	0.68	1.14	0.31	2.60	5.19	1.04	12.98	0.73	0.05	30.48	0.26	0.26
NP2-Low Li	28.80	17.38	0.66	0.66	1.11	0.30	2.53	4.00	1.01	12.63	0.71	0.05	29.66	0.25	0.25
NP2-Low Na	29.48	17.79	0.67	0.67	1.14	0.31	2.59	5.17	1.03	9.50	0.72	0.05	30.36	0.26	0.26
NP2-Low Si	29.85	18.02	0.68	0.68	1.15	0.31	2.62	5.24	1.05	13.09	0.73	0.05	26.00	0.26	0.26
NP2-Very High Al	34.50	15.76	0.60	0.60	1.01	0.27	2.29	4.58	0.92	11.45	0.64	0.05	26.89	0.23	0.23
NP2-Very High Si	25.41	15.34	0.58	0.58	0.98	0.27	2.23	4.46	0.89	11.15	0.62	0.04	37.00	0.22	0.22
NP2-Very Low Si	31.46	18.99	0.72	0.72	1.21	0.33	2.76	5.52	1.10	13.80	0.77	0.06	22.00	0.28	0.28
NP-MC-AlB-1	25.50	22.00	0.63	0.63	1.06	0.29	2.42	4.83	0.97	12.09	0.68	0.05	28.38	0.24	0.24
NP-MC-AlBNa-1	26.80	15.39	0.73	0.73	1.24	0.34	2.81	5.62	1.12	10.80	0.79	0.06	33.00	0.28	0.28
NP-MC-AlBNa-2	29.83	19.32	0.58	0.58	0.97	0.27	2.21	4.43	0.89	13.83	0.62	0.04	26.00	0.22	0.22
NP-MC-AlBSi-1	31.50	22.00	0.53	0.53	0.90	0.25	2.05	4.11	0.82	10.27	0.58	0.04	26.00	0.21	0.21
NP-MC-AlLi-1	25.50	18.23	0.69	0.69	1.17	0.32	2.65	4.00	1.06	13.25	0.74	0.05	31.12	0.27	0.27
NP-MC-AlLi-2	31.50	16.68	0.63	0.63	1.07	0.29	2.42	4.00	0.97	12.12	0.68	0.05	28.47	0.24	0.24
NP-MC-AlLiNa-1	31.14	15.24	0.58	0.58	0.97	0.27	2.21	5.88	0.89	15.14	0.62	0.04	26.00	0.22	0.22
NP-MC-AlLiSi-1	25.50	20.60	0.78	0.78	1.32	0.36	2.99	4.00	1.20	14.97	0.84	0.06	26.00	0.30	0.30
NP-MC-AlNa-1	25.50	17.20	0.65	0.65	1.10	0.30	2.50	5.00	1.00	15.50	0.70	0.05	29.35	0.25	0.25
NP-MC-AlSi-1	25.50	19.79	0.75	0.75	1.27	0.35	2.88	5.75	1.15	14.38	0.81	0.06	26.00	0.29	0.29
NP-MC-AlSi-2	31.50	14.49	0.55	0.55	0.93	0.25	2.11	4.21	0.84	10.53	0.59	0.04	33.00	0.21	0.21
NP-MC-BLiSi-1	27.65	14.00	0.63	0.63	1.07	0.29	2.43	6.00	0.97	12.13	0.68	0.05	33.00	0.24	0.24
NP-MC-BLiSi-2	27.06	22.00	0.62	0.62	1.04	0.28	2.37	6.00	0.95	11.87	0.66	0.05	26.00	0.24	0.24
NP-MC-BNa-1	28.58	14.00	0.65	0.65	1.10	0.30	2.51	5.01	1.00	15.50	0.70	0.05	29.43	0.25	0.25
NP-MC-BNaSi-1	30.97	14.00	0.71	0.71	1.20	0.33	2.72	5.43	1.09	15.50	0.76	0.05	26.00	0.27	0.27
NP-MC-BSi-1	25.50	20.40	0.58	0.58	0.98	0.27	2.24	4.47	0.89	11.18	0.63	0.05	31.78	0.22	0.22

In Table 1, the OCAT glasses have glass IDs beginning with NP2-, and the 2,3CAT glasses have glass IDs beginning with NP-MC-.

MATERIALS AND METHODS

All of the glasses were prepared in 250 g batches and melted in Pt-10%Rh crucibles. Oxides (Bi_2O_3, Cr_2O_3, Fe_2O_3, MnO, SiO_2, and ZrO_2) and carbonates ($CaCO_3$, Li_2CO_3, and Na_2CO_3) comprised most of the batching material. NaF, $NaPO_3$, and Na_2SO_4 were used as the sources of F, P_2O_5, and SO_3, respectively. Boric acid was used as the B_2O_3 source, $Al(OH)_3$ was chosen as the aluminum additive, and $RuNO(NO_3)_3$ was the source of RuO_2.

First, the appropriate amount of 1.5% $RuNO(NO_3)_3$ solution was added drop wise to pre-weighed SiO_2 on a large watch glass. The SiO_2 + $RuNO(NO_3)_3$ was placed in a drying oven at 90 °C for at least 1 hour. Once dry, the appropriate amount of each source was combined, the mixture was placed in an agate mill with an agate puck on a vibratory fixture for 4 minutes to obtain homogeneity. The batch was loaded into a Pt-10%Rh crucible, covered with a lid and placed in a furnace preheated to the melt temperature (T_M = 1150-1350°C), and the sample was allowed to dwell for a period of ~1 hour. The glass samples were quenched by pouring the melt onto an Inconel plate and allowed to cool. The glass was then placed in a tungsten carbide mill on a vibratory fixture for 4 minutes and ground to a powder. Finally, the powder was re-melted, poured, and quenched using the conventions previously mentioned.

Sub-samples of the quenched glasses were taken to perform the CCC heat treatment. The CCC samples were prepared by placing pieces of quenched glass in 1.2 cm^3 crucibles made from Pt-10%Rh foil. The heat treatment schedule is presented in Table 2.

Table 2 - CCC heat treatment schedule

Start Temp (°C)	Rate (°C/min.)
1050	-1.556
980	-0.806
930	-0.591
875	-0.388
825	-0.253
775	-0.278
725	-0.304

Before proceeding with the heat treatment schedule in Table 2, the glasses were held at their respective T_M for 30 minutes and ramped down to 1050 °C at 12.5 °C/minute. Once the furnace temperature reached 400 °C (below the glass transition temperature), the furnace was powered off and allowed to cool naturally. The samples were removed from the furnace at temperatures between 400 °C and 25 °C.

Quantitative X-ray diffraction (XRD) was performed to identify and quantify the major crystalline phases in the CCC heat-treated samples. A 1 – 3 g portion of the CCC sample was ground for ~15 s in a tungsten carbide mill, spiked with ~5.0 mass% CaF_2 or ZnO, and then milled for another ~45 s. The powdered samples were then loaded into plastic holders for analysis. The analysis was performed with a Bruker D8 Advance (Bruker AXS Inc., Madison, WI) with Cu Kα X-rays. The detector used was a LynxEyeTM position-sensitive detector with a collection window of 3° 2θ. Scan parameters were 5 - 75° 2θ with a step of 0.015° 2θ and a 0.6 s dwell at each step. Bruker AXS DIFFRACplus EVA software was used to identify crystalline phases. Topas software was used to quantify the fractions of each crystal using a whole pattern fitting technique.[11]

RESULTS

The baseline glass precipitated ~12 mass% nepheline after CCC heat treatment. Eleven of the heat-treated OCAT glasses precipitated nepheline with concentrations ranging from 1.8 to 58.5 mass%, and one precipitated 8.0 mass% eucryptite. Eleven of the heat-treated 2,3CAT glasses precipitated nepheline with concentrations ranging from 0.7 to 63.5 mass%, two of these glasses also precipitated eucryptite with concentrations of 4.9 and 13.0 mass%. Table 3 summarizes the XRD results for all of the CCC heat-treated glasses.

Table 3 - Quantitative XRD results for CCC heat-treated glasses (mass%)

Glass ID	Nepheline	Eucryptite	Other Crystalline Phases
BL3	12.3	0.0	(1.5) Lithium Iron Oxide (2.0) Lithium Iron Manganese Oxide
NP2-High Al	9.0	0.0	(1.7) Lithium Iron Manganese Oxide (2.8) Lithium Iron Chromium Oxide
NP2-High B	0.9	0.0	(2.9) Lithium Iron Oxide
NP2-High Li	42.2	0.0	(1.7) Maghemite (2.0) Lithium Iron Manganese Oxide
NP2-High Na	52.7	0.0	(2.3) Maghemite
NP2-High Si	27.4	0.0	(1.0) Lithium Iron Oxide (1.1) Lithium Iron Manganese Oxide
NP2-Low Al	4.0	0.0	(3.1) Lithium Iron Manganese Oxide
NP2-Low B	58.5	0.0	(3.2) Maghemite
NP2-Low Li	1.8	0.0	(2.7) Lithium Iron Oxide
NP2-Low Na	0.0	8.0	(3.7) Lithium Iron Manganese Oxide
NP2-Low Si	5.0	0.0	(3.2) Lithium Iron Manganese Oxide
NP2-Very High Al	28.8	0.0	(3.0) Aluminum Oxide
NP2-Very High Si	0.0	0.0	(2.0) Chromite
NP2-Very Low Si	22.4	0.0	(1.5) Lithium Iron Manganese Oxide (3.8) Lithium Iron Chromium Oxide
NP-MC-AlB-1	0.0	0.0	(3.1) Lithium Iron Manganese Oxide
NP-MC-AlBNa-1	0.0	0.0	(1.6) Lithium Iron Manganese Oxide (1.9) Lithium Iron Oxide
NP-MC-AlBNa-2	5.1	0.0	(2.2) Maghemite
NP-MC-AlBSi-1	0.0	0.0	(1.4) Iron Chromium Oxide
NP-MC-AlLi-1	0.7	0.0	(3.1) Iron Silicon Oxide
NP-MC-AlLi-2	35.5	0.0	(2.5) Lithium Iron Oxide
NP-MC-AlLiNa-1	59.9	0.0	(3.8) Lithium Iron Manganese Oxide (11.3) Lithium Aluminum Borate
NP-MC-AlLiSi-1	5.3	0.0	(2.0) Iron Silicon Oxide (2.0) Lithium Iron Manganese Oxide
NP-MC-AlNa-1	57.0	0.0	(1.1) Maghemite (1.6) Nosean (1.9) Lithium Iron Manganese Oxide
NP-MC-AlSi-1	49.5	0.0	(1.3) Lithium Iron Manganese Oxide (3.1) Lithium Iron Oxide

Glass ID	Nepheline	Eucryptite	Other Crystalline Phases
NP-MC-AlSi-2	9.4	13.0	(1.1) Eskolaite (1.2) Iron Chromium Oxide
NP-MC-BLiSi-1	56.1	4.9	(1.2) Silicon Oxide (2.7) Lithium Iron Manganese Oxide
NP-MC-BLiSi-2	0.0	0.0	(1.2) Maghemite (1.7) Lithium Iron Manganese Oxide
NP-MC-BNa-1	63.5	0.0	(1.1) Lithium Iron Manganese Oxide (2.8) Lithium Iron Oxide
NP-MC-BNaSi-1	51.5	0.0	(6.4) Lithium Iron Manganese Oxide
NP-MC-BSi-1	0.0	0.0	(1.1) Eskolaite

DISCUSSION

Findings from OCAT study

From the OCAT matrix results, it appears that increasing B_2O_3 is the most effective method of suppressing nepheline formation and increasing alkali promotes nepheline formation. Both Al_2O_3 and SiO_2 appear to have highly non-linear effects on the fraction of nepheline formed on CCC. Figure 4 summarizes the results of the OCAT matrix by plotting the difference between the adjusted component and the baseline against the measured mass fraction of nepheline. The single eucryptite point is plotted as well.

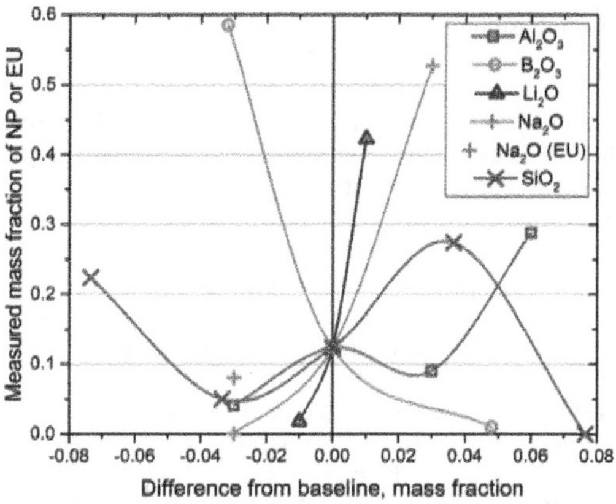

Figure 4 - Component difference from baseline vs measured mass fraction. NP = nepheline, EU = eucryptite

Findings from 2,3CAT study

In the 2,3CAT matrix it was observed that increasing Na_2O/Al_2O_3 in glass NP-MC-AlNa-1 increased the nepheline fraction at a higher rate compared to increasing the Na_2O/B_2O_3 in glass NP-MC-BNa-1. This can be seen in Figure 5, where the Na_2O/Al_2O_3 slope is steeper than the Na_2O/B_2O_3 slope. Figure 5 plots the difference from the baseline glass against the measured mass fraction for some of the 2,3CAT glasses. This plot also shows that increasing B_2O_3/Al_2O_3 in glass NP-MC-AlB-1 decreases the nepheline fraction.

The ineffectiveness of SiO_2 on reducing the fraction of nepheline was seen with glasses NP-MC-BLiSi-1 and NP-MC-BLiSi-2, where increasing $B_2O_3/(B_2O_3 + SiO_2)$ reduced the nepheline fraction and increasing $SiO_2/(B_2O_3 + SiO_2)$ increased the nepheline fraction. These are also plotted in Figure 5 along with nepheline + eucryptite fraction for glass NP-MC-BLiSi-1.

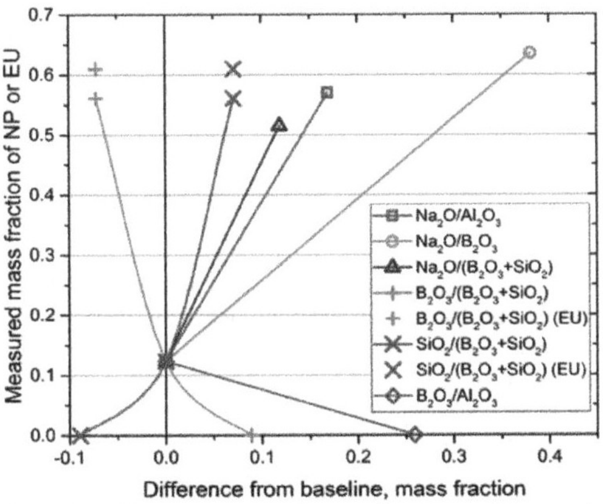

Figure 5 - Difference from baseline vs measured mass fraction. NP = nepheline, EU = eucryptite

CONCLUSIONS

In this study, 30 unique compositions (including the baseline glass) were melted, heat-treated, and the nepheline fraction was measured. Twenty three of the glasses precipitated nepheline upon CCC heat treatment, two of which also precipitated eucryptite. One of the 30 glasses formed only eucryptite.

It was observed in the OCAT study that B_2O_3 was the most effective component for suppressing nepheline formation, and that increasing alkali promoted nepheline formation. The effects of SiO_2 and Al_2O_3 were non-linear across the composition region investigated here. Also, it seems that eucryptite formed with lower concentrations of Na_2O rather than higher concentrations of Li_2O.

The ND was observed to be conservative in this study since all of the glasses have $N_{Si} < 0.62$, but not all formed nepheline on CCC. The observation of the effects of SiO_2 on nepheline formation were non-linear may explain the conservatism seen in the ND.

Future work is planned to refine a ternary model that takes advantage of the success of the ND, but includes the effects of other components.[10] Also, a model will be developed to predict PCT response of CCC glasses.

ACKNOWLEDGEMENTS

Pacific Northwest National Laboratory is operated by Battelle under Contract Number DE-AC05-76RL01830. The authors are grateful for the financial support provided by the US DOE Office of River Protection's Waste Treatment and Immobilization Plant Federal Project office under the direction of Dr. Albert Kruger. The scientific discussions and guidance of Brian

Riley, Jarrod Crum, and Jesse Lang are greatly appreciated. Additional analytical laboratory support was provided by Savannah River National Laboratory with Kevin Fox as the facilitator.

REFERENCES

[1]J. Perez, S. Bames, S. Kelly, Vitrification testing and demonstration for the Hanford waste treatment and immobilization plant, *Ceram. Trans.*, **168**, 3 – 19 (2005).

[2]J. Vienna, D. Kim, D. Skorski, J. Matyas, Glass Property Models and Constraints for Estimating the Glass to Be Produced at Hanford by Implementing Current Advanced Glass Formulation Efforts, *PNNL-22631, Rev. 1, ORP-58289*, Pacific Northwest National Laboratory, Richland, WA (2013).

[3]D. Peeler, J. Vienna, M. Schweiger, K. Fox, Advanced High-Level Waste Glass Research and Development Plan, *PNNL-24450*, Pacific Northwest National Laboratory, Richland, WA (2015).

[4]D. Kim, D. Peeler, P. Hrma, Effect of Crystallization on the Chemical Durability of Simulated Nuclear Waste Glasses, *Ceram. Trans.*, **61**, 177 – 185 (1995).

[5]ASTM, Standard Test Methods for Determining Chemical Durability of Nuclear, Hazardous, and Mixed Waste Glasses and Multiphase Glass Ceramics: The Product Consistency Test (PCT), *ASTM C1285-14*, ASTM International, West Conshohocken, PA (2008).

[6]DOE, Waste Acceptance Product Specifications for Vitrified High-Level Waste Forms (WAPS), *DOE/EM-0093*, U.S. Department of Energy, Office of Environmental Management, Washington, D.C. (1996)

[7]J. McCloy, J. Vienna, Glass Composition Constraint Recommendations for Use in Life-Cycle Mission Modeling, *PNNL-19372*, Pacific Northwest National Laboratory, Richland, WA (2010).

[8]J. Vienna, M. Schweiger, Y. Chou, J. Kroll, S. Cooley, K. Fox, A. Kruger, High-Alumina Glass Formulation and Testing for U.S. High-Level Tank Wastes, *PNNL-SA-113454*, Pacific Northwest National Laboratory, Richland, WA (2015).

[9]H. Li, J. Vienna, P. Hrma, D. Smith, M. Schweiger, Nepheline Precipitation in High-Level Waste Glasses: Compositional Effects and Impact on the Waste Form Acceptability, *Mater. Res. S. C.*, **465**, 261 – 268 (1997).

[10]J. Vienna, J. Kroll, P. Hrma, J. Lang, J. Crum, D. Skorski, High-Level Waste Glass Composition Region Prone to Nepheline Precipitation, Submitted to *Int. J. Appl. Glass. Sci.*, (2015).

[11]R. Cheary, A. Coelho, J. Cline, Fundamental Parameters Line Profile Fitting in Laboratory Diffractometers, *J. Res. Natl. Inst. Stand. Technol.*, **109**, 1 – 25 (2004).

EVOLUTION OF REPOSITORY, CONTAINER, WASTE FORM CHARACTERIZATION AND DESIGN AT THE PROPOSED US DISPOSAL SYSTEM IN VOLCANIC TUFF

Rob P. Rechard
Nuclear Waste Disposal Research & Analysis
Sandia National Laboratories
Albuquerque, NM 87185-0747, USA

ABSTRACT

Over 25 years, scientists and engineers designed engineered features to complement attributes of the natural barrier of volcanic tuff at Yucca Mountain in southern Nevada such that a proposed repository in the unsaturated zone would safely isolate spent nuclear fuel and high-level radioactive waste over 10^6 years. Initially in 1983, an engineered barrier design applicable to several geologic media was used. With the Congressional direction to characterize Yucca Mountain, the engineered design gradually adapted to conditions in unsaturated tuff in the 1990s. The repository switched from floor emplacement of waste in small, single-walled stainless steel canisters to in-drift emplacement in large, double-layered containers. By 2000, the outer layer was high-nickel alloy to resist corrosion and the inner layer was stainless steel for strength. To avoid localized corrosion during the ~1000-yr thermal period, titanium drip shields were also added above the containers. By 2008, a modular design of the repository was used for flexibility.

In general, flexibility in accommodating various waste forms has been an intended attribute of geologic disposal system designs, rather than tuning the disposal system to specific characteristics of waste durability. The degradation rate of the radioactive waste matrix was an important parameter of the source-term in early modeling analysis. However, by the mid1990s, analyses used fairly rapid degradation rates within the oxygenated environment of the unsaturated zone. Other components of the multiple barrier disposal system compensated for high degradation rates.

INTRODUCTION

Since 1978, the US Department of Energy (DOE) has explored developing a geologic repository at Yucca Mountain (YM), located at the boundary of the Nellis Air Force Range and the Nevada Nuclear Security Site (NNSS and formally known as Nevada Test Site) (Fig. 1).[1] The purpose was to dispose commercial spent nuclear fuel (CSNF), DOE-managed spent nuclear fuel (DSNF), and high-level radioactive waste (HLW), which is an engineered waste form produced when reprocessing a small amount of CSNF, prior to 1972, and DSNF until 1992. The design of the engineered barrier to accommodate this waste varied considerably over the life of the Yucca Mountain Project (YMP) as understanding of the natural barrier increased, techniques for excavation advanced, and desires for large containers for high throughput were expressed.[2-4]

To provide perspective on future material issues for radioactive waste management, this paper describes the evolution of the engineered barrier system between 1984, when the YM disposal system was first described, and 2008, when the safety analysis report for the license application (SAR/LA) to construct a repository was submitted to the US Nuclear Regulatory Commission (NRC).

Geologic Disposal System

Similar to all geologic disposal systems, the YM disposal system consists of two main systems: (1) the natural barrier system (NBS) or the geologic setting that hosts the repository, and (2) the engineered barrier system (EBS). The natural barrier must be characterized through scientific study; whereas, parts of the engineered barrier are designed.

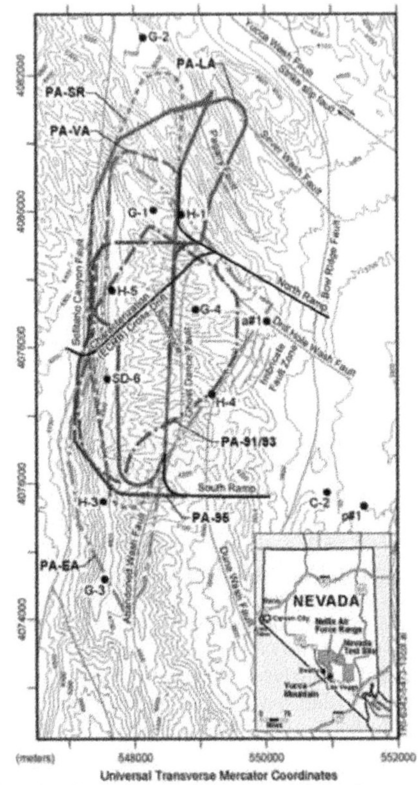

Fig. 1. Location and extent of various proposed layouts for YM repository.[2, Fig. 1]

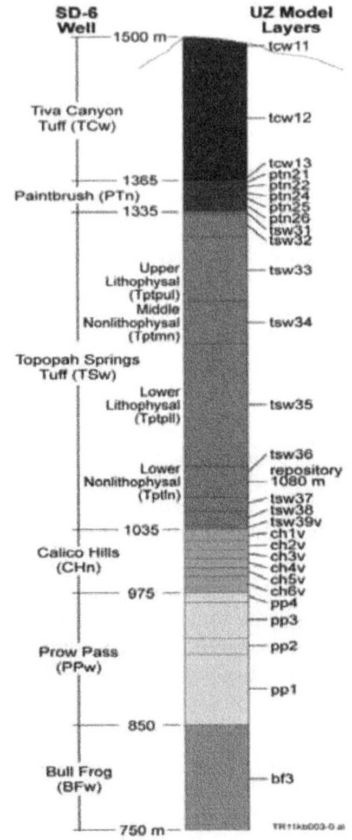

Fig. 2. Stratigraphy of Yucca Mountain volcanic tuff at SD-6 borehole.[2, Fig. 2]

As NRC defined in the 2001 site-specific regulation for the YM repository—10 CFR 63 (and similar to definitions in the earlier 1983 generic regulation—10 CFR 60), the "*engineered barrier system* means the waste packages, including engineered components and systems other than the waste package (e.g., drip shields), and the underground facility" where the "*underground facility* means the underground structure, backfill materials, if any, and openings that penetrate the underground structure (e.g., ramps, shafts, and boreholes, including their seals);" "*waste package* [WP] means the waste form and any containers, shielding, packing, and other absorbent materials immediately surrounding an individual waste container;" and "*waste form* means the radioactive waste materials and any encapsulating or stabilizing matrix." Although much of the engineered barrier is designed, its long-term behavior over 10^4 or 10^6 years must still be determined. For example, waste form designs have not emphasized long-term durability after disposal (e.g., reactor operation dominates design of fuel assemblies and desire for high waste loading dominates design for HLW); hence, waste characterization is also an important activity.

Regulatory Performance Criteria

In the site-specific radiation protection standard for the YM disposal system, first promulgated in 2001, the US Environmental Protection Agency (EPA) requires a performance assessment (PA), which is a quantitative stochastic estimate of dose to a hypothetical individual.[1] EPA limits the arithmetic mean of the various estimates of dose from the highest concentration in a hypothetical contaminant plume no further south than the southern edge of NNSS (~18 km from the repository) to 0.15 mSv/yr for 10^4 years after closure ($t \leq 10^4$ yr) and to 1 mSv/yr for $10^4 < t \leq 10^6$ yr. The earlier generic EPA standard, promulgated in 1985, had limited the cumulative release, rather than dose.[1] The use of dose as a health indicator (in contrast to cumulative release) increases the necessary fidelity of estimating the rate of radionuclide release to the accessible environment and, thereby, the importance of the EBS, which can presumably be more readily and accurately characterized for analysis.

EBS Modeling and Analysis Iterations

Two types of models are typically developed for analyzing a disposal system, in general, and the EBS, in particular: (1) a detailed mechanistic model to develop better designs through understanding of EBS behavior, and (2) an EBS model as a subsystem in the PA to understand the EBS role in overall performance. Whether the same model is used for both roles depends upon computational capabilities and the influence of the EBS in the overall PA performance. This paper describes the simplified EBS model used in the PA of the YM disposal system.[3, 4] The YM EBS model evolved to include 3 primary modeling parts: EBS chemistry, container, and waste form. The waste form eventually consisted of 12 components:[4, Figure 18] EBS flow, EBS transport, waste water chemistry, radioelement solubility, sorption on container rust, reversible colloid formation, irreversible colloid formation, inventory decay, DSNF degradation, CSNF cladding failure, CSNF degradation, and HLW degradation. Only the modeling evolution of the container, CSNF cladding, CSNF degradation, and HLW degradation is summarized here.

Seven PA iterations, of the many performed by YMP, provide convenient points to discuss evolution of the EBS.[1-4] The 1984 deterministic analyzes, conducted for Environmental Assessment (EA) required by the *Nuclear Waste Policy Act of 1982* (NWPA) and designated herein as PA-EA, provides the initial marker for the paper. The first stochastic PA, conducted in 1991 (PA-91), serves as the second marker. PA-93 and PA-95, which serve as the third and fourth markers, respectively, provided preliminary guidance on site characterization and EBS design. The congressionally requested viability assessment (PA-VA), completed in 1998, serves as the fifth marker. The conclusion of natural barrier characterization culminated with an analysis in late 2000 for the site recommendation—PA-SR, which serves as the sixth marker. PA-LA, which forms the basis of SAR/LA, serves as the final marker.

ENGINEERED BARRIER FOR 1984 PA-EA

Repository Design

A repository concept was developed for the 1984 PA-EA with a stair-step design to remain horizontal in the welded tuff of the Topopah Spring Formation (TSw of Fig. 2).[2] The YM repository was set >200 m below the surface and >100 m above the water table. These vertical limits remained a design goal for the repository through 2008 PA-LA, 24 years later.

A repository for 70,000 metric tons of heavy metal (MTHM), the maximum size legally allowed in NWPA for the first repository (until a second repository was constructed) required 6 km^2, based on an initial areal power loading of 12 W/m^2. The design criteria was to keep the tuff temperature in the repository < 200 °C and, thereby, avoid altering zeolitic tuff in layers under the repository.[2] The latter criterion, expressed as a drift wall temperature <200 °C, remained a design goal through PA-LA.

In PA-EA, the repository disposed 3000 MTHM/yr, a rate ~50% higher than the annual rate of generation of CSNF in the US. This rate would guide repository designs for the next 25 years to ensure that CSNF storage at reactors would decrease once the repository was operating.[2]

Both vertical and horizontal emplacement were considered during the mid-1980s.[2] Vertical emplacement was in the floor of a 6.7 m tall by 6.1 m wide disposal drift with one container placed in a 0.7-m diameter borehole spaced every 9 m. Horizontal emplacement was in the pillars of a 3.7 m tall by 6.1 m wide drift with 35 containers per borehole spaced every 14 m.

Package Design

Initially, a generic small container was used for SNF and HLW, which maintained flexibility for disposal in various geologic media being considered in the early 1980s (Fig. 3). A small container with a low thermal load can be disposed in many media. By 1983, stainless steel and high-nickel alloys were candidate container materials for a container in salt, basalt, or tuff repositories.[2, 3] Copper alloys were also evaluated for the anoxic, saturated zone (SZ) of a basalt repository but were not necessary in salt and not effective in the oxidizing, unsaturated zone (UZ) for the tuff repository.

YMP anticipated both CSNF and HLW in the repository. However, only a 9.5-mm thick 304 stainless-steel container of CSNF was modeled in PA-EA, although other materials and designs were under consideration. At this time, the EBS did not include other components of the EBS that would delay release of radionuclides, such as a corrosion-resistant overpack, borehole liner, or adsorptive backfill. The containers for CSNF from either pressurized water reactors (PWRs) or boiling water reactors (BWRs) were assumed to be 0.65 m in diameter and between 4.0 and 4.75 m long (Fig. 4). About 33,000 containers were anticipated, if CSNF was not consolidated, or ~18,000, if CSNF was consolidated at the repository by removing hardware surrounding the fuel rods.[2, Table 2]

Although HLW was not modeled in PA-EA, designs were available. The pour canisters for defense HLW were to be 0.61 m in diameter and 3 m long (the standard canister used in the vitrification plant at the Savannah River Site); the pour canisters for commercial HLW were to be 0.32 m in diameter and 3 m long (Fig. 3).

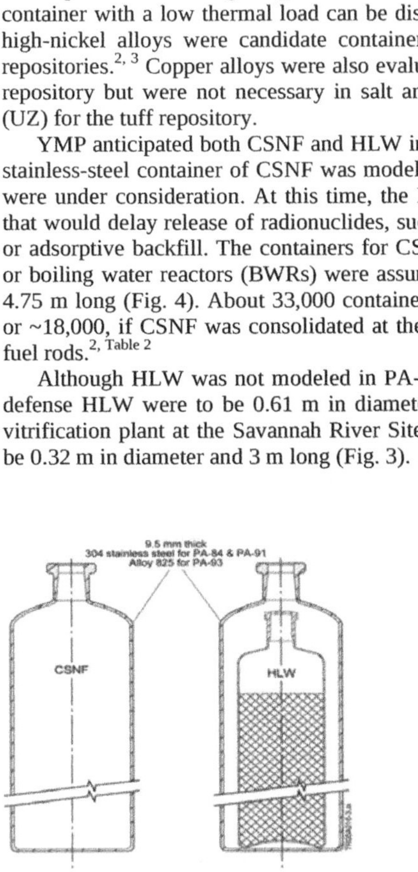

Fig. 3. Small, thin-walled container design of stainless steel or high-nickel Alloy 825 considered for 1984 PA-EA, PA-91, and PA-93 can easily adapt to geologic media that cannot accomodate high heat loads.[3,Fig.2]

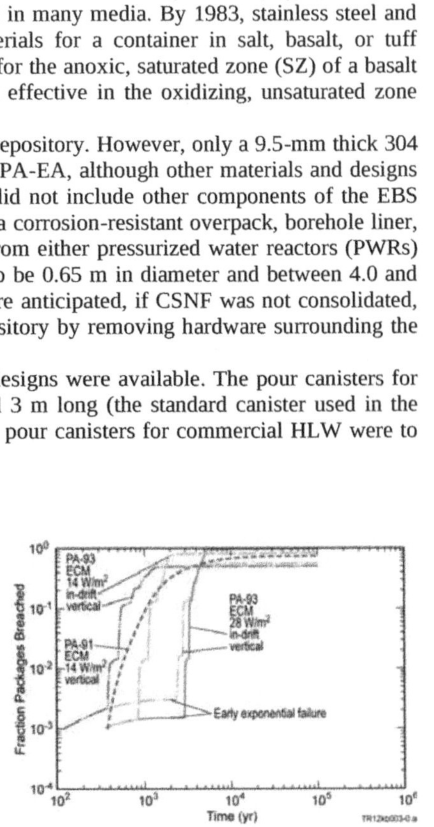

Fig. 4. Fraction of packages breached ($F^{WP}(t)$) for vertically emplaced containers with single Alloy 825 layer (using defined distribution for pitting) and in-drift container with double layer of Alloy 825 and carbon steel (using parabolic equation for general corrosion) at two heat loads in PA-93; comparable release curve shown for PA-91.[3, Fig. 4]

The cumulative fraction of containers breached ($F^{WP}(t)$) was modeled either as (1) instantaneous failure of all containers, or (2) exponential degradation. The time of instantaneous failure was either 300 yr or 1000 yr, which corresponded to the range for the minimum lifetime required in the 1983 NRC implementing regulations (10 CFR 60, §60.113).[1] Instantaneous failure accounted for the susceptibility of stainless steel to stress corrosion cracking.

For exponential degradation, the cumulative fraction of containers breached was

$$F^{WP}(t) = 1 - e^{-\lambda_{WP} t} \tag{1}$$

The mean time to failure ($1/\lambda_{WP}$) was 10^4 years. The λ_{WP} was assumed to include both the degradation of the stainless steel container and the degradation of CSNF cladding (i.e., $\lambda_{WP} = \lambda_{can} + \lambda_{clad}$). Exponential degradation accounted for a portion of containers that failed early and the limited ability of water to initially enter containers through stress corrosion cracks.

Waste Characterization

In PA-EA, only advective release of radionuclides from waste ($\Gamma_{CSNF,r}$) (kg/yr-WP) was modeled and assumed to be controlled by the fractional degradation rate of the CSNF fuel matrix ($\Lambda_{matrix}(t)$) (yr^{-1}) and the radionuclide mass per container ($\bar{M}_{CSNF,r}(t)$) (kg/WP). The $\Lambda_{matrix}(t)$ was modeled as a function of uranium solubility (S_U) (kg/m^3), U mass per container (M_U), $F^{WP}(t)$ from Eq. (1)), and quantity of water entering a package (Q^{WP}) (m^3/yr-WP). That is,[4]

$$\Gamma_{CSNF,r}(t) = \Lambda_{matrix}(t)\bar{M}_{CSNF,r}(t) \tag{2}$$

where

$$\Lambda_{matrix}(t) = \frac{S_U}{M_U} Q^{WP} F^{WP}(t) \tag{3}$$

The Q^{WP} was limited to a fraction of the percolation ($f^{perc} q^{perc}$) (m/yr) passing through Yucca Mountain and the intercept area of a package (repository area/number of WP), yet, all containers experienced some seepage. PA-EA did not account for the short time temperatures might exceed 100 °C near the boreholes, thus, Q^{WP} was not a function of temperature. The source-term for CSNF did not consider sorption and transport within the EBS domain; hence, $\Gamma_{CSNF,r}$ was used directly in the transport model outside the EBS. The equivalent dissolution rate was quite low (Fig. 5a).

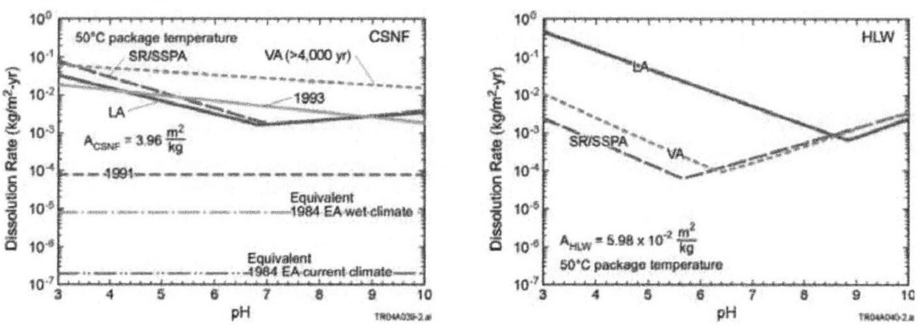

(a) CSNF (b) HLW

Fig. 5. Dissolution of CSNF and HLW matricies eventually set at similar rapid rates in YMP PAs, but alteration rates differed because of differences in specific surface areas.[4, Fig. 20]

ENGINEERED BARRIER FOR PA-91

Repository Design

Similar to PA-EA, the repository design for PA-91 considered either vertical floor emplacement or horizontal pillar emplacement using small waste containers. While ramps were to be constructed using a tunnel boring machine, most of the underground facility was to be constructed with conventional drill and blast methods.

Waste Characterization

Although only degradation of the CSNF matrix was again modeled in PA-91, the conceptual model of CSNF degradation had changed. The radionuclide source rate was dependent on either (1) the solubility of the radioelement, as in PA-EA, or (2) the potentially rapid degradation of CSNF matrix in the UZ. Each radionuclide was assigned, based on analyst judgment, to either the first or second situation in PA-91. In later PAs, however, the release rates from both situations were evaluated and the minimum rate chosen (i.e., $\Gamma_{CSNF,r}(t) = \min\{\Gamma^{\deg}_{CSNF,r}(t), \Gamma^{sol}_{CSNF,r}(t)\}$).

For 5 of the 9 radionuclides tracked, the source rate (${}^{sol}\Gamma_{CSNF,r}(t)$) was the product of the solubility of the radioelement e (S_e) and the $Q^{WP}(t)$ amount of water entering a package:

$$ {}^{sol}\Gamma_{CSNF,r}(t) = S_e Q^{WP}(t) \text{ for } e \sim {}^{243}\text{Am}, {}^{237}\text{Np}, {}^{239}\text{Pu}, {}^{126}\text{Sn}, {}^{234}\text{U}. \quad (4) $$

For the other 4 radioelements, the source rate $\Gamma_{CSNF,r}(t)$ was a combination of the degradation release from the CSNF matrix (Λ_{matrix}) and instantaneous release (Λ_{prompt}) of a small portion of radionuclides trapped between the CSNF matrix and cladding or trapped in grain boundaries within the CSNF matrix:

$$ \Gamma^{\deg}_{CSNF,r}(t) = \bar{M}_{CSNF,r}(t)\,[(1 - f_r^{prompt})\Lambda^{matrix}_{CSNF}(t) + f_r^{prompt}\Lambda^{prompt}_{CSNF}(t)] \quad (5) $$

for $r \sim {}^{135}$Cs, ^{129}I, ^{99}Tc, ^{79}Se where $\bar{M}_r(t)$ was the time varying mass of radionuclide r in a single CSNF package and f_r^{prompt} was instantaneous release fraction. The f_r^{prompt} was not an important parameter even when increased and varied in later PAs.

ENGINEERED BARRIER FOR PA-93

Repository Design

PA-93 considered vertical floor emplacement but also considered in-drift emplacement with large packages. Layout of the underground facility was redesigned to facilitate construction with a tunnel boring machine and, thereby, reduce disturbance of the tuff near the disposal drifts.

Package Design

Both vertical and in-drift configurations in PA-93 and thereafter considered corrosion resistant material for the containers (Fig. 6). General corrosion of carbon steel container under wet conditions (outside the dry-out zone for <100 °C) for the in-drift package was represented by a parabolic equation. PA-93 also included early failure of containers where a container could fail at early times from manufacturing defects, as represented by an exponential function (Fig. 4).[3]

PA-93 examined the implications of higher temperatures on container degradation for more closely packed packages and higher heat loads, which necessitated introducing a thermal module. That is, PA-93 made the first steps toward evaluating the interaction between the EBS and NBS.

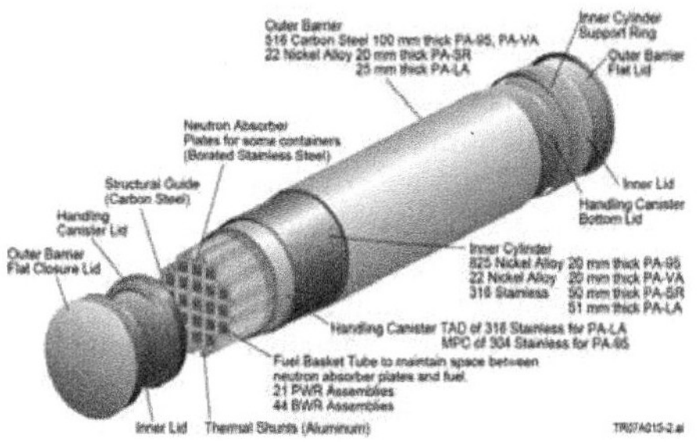

Fig. 6. In-drift container considered for PA-93 and thereafter included nner handling canister for PA-95 (multipurpose canister or MPC) and for PA-LA (transportation, aging, and storage or TAD).[3, Fig. 3]

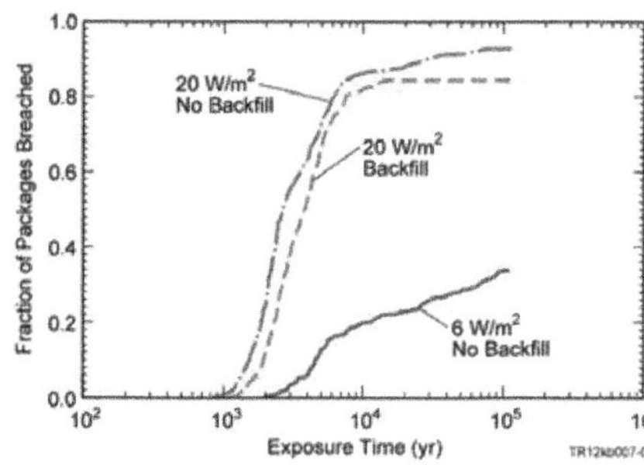

Fig. 7. Fraction of in-drift containers breached ($F^{WP}(t)$) via pits through Alloy 825 layer (using exponential function of temperature) for hot and cool repository with and without backfill in PA-95.[3, Fig. 5]

Waste Characterization

The degradation release rate $\Gamma^{\deg}_{CSNF,r}(t)$ was evaluated with Eq. (5). However, the time dependent evaluation of the matrix fractional releases rates ($\Lambda^{matrix}_{CSNF}(t)$) became more complicated because of the need to account for the breach time of the waste container (τ_{fail}) in addition to $\dot{r}_{CSNF}$ (and the breach time of CSNF cladding in PA-VA and PA-SR). These breach times and degradation rates, which varied with temperature, were estimated at each analysis time step.

The CSNF matrix dissolution rate ($\dot{r}_{CSNF}$) (or fractional alteration rate–yr^{-1} $\dot{a}_{CSNF} = \dot{r}_{CSNF}\mathcal{A}_{surf}$ where $\mathcal{A}_{surf}$ is the specific surface area) was an important input to the analysis algorithm. The dissolution rate $\dot{r}_{CSNF}(t)$ (kg/m^2-yr) was evaluated using an Arrhenius relationship except that the rate was also a function of the total carbonate concentration ($C^{CO_3}_{CSNF}$) and pH (Fig. 5a):

$$\log \dot{r}_{CSNF}(t) = \kappa_0 - \frac{\kappa_1}{T^{WP}_{CSNF}(t)} + \kappa_2 \log C^{CO_3}_{CSNF} - \kappa_3 pH_{CSNF} \tag{6}$$

Experimental data were used to evaluate the regression coefficients κ_0, κ_1 (activation energy), κ_2, and κ_3 for PA-93 and thereafter.

ENGINEERED BARRIER FOR PA-95

YMP conducted PA-95 to again provide guidance on (a) characterizing the site, (b) two options for the heat load, and (c) two options for package placement in the repository.[3, 4, 5, Table 1]

Repository Design

In 1994, DOE formally adopted the use of in-drift emplacement of large packages. The appeal was the ability to more easily receive 3000 MTHM/yr of SNF and HLW since ~⅔ fewer packages were emplaced. A mountain site facilitated use of large packages since gently inclined ramps could be used to move large, heavy packages by rail into the repository.

The PA-95 design closely spaced the drifts 28 m apart such that decay heat would dry out the UZ for ~1000 yr.[2] The moderately hot and more compact repository fit west of the Ghost Dance Fault (Fig. 1). The design saved tunneling costs by reducing the repository to ~200 km of drifts (160 drifts up to 1200 m long).

Package Design

For PA-95, the container consisted of an inner layer of 20-mm nickel Alloy 825 for corrosion-resistance, and an outer layer of 100-mm thick carbon steel for structural strength and corrosion sacrificial allowance. The package also included a 25-mm thick multipurpose handling canister (MPC) of 304 stainless steel (Fig. 6).[2, Table 3] The MPC was not considered as a barrier.

PA-95 examined the influence of percolation, seepage, and relative humidity on the thermal regime in an expanded thermal-hydrologic module. With the proposed change to an individual dose performance measure, PA-95 also took the major step of modeling corrosion stochastically to evaluate the diffusive and advective barrier capability of the container.[3] Radionuclides sorption on rust as the container degraded was not include but would be for PA-SR and PA-LA.

Two modes of corrosion were considered for the outer carbon steel corrosion allowance layer: humid air corrosion and wet (aqueous) corrosion. For humid air corrosion, the layer removal was an exponential function:

$$s_{steel}^{air}(t) = e^{\kappa_0 + \frac{\kappa_1}{T_{CSNF}^{WP}(t)} + \frac{\kappa_2}{RH_{CSNF}^{drift}(t)} + \kappa_3 \ln t} \quad (\mu\text{m}) \tag{7}$$

where $RH_{CSNF}^{drift}(t)$ was the relative humidity (RH) in the drift, and $T_{CSNF}^{WP}(t)$ was the temperature of a representative CSNF package (evaluated for two heat load cases with and without backfill and two infiltration cases). The parameters κ_0, κ_1, and κ_2 were determined from published literature.

Wet corrosion of the carbon steel outer layer was modeled as

$$s_{wet}^{steel}(t) = e^{\kappa_0 + \frac{\kappa_1}{T_{CSNF}^{WP}(t)} + \kappa_3 \ln t + \kappa_4 T_{CSNF}^{WP}(t)^2} \quad (\mu\text{m}) \tag{8}$$

where the parameters κ_0, κ_1, κ_3, and κ_4 were determined from published literature.

Only wet, pitting corrosion was modeled for the Alloy 825 layer. The pit penetration rate was an exponential function of temperature with parameters based on scientific judgment (Fig. 7).[3]

$$\dot{s}_{wet}^{A825\,pit}(t) = e^{\kappa_0 - \frac{\kappa_1}{T_{CSNF}^{WP}(t)}} \quad (\text{mm/yr}) \tag{9}$$

Waste Characterization

The degradation rate for CSNF was of the same functional form as PA-93 (Eq. (6)) with a generic CSNF package temperature ($T^{WP}(t)$) calculated for 2 heat loads, 2 backfill options, and 2 moisture infiltration cases.[3]

The degradation rate of HLW encapsulated in borosilicate glass ($\dot{r}_{HLW}(t)$) was derived from transition state theory to obtain a chemical affinity reaction rate equation (Fig. 5b):[3]

$$\dot{r}_{HLW}(t) = \kappa^0(pH(t), T(t)) \cdot \left(1 - \frac{C_{SiO_2}(t)}{K_{SiO_2}}\right) \tag{10}$$

where κ^0 was the forward or intrinsic dissolution rate and an empirical function of time varying temperature and pH, obtained from regression analysis of experimental results.[4]

The second term of Eq. (10) is the chemical affinity, which accounts for reduced glass dissolution as the silica concentration in the solution surrounding the glass ($C_{SiO_2}(t)$) gradually increases to the solubility equilibrium constant for amorphous silica dissolution (K_{SiO_2}). For PA-95, however, the chemical affinity term was a function of temperature (i.e., $C_{SiO_2} / K_{SiO_2} = f\{T_{CoWP}^{WP}\}$), because $C_{SiO_2}(t)$ was not yet available during the calculation; hence, Eq. (10) became a first-order reaction.

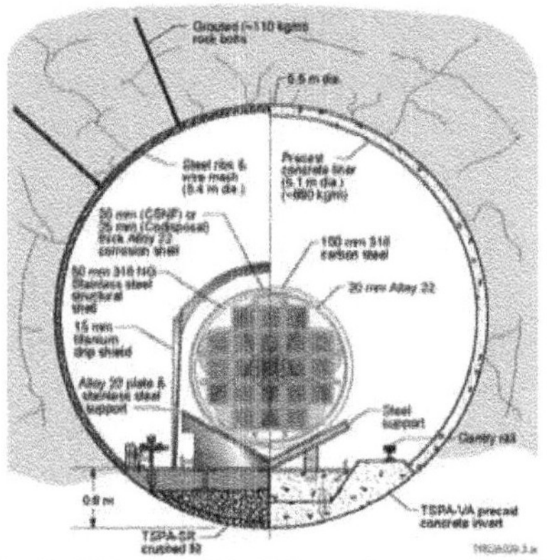

Fig. 8. Although PA-VA used concrete; PA-SR used steel and titanium drip shields to limit altering water chemistry and influence of water drips during thermal period.[2, Fig. 9]

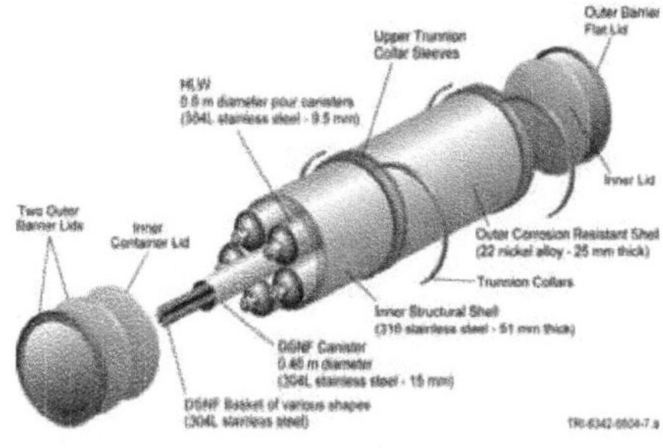

Fig. 9. Co-disposing DSNF with HLW limited fissile content in package yet made efficient use of space for PA-VA and thereafter.[3, Fig. 6]

ENGINEERED BARRIER FOR 1998 PA-VA

PA-VA was conducted to demonstrate the viability of the YM disposal system to Congress.[3, 4, 5, Table 1] YMP used the most current information as interpreted by expert panels prior to completion of experiments. For example, a Waste Package Degradation Expert Elicitation Panel, which consisted of 6 experts with three outside DOE, evaluated the status of current waste package models or proposed for PA-VA, and aggregated disparate data available in the literature.

Repository Design

For the PA-VA, the repository was extended farther north than in PA-95 (Fig. 1). The repository was roughly horizontal and primarily in the lower lithophysal unit (Tptpll) with a small fraction in the middle non-lithophysal unit (Tptpmn) and lower non-lithophysal unit (Tptpln) (Fig. 2). A moderate temperature repository concept with 28-m drift space was used for PA-VA. Spacing down the drift between hot CSNF packages was large and varied between 10.6 and 8.9 m. A cool HLW/DSNF package might be placed between two CSNF packages and in this case spacing between packages was reduced to 2.3 m.

Backfill to create a capillary barrier was studied as an option, but its use was not recommended. Backfill caused operational difficulties in filling the long drift. Also, fine backfill would insulate the waste containers and promote internal temperatures above 350 °C for the large containers during the first 1000 yr after emplacement, which could prematurely damage CSNF cladding. Steps taken to maintain operational safety and allow package retrieval included pre-cast concrete lining and invert for the 5.5 m diameter disposal drifts (Fig. 8).

Package Design

For PA-VA, the inner 20-mm Alloy 825 layer was replaced with another high-nickel Alloy 22 to more reliably increase life. Also, DSNF was included in PA-VA and envisioned to be mostly co-disposed with HLW. Co-disposal of 5 HLW canisters with usually one DSNF canister reduced the need for criticality control inside the package since the DSNF fissile mass was limited (Fig. 9). Generalized corrosion of the Alloy 22 was added for PA-VA and represented by probability density functions (PDF) at 3 different temperatures at humid and wet conditions, as assigned by expert judgment, though humid corrosion was unimportant (Fig. 10).

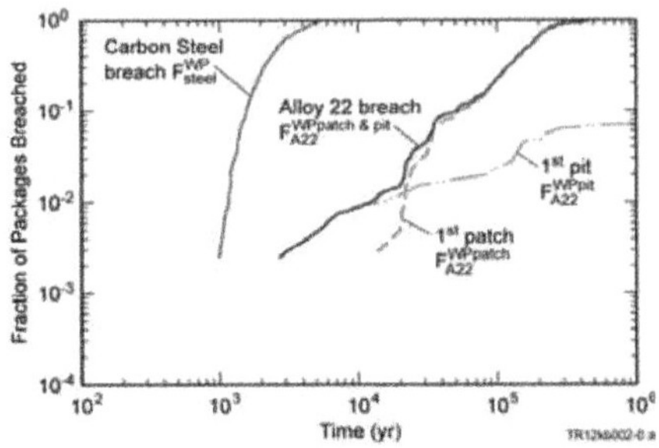

Fig. 10. Fraction of in-drift CSNF containers breached ($F_{CSNF}^{WP}(t)$) for PA-VA; general corrosion of Alloy-22 patches used PDFs set at 3 temperatures and pitting corrosion used exponential function of temperature.[3, Fig. 9]

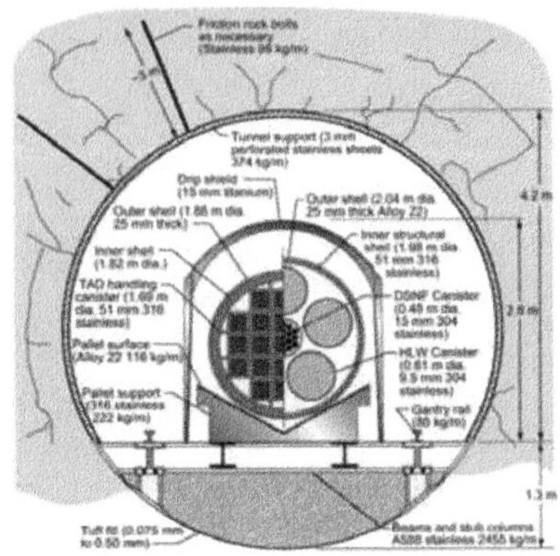

Fig. 11. Ti drip shield and stainless steel invert and ground support for PA-LA.[2, Fig. 13]

Waste Characterization

Since the 1950s, most CSNF has been clad with between 600 through 900 μm of corrosion resistant Zircaloy, an alloy that is ~98% zirconium with small amounts of tin, iron, nickel, and chromium. The Zircaloy cladding was not a designed engineered barrier for disposal, but rather an existing characteristic of CSNF that could influence the release rate of radionuclides once the container breached. YMP was generally reluctant to account for its presence because of the perceived difficulty in defending its condition at disposal and uncertainty in degradation performance over time. Thus, cladding performance was not considered in early PAs. However, NRC used cladding data as the basis for extending the period of wet storage, for licensing dry storage facilities, and for licensing shipping casks for CSNF. Hence for PA-VA, YMP scientists added a rudimentary component for cladding degradation to explore the impact. At 10^6 years, the total fraction of fuel exposed was between 1.7 and 52%. Including cladding had a major effect on dose because the cladding prevented release of 48% to 97.3% of the inventory,[4, 5] because degradation of the CSNF matrix could not proceed until the cladding had perforated.

Once perforated, the exposed CSNF matrix degraded at the rate ($\dot{r}_{CSNF}$) as determined from an 11-term regression equation dependent on temperature, water chemistry (pH, total carbonate concentration, and oxygen partial pressure) and several cross-products. The coefficients were based on batch experiments with irradiated CSNF.[3]

For HLW, Eq. (10) was used in PA-VA, except the model added a long-term residual reaction rate (k_{long}) when the surrounding water film solution was saturated with silica (Fig. 5b). The wet dissolution rate ($\dot{r}_{HLW}^{Wet}(t)$) was used for all environmental conditions because it was surmised that the usually high relative humidity in the drifts would ensure adsorbed water films on the glass.[4] The dissolution rate varied over 5 orders of magnitude during the simulations; yet, the HLW dissolution rate was not a sensitive parameter in PA-VA.[5, Table 2]

ENGINEERED BARRIER FOR 2000 PA-SR

By December 2000, YMP completed PA-SR to support the decision of the Secretary of Energy and President to recommend the YM site for a repository.[1] PA-SR made liberal use of conservatisms in model choices and parameter assignments, and in July 2001, YMP conducted a more realistic perspective of disposal system performance (along with an evaluation of an alternative cooler repository design): the Supplemental Science Performance Analysis (SSPA).

Another analysis was completed for the 2002 final Environmental Impact Statement (EIS) on site-suitability (PA-EIS), which built upon SSPA.

Repository Design

For the 2000 PA-SR and 2008 PA-LA, spacing between drifts was increased to 81 m to allow water to percolate through cool pillars. However, some flow percolating through fractures could become drips, even though a large drift cavity provided a capillary barrier. Titanium drip shields were added to avoid drips on the package and, thereby, reduce the potential for localized corrosion of Alloy 22 during the ~1000-yr thermal period (Figs. 8 and 11).

Package Design

Because the repository was in the UZ, the drifts did not need to be backfilled (at least for an extended period); hence, convection and radiative heat transfer in the open drifts lowered temperatures on the large package surface. Yet, temperatures in the drift exceeded boiling, which prompted questions about the coupling of thermal, hydrologic, and chemical processes during the ~1000 yr thermal period. Hence, YMP conducted much experimental work and code development to advance coupled thermal-hydrologic-chemical modeling. Eventually, concepts from the coupled process modeling were incorporated into a simplified EBS chemistry module for PA-SR and PA-LA to evaluate the possibility of localized corrosion of the Alloy 22 outer layer. However, in both PA-SR and PA-LA, the range of chemical conditions was generally mild and not conducive to localized corrosion; thus, the containers were quite long lived (Fig. 12).

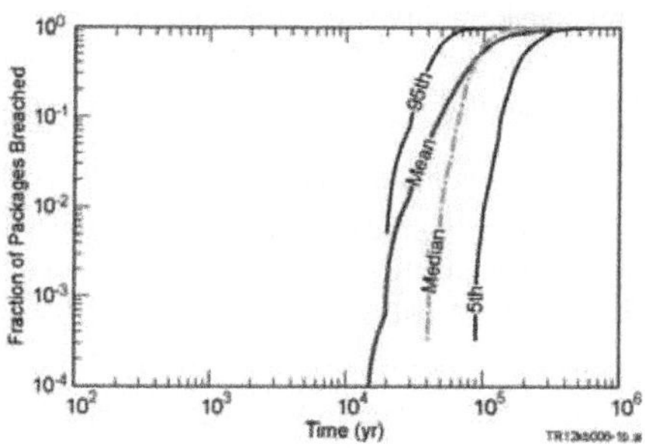

Fig. 12. Calculated distribution for fraction of Alloy 22 outer layer breached ($\mathcal{G}\{F^{WP}(t)\}$) from generalized corrosion (since pitting omitted by drip sheilds) in PA-SR.[3, Fig. 11]

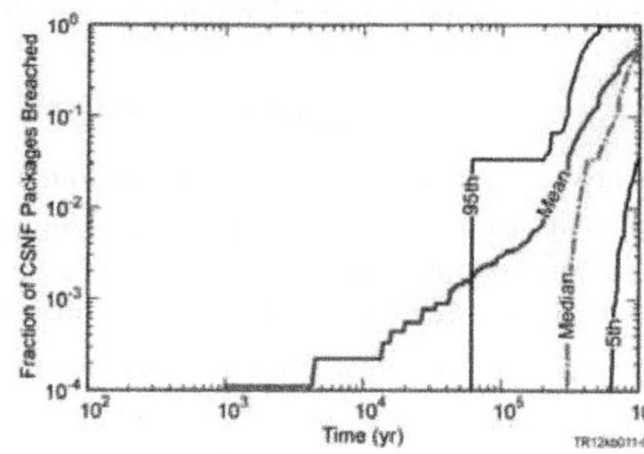

Fig. 13. Wide distribution for fraction of CSNF containers breached was calculated when possibility of seismic damage added in PA-LA ($\mathcal{G}\{F^{WP}_{CSNF}(t)\}$).[3, Fig. 13]

Waste Characterization

The cladding component was refined for PA-SR .Two steps for cladding degradation were included, perforation and unzipping, rather than just perforation as in PA-VA. Three perforation modes were defined:

$$F^{clad}(t) = F^{clad}_{initial}(0) + F^{clad}_{creep}({}^{\max}T^{WP}_{CSNF}) + F^{clad}_{corr}(t; \tau^{fail}_{CSNF}(t)) \quad (11)$$

where $F^{clad}_{initial}(0)$ was the fraction of cladding initially failed when CSNF was placed in containers; $F^{clad}_{creep}({}^{\max}T^{WP}_{CSNF})$ was the fraction of cladding failed by creep and stress corrosion cracking during storage, transportation, and emplacement (~8%);[4] F^{clad}_{corr} $(t;$ $\tau^{fail}_{CSNF}(t))$ was the fraction of cladding failed by localized corrosion from some unknown chemical species such as iodide or fluoride in water entering a failed container much later in time. F^{clad} was set to 1 when a seismic event caused sufficient ground motion to perforate all the cladding.

Once cladding was perforated, unzipping of the cladding could occur as the UO_2 of CSNF formed schoepite with its larger volume. The alteration rate was assumed proportional to the degradation rate of the UO_2 matrix. The unzipping of the cladding was quite rapid in an oxic environment. Hence, CSNF cladding was not important in PA-SR and not considered for PA-LA.[5, 13] While the container was structurally intact, the container provided substantial diffusive and advective resistance to flow even without the presence of the cladding.[4, 5]

For PA-SR and PA-LA, the generic rate of CSNF degradation in a container that failed at time τ^{fail} was expressed similar to PA-95 (Eq. (6)) (Fig 5a):[4]

$$\log\{\dot{r}_{CSNF}(t;\tau^{fail}_{CSNF}(t))\} = (\kappa_0 U^{a_0} + \kappa_1 / T^{WP}_{CSNF}(t) + \kappa_2 \log C^{CO_3}(t;\tau^{fail}_{CSNF}(t)) + \kappa_3 pH(t;\tau^{fail}_{CSNF}(t)) + \kappa_4 \log P^{O_2}) \quad (12)$$

where U^{a_o} was a scale factor to account for overall uncertainty, P^{O_2} was the partial pressure of oxygen, and κ_0, κ_1, κ_2, κ_3 and κ_4 were pH-dependent regression coefficients determined from experimental results.[3] For PA-LA, more data was available, especially at low pH, and the 5 coefficients were adjusted slightly.

For PA-SR and PA-LA, the HLW degradation model was simplified to an Arrhenius rate equation.[4] The model of glass dissolution rate ($\dot{r}_{HLW,b}(t)$) was purposefully biased to bound the initial rate when immersed in liquid water when the relative humidity was > 5% (Fig. 5b):

$$\dot{r}_{HLW}(t) = \kappa^0_{pHseg} \bullet 10^{\eta_{pHseg} pH_{CoWP}(t)} \bullet e^{-\frac{E^a_{pHseg}}{\Re T_{CoWP}(t)}} \quad (13)$$

where $\dot{r}_{HLW}(t)$ was a piece-wise linear function with two pH segments (pH_{low}, pH_{high}); κ^0_{pHseg} was the intrinsic glass dissolution rate constant in either the low or high pH segment and no longer a function of temperature. It was dependent on pH only as a segment switch. The 2nd power term represented the catalyzing influence of either H^+ of OH^- where η_{pHseg} was a coefficient dependent upon either the low or high pH segment, and $pH_{CoWP,b}(t)$ was the time-varying pH in a co-disposal package. The 3rd exponential term represented the increase in SiO_2 hydrolysis as the temperature increased where E^a_{pHseg} was the activation energy in either low or high pH segment, $T_{CoWP}(t)$ was the time-varying temperature in a co-disposal package, and $\Re$ was the universal gas constant. As a bounding equation, the reaction affinity term (Eq. (10)) was omitted since its influence was to decrease the dissolution rate. Also, a long-term residual reaction rate (k_{long}) was not necessary with the omission of the affinity term. Parameter uncertainty was still large for PA-SR but reduced for PA-LA.

ENGINEERED BARRIER FOR 2008 PA-LA

A compliance analysis, to support an NRC license application to construct the repository, was started after Congressional authorization in July 2002.[1, 5, Table 1] Following two interim PAs in 2004 and 2005, PA-LA was completed in June 2008.

Repository Design

For PA-LA, YMP embraced a staged construction approach for the repository and surface facilities as tentatively proposed in the 1986 EA. Also, YMP used a modular design strategy for the repository underground and surface facilities to be (a) consistent with the annual funding available, and (b) able to flexibly adapt to design changes in the container and waste form.

Drift disposal without backfill increased the importance of the disruptive events such as seismic and igneous intrusion. The effects of seismic ground motion on in-drift disposal, as modeled in the 2008 PA-LA, included fault displacement; rockfall damage from a collapsed drift; degraded internal supports, which allowed spent fuel assemblies to puncture the container; and containers hitting other containers and the invert supports. The possibility of seismicity broadened the time when containers might breach (Fig. 13).

Package Design

For PA-LA, the temperature dependence of general corrosion and early container failure was reintroduced and the stress corrosion cracking model for the Alloy 22 outer layer of the container updated. PA-LA added a 25-mm thick transport, aging, and disposal (TAD) handling canister loaded and sealed at the utility, which eliminated handling assemblies whenever transporting, storing, or disposing CSNF (Fig. 6). The TAD was functionally similar to the proposed MPC in PA-95. The TAD handling canister made CSNF containers less susceptible to seismic damage.

GUIDANCE FOR FUTURE DIRECTION OF EBS STUDY

The influence of waste form behavior and its uncertainty is diminished in geologic disposal because other barriers often control the release, whether by design in the case of a robust container or by existing conditions in the natural barrier. Indeed, a benefit of geologic disposal is to provide sufficient flexibility to accommodate a variety of SNF from commercial reactors, experimental reactors, and HLW from reprocessed SNF from current and future fuel cycles.

The role of the EBS differs with the host geologic media. For all repositories, the container provides important short-term radionuclide confinement for operations and when retrieval might be necessary. However, long-term container and waste form performance is less important for a mined repository in salt and clay/shale because the NBS provides substantial long-term isolation. In these geologic media, the long-term EBS function consists primarily of drift and shaft seals.

Long-term radionuclide isolation is important for a mined repository in crystalline rock and volcanic tuff, when using a dose standard. In these media, the NBS functions to establish a long-term stable environment for the important EBS barrier. In the YM disposal system, for example, the container provides substantial diffusive and advective resistance to radionuclide transport while the container is structurally intact; thus, disposal performance relies less on degradation rates of HLW glass, CSNF cladding, and CSNF/DSNF matrix. To fully account for container performance in the YM disposal system, container modeling evolved from assuming complete failure between 300 and 1000 years in early PAs to stochastically describing degradation of the Alloy-22 corrosion-resistant layer in PAs after 1993.[5, Table 2]

The interaction between the EBS and NBS is also important when long-term EBS performance is important. PA-93 introduced a thermal model to evaluate this interaction. Each successive PA would evaluate this EBS-NBS interaction in more detail. In the later PAs, container degradation was a complex function of both the thermal and chemical history and provides a motivation for developing efficient EBS mechanistic models and using high performance computing in the future.[4]

Although disposal performance influenced EBS design, repository operations did also. Flexibility in the size of the handling container is useful in order to accommodate various thermal loads when searching for a new geologic disposal site (as is the current situation in the US). Specifically, the thermal load of the sealed waste container determines when it can be sent to the repository and the requirements for (a) drift spacing, (b) thermal capability of backfill material, and (c) long-term ventilation in various geologic media. A small container with a low thermal load can be disposed in many media, while a large container with a high thermal load cannot (such as crystalline rock with a surrounding clay engineered barrier) unless the container is cooled in long-term storage (either on the surface or in the repository with ventilation). For applications in saturated clay or crystalline environments, enhancing natural materials to withstand high heat loads from large containers would facilitate operations and increase disposal options.

After the host medium has been selected, a waste management organization can optimize the EBS design. For example, in-drift disposal of large containers with a high heat load improved the operation of the YM repository by reducing repository size and number of containers emplaced. The YMP could, thereby, more easily accommodate a high 3000 MTHM/yr rate of disposal for the large amounts of CSNF in the US. Similarly, a TAD handling canister eased surface loading operations.

Crystalline and tuff host rocks, currently depend on the long-term isolation provided by the waste package to flexibly dispose of a wide variety of waste forms without fully modeling waste behavior. However, thorough knowledge of waste form behavior in a variety of environments, conditioning the EBS environment to lower radioelement solubility, and/or developing waste forms with an emphasis on long-term disposal durability may make container longevity using expensive alloys less important in the future. Examples of a durable waste forms include silicon-carbide coatings on reactor fuels, such as on pellet fuel for high-temperature gas reactors or development of robust engineered HLW matrices with slow degradation rates.

ACKNOWLEDGEMENTS

Sandia National Laboratories (SNL) is a multi-program laboratory operated by Sandia Corporation, a wholly owned subsidiary of Lockheed Martin Corporation, for the US Department of Energy (DOE) National Nuclear Security Administration under contract DE-AC04-94AL85000. SAND2016-4237C. As a historical review, the author is reporting on the work of others; however, the perspective presented is that of the author and is not necessarily held by DOE, SNL, or the many other institutions and persons involved at YMP. Unfortunately, directly citing the vast work related to this summary is not possible here. Indeed, the NRC Licensing Support Network, which supports SAR/LA, contains over 3.4 million documents. The 5 articles cited here list references and directly acknowledge many persons contributing to the EBS work. In general, the institutions gathering data and modeling long-term degradation of CSNF and HLW were Argonne National Laboratory (ANL) and Pacific Northwest National Laboratory (PNNL). Idaho National Laboratory was responsible for DSNF. Lawrence Livermore National Laboratory was responsible for EBS performance and, thus, involved with evaluating container performance and the experimental work from ANL and PNNL. SNL led the effort of incorporating EBS performance into PA-EA, PA-91, and PA-LA. SNL and the DOE management contractor (primarily INTERA/Duke/Areva) jointly led PA-95, PA-VA, PA-SR.

REFERENCES

[1]R.P. Rechard, Milestones for Selection, Characterization, and Analysis of the Performance of a Repository at Yucca Mountain, SAND2015-1060 Sandia National Laboratories (2015).(465 refs)

[2]R. Rechard, and M.D. Voegele, Evolution of Repository and Package Designs for Yucca Mountain Disposal System, Reliability Eng Sys Safety, 122(2), 53-73 (2014). (150 refs)

[3]R.P. Rechard, J.H. Lee, E. Hardin, and C.R. Bryan, Waste Package Degradation from Thermal and Chemical Processes in Performance Assessments of the Yucca Mountain Disposal System, Reliability Eng Sys Safety, 122(2), 145-64 (2014). (71 refs)

[4]R.P. Rechard, and C.T. Stockman, Waste Degradation and Mobilization in Assessments of Yucca Mountain Disposal System, Reliability Eng Sys Safety, 122(2), 165-88 (2014). (91 refs)

[5]R.P. Rechard, Results from Past Performance Assessments of the Yucca Mountain Disposal System, Reliability Eng Sys Safety, 122(2), 207-22 (2014). (62 refs)

EFFECT OF HYDRATION HEAT ON IODINE DISTRIBUTION IN GYPSUM-ADDITIVE CALCIUM ALUMINATE CEMENT

Tomofumi Sakuragi, Yu Yamashita[1], and Shigeto Kikuchi[1]
Radioactive Waste Management Funding and Research Center, Tsukishima 1-15-7, Chuo City, Tokyo, Japan
[1] TOSHIBA Corporation, Ukishima Cho 4-1, Kawasaki Ward, Kawasaki, 210-0862, Japan

ABSTRACT

Radioactive iodine (^{129}I) generated during the reprocessing of spent nuclear fuel presents a major challenge to radioactive waste disposal strategies. Cementitious materials are widely used in disposal facilities and are expected to perform iodine fixation. Recent studies indicate that iodate is confined in cement hydrates enriched by adding gypsum to calcium aluminate cement (IO_3-cement). To ensure that thermal affects are avoided when scaling up, the characteristics of IO_3-cement containing 2wt% of iodate were demonstrated for two different sizes (200 mL and 20 L). The mineral composition in both sample sizes was equal, whereas the dominant iodine-bearing mineral was changed from ettringite to hydrogarnet for the larger sample. This result can be attributed to the temperature difference in the interior of the IO_3-cement during curing. A solubility equilibrium model is applied to evaluate the iodine fixation property based on the degradation of iodine-bearing minerals.

INTRODUCTION

Iodine 129 (^{129}I) is a typical fission product (fission yield of ^{235}U is 0.543%) with a long half-life of 1.57×10^7 years. For instance, the Rokkasho commercial reprocessing plant in Japan, which has not yet been in full-scale operation, has been estimated to have a total ^{129}I inventory of 5.1×10^{13} Bq from 40 years of operation with a total of 32,000 MTU (burn-up of 45 GWd/MTU).[1] This plant is equipped with an iodine absorbing system that captures volatile iodine on an alumina-based filter (the silver-sorbent) in the form of silver iodide. The spent silver-sorbent, which is estimated at 137 tons, is planned to be disposed of in a deep underground repository. The safety aspects of the radioiodine after disposal are important due to the long half-life and mobility of the substance within the engineered and geological media.

The challenge of iodine immobilization has been studied since the 1980s, when immobilization forms such as simple iodine compounds, glasses, and iodine-bearing minerals were suggested.[2] Since then, the synthesis of certain iodine-bearing materials such as apatite[3, 4], Synroc[5], and zeolite[6] has been demonstrated. Recently, consolidation of a mordenite-base iodine filter has been carried out by hot isostatic pressing[7], but these studies have focused on elucidating the behavior of iodine uptake in the matrix and solids. However, other studies are evaluating durability and iodine leaching behavior from novel waste-forms under conditions that are expected to be encountered during disposal.[8–11]

Cementitious materials are widely used in disposal facilities for both construction and backfill, and are expected to perform a safety function as engineered barriers for radionuclides. One reason for their selection is that cement hydrate minerals are considered to have good potential for radionuclides retention even for iodine fixation. Several studies have focused on iodine sorption by cement paste and cement hydrate minerals such as calcium silicate hydrate (CSH) and ettringite (AFt).[12–20] Tomita et al. studied the sorption behavior of anion species in synthesized AFt and monosulfate (AFm) and found that iodate (IO_3^-) is absorbed more selectively than iodide (I^-).[14] Toyohara et al. have succeeded in the synthesis of novel cement enriched with AFt by adding gypsum to alumina calcium cement.[15] The substitution of SO_4^{2-} for IO_3^- in an anion channel of AFt has been suggested by Iwaida et al.[16] More recent work by Idemitsu et al. has clarified the detailed structural formula of iodine-bearing AFt by using measurement by X-ray absorption near-edge structure (XANES).[20] Following these studies, Haruguchi et al. synthesized gypsum-additive calcium aluminate cement within IO_3^- (IO_3-cement) at laboratory scale (200 mL) and investigated the iodine fixation and release properties of the IO_3-cement solid based on degradation of the iodine-bearing minerals through interaction with water.[10]

Scaling-up is the next challenge to address for IO_3-cement. Cementitious materials generate hydration heat during the hardening process. Generally, calcium aluminate cement is rapidly hardened within a day. Damidot et al. suggested through use of a phase diagram that the stable region of AFt shrinks in a CaO-Al_2O_3-$CaSO_4$-H_2O system when the temperature rises from 50°C to 85°C.[21] It is also known that hydrate composition in calcium aluminate systems can be altered by temperature.[22] The objective of the current study is to evaluate the scale effects upon the mineral composition and iodine-bearing minerals in IO_3-cement, under condition in which the thermal influence from hydration heat is raised. To evaluate these effects, two different size samples of IO_3-cement were prepared. The proportion of cement minerals in the hardened IO_3-cement and the distribution of iodine to these minerals were obtained by chemical analysis using modification of the sequential extraction technique proposed by Suzuki et al. for obtaining the mineral composition of ordinary Portland cement paste.[23] The iodine fixation and release property of the IO_3-cement samples with different thermal history were then evaluated by using a solubility equilibrium model that calculates the degradation of iodine-bearing minerals.

EXPERIMENT

Sample Preparation and Thermal Measurement

The gypsum-added calcium aluminate cement including iodate (IO_3-cement) was prepared by mixing calcium aluminate cement from Covalent Materials Corporation, gypsum ($CaSO_4 \cdot 2H_2O$), and a 0.4 mol L^{-1} $NaIO_3$ solution in a ratio of 54.0:8.5:37.5 by weight. The components of the calcium aluminate cement are shown in Table I. For the smaller samples, the IO_3-cement was cast into plastics vessels of 100 mL (50 mm diameter × 50 mm height cylinder) and 200 mL (50 mm diameter × 100 mm height cylinder). For the larger sample of 20 L, the IO_3-cement was mixed in a tin vessel (286 mm diameter × 311 mm height cylinder). One thermocouple was promptly inserted in each of the 100 mL and 200 mL samples, and several were promptly inserted in the 20 L mixtures. After curing and hardening under atmosphere conditions at approximately 22°C for 100 hours, the IO_3-cement samples were obtained and dried for chemical analysis.

Table I. Components of calcium aluminate cement

Al_2O_3	CaO	SiO_2	Fe_2O_3	MgO	$K_2O + Na_2O$	TiO_2	SO_3
69.5 - 71.5	27 - 29	< 0.8	< 0.3	< 0.3	< 0.5	< 0.1	< 0.3

Chemical Analysis of Mineral Composition and Iodine Distribution

As shown in Figure 1, five minerals were identified in the IO_3-cement by X-ray diffraction (XRD): ettringite (AFt), monosulfate (AFm), hydrogarnet (HG), unhydrated calcium aluminate (CA), and aluminum hydroxide (AH). This result was then used to set assumptions for the mineral composition and iodine-bearing minerals in the IO_3-cement. However, the analysis for HG, CA, and AH are not included in the Suzuki method[23] referred to above for analyzing the proportion of cement minerals in the hardened IO_3-cement paste. We therefore modified Suzuki's method to include these minerals. Figure 1 shows the change in XRD patterns after treatment with a selected sequential extraction carried out in the following order: 1) ethylene glycol, 2) salicylic acid, 3)0.1 mol L^{-1} HCl, and 4) 1.0 mol L^{-1} HCl. After each treatment, the peaks corresponding to the target minerals of AFt, AFm, HG, and CA are disappeared whereas AH remains. Therefore, the minerals and the distribution of iodine to these chemicals were measured by sequential extraction as shown in Figure 2 (details are described in the following paragraph) Calcium carbonate (calcite) was not found in the IO_3-cement, but was checked according to Suzuki's method[23] because carbonate from the atmosphere can contaminate the cement.

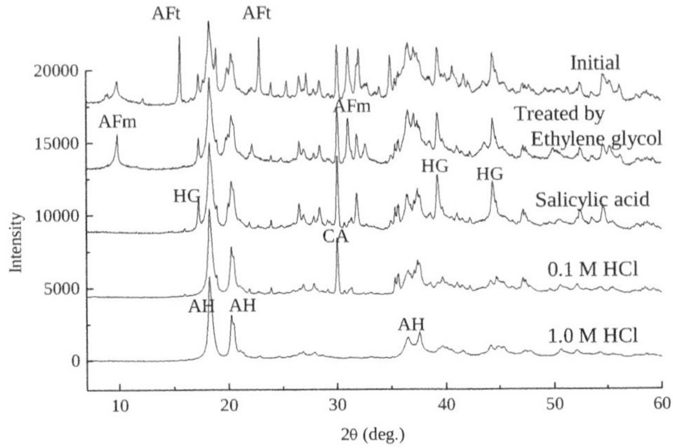

Figure 1. X-ray diffraction patterns of IO_3-cement powder before and after sequential dissolution treatments.

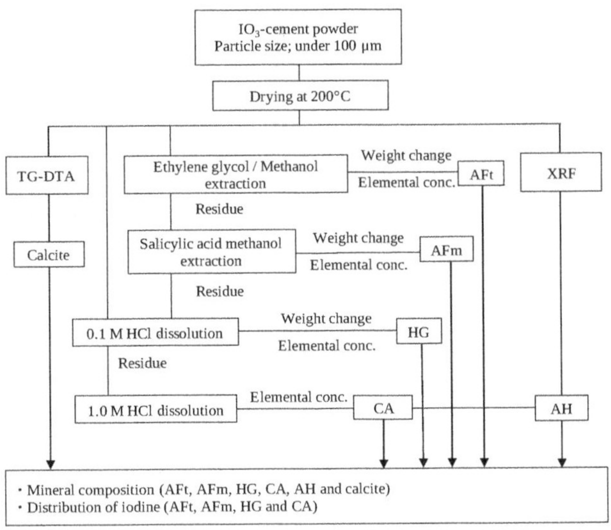

Figure 2. Overview of analytical procedure for mineral composition of IO_3-cement using sequential extraction treatments.

Ettringite (AFt; $Ca_6Al_2(SO_4)_3(OH)_{12}\cdot 26H_2O$) was first extracted from IO_3-cement (0.2 g) with an ethylene glycol and methanol mixture (mixing ratio of 1:3, 200 mL) by stirring for 3 hours at room temperature. The amount of AFt was obtained from the dry weight difference before and after the extraction. Because the direct determination of iodine in the ethylene glycol extract would have resulted in loss of iodine during the drying process, the iodine included in the AFt phase was calculated from the difference in the iodine content within and outside of the AFt phase.

Monosulfate (AFm; $Ca_4Al_2SO_4(OH)_{12}\cdot 6H_2O$) was secondarily extracted from the residue described above (0.2 μm filtered) by a solvent containing salicylic acid (6.0 g) in methanol (40 mL) by stirring for 2 hours at room temperature. The amount of AFm was obtained from the dry weight difference before and after the extraction. Iodine included in the AFm phase was also calculated from the difference of iodine content within and outside of the AFm phase.

Hydrogarnet (HG; $Ca_3Al_2(OH)_{12}$) was dissolved using a 0.1 mol L^{-1} HCl solution (200 mL) from the residue described above by stirring at room temperature for 30 minutes. The amount of HG was obtained from the dry weight difference in the same manner as for AFt and AFm. Iodine in the HG phase was directly measured in the dissolved solution by ICP-AES. It should be noted that a small amount of residual elements from the previous AFt and AFm extraction process might be included in this fraction.

Unhydrated calcium aluminate (CA; $CaO\cdot Al_2O_3$) was dissolved in a 1.0 mol L^{-1} HCl solution (100 mL) by stirring at room temperature for 10 minutes after simultaneously removing AFt, AFm and HG from 0.1 g of IO_3-cement by use of 0.1 mol L^{-1} HCl. The amount of CA was derived from the Ca quantity that corresponds to its composition formula. Iodine in the 1.0 mol L^{-1} HCl dissolution was directly measured.

Aluminum hydroxide (AH; $Al(OH)_3$) is assumed as insoluble fraction of IO_3-cement. The amount of AH was derived from the difference between the total amount of Al in IO_3-cement, which was obtained by X-ray Fluorescence Spectrometry (XRF), and Al in the 1.0 mol L^{-1} HCl dissolution of the CA phase described above.

Calcium carbonate (Calcite; $CaCO_3$) was determined by thermogravimetry and differential thermal analysis (TG-DTA). Figure 3 shows the TG-DTA data for the IO_3-cement. The temperature range from 200°C to 400°C corresponds to the dehydration of aluminum hydroxide. In the range from 500°C to 800°C, there is a small change in DTA associated with the loss of carbonate in calcite.

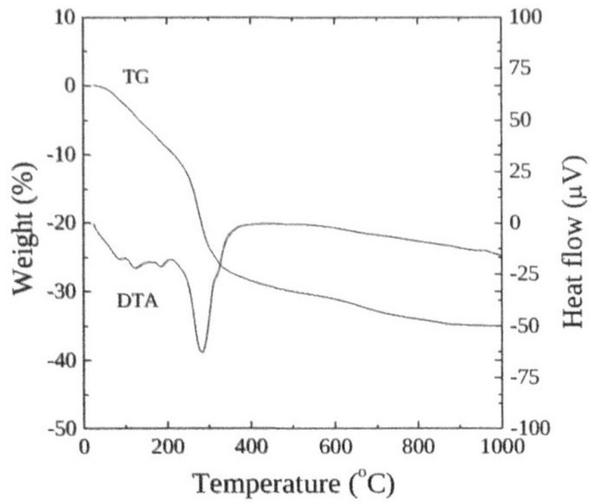

Figure 3. Thermogravimetry and differential thermal analysis (TG-DTA) of IO_3-cement for calcium carbonate measurement.

RESULTS AND DISCUSSION

Thermal Measurement

The internal temperature for cement material increases during the hardening process due to the hydration reactions of its component minerals. Figure 4 shows the temperature change in the different size samples of gypsum-additive calcium aluminate cement containing iodate (IO_3-cement) measured at the sample center. Data was collected for one 100 mL, four 200 mL, and three 20 L IO_3-cement samples. With the exception of one 20 L sample, the internal temperature rose after 12 hours of mixing and reaches its peak within 24 hours. The maximum temperatures reached were around 30°C for small cement samples. However, for the 20 L cement samples, the temperature rose rapidly to 150°C. This result is simply understood to mean that the smaller body dissipates heat easily and the larger body stores the heat. The maximum temperatures reached during the hardening process for each chemically analyzed sample are summarized in Table II. Even in the same 20 L sample, the maximum temperatures were different depending upon whether the location was close to the surface, where the temperature was low.

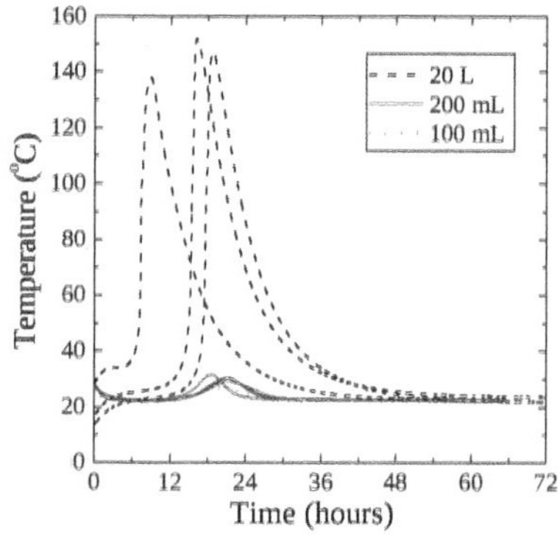

Figure 4. Thermal measurement during cure and hardened process at the center of different IO_3-cement sizes.

Mineral Composition and Iodine Distribution

Figure 5 (left half) shows the mineral composition in the different IO_3-cement sizes of 200 mL and 20 L. Additional data for 200 mL samples (sample 200 mL-b) was obtained to confirm the reproducibility of the previous data for 200 mL-a.[10] For the 20 L samples, data were collected from the sample top, center, and bottom. Although, each sample has a different thermal history, the mineral composition is largely constant. Aluminum hydroxide (AH) is dominant at about 45wt%; monosulfate (AFm) and hydrogarnet (HG) have comparable amounts at 20 wt% to 25wt%; ettringite (AFt) is around 10wt%, and calcium aluminate (CA) is less than 2wt%. The amount of calcium carbonate (calcite) was nearly undetectable but the detectable calcite found in the top of the cement may have been formed from atmospheric contaminant. We did not find a change in mineral composition by scaling up the cement as suggested in the studies on similar cementitious materials.[21, 22] Table II summarizes the chemical analysis results.

Figure 5 (right half) shows the content of the iodine-bearing minerals. Iodine in the IO_3-cement is distributed among three hydrates: AFt, AFm, and HG. In contrast to mineral composition, the distribution of iodine to the hydrates was significantly affected by the thermal effects as a consequence of the scaled-up of the IO_3-cement. AFt is the most abundant hydrate, containing about 80% iodine for laboratory-scale cement of 200 mL as was postulated. At this small size, AFm and HG are minor components that hold 5% to 15% of the iodine. However, as the temperature during curing rose to the maximum value of 150°C due to hydration heat, the dominant iodine-bearing mineral was no longer AFt but alter to HG containing about 70% iodine. The relationship between the amount of iodine distributed to AFt and HG and the maximum temperature to which the samples was exposed is depicted in Figure 6, which shows that as temperature rises less iodine is distributed to AFt and more is distributed to HG.

The sorption and incorporation of iodine in calcium silicate hydrate (CSH)[13, 18, 19], AFt[13, 14, 16], and AFm[14] were observed in the previous studies. Although iodate-bearing AFt and AFm have been synthesized in other studies[20], only material in which interaction of iodine with HG is suggested is the IO_3-cement. It is possible that some of the hydroxyls in the HG structure substitute by IO_3 ions. However, the formation mechanism of iodine-bearing hydrate in cement is not understood at present.

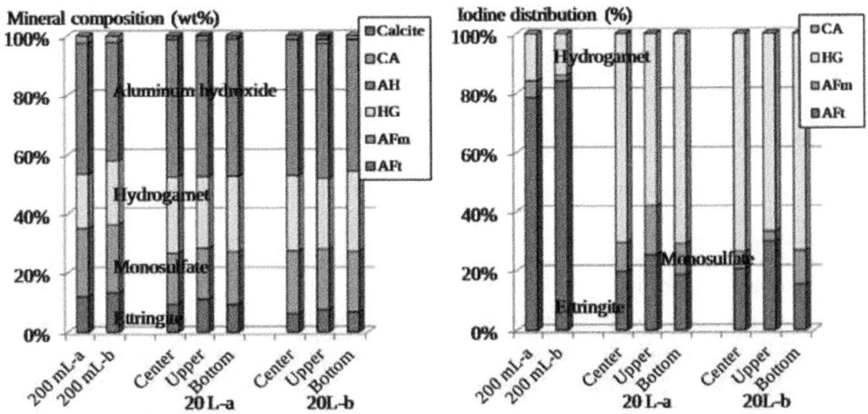

Figure 5. Mineral compositions (left) and iodine-bearing minerals (right) obtained from different IO_3-cement size and location. Data for 200mL-a are obtained from published studies.[10]

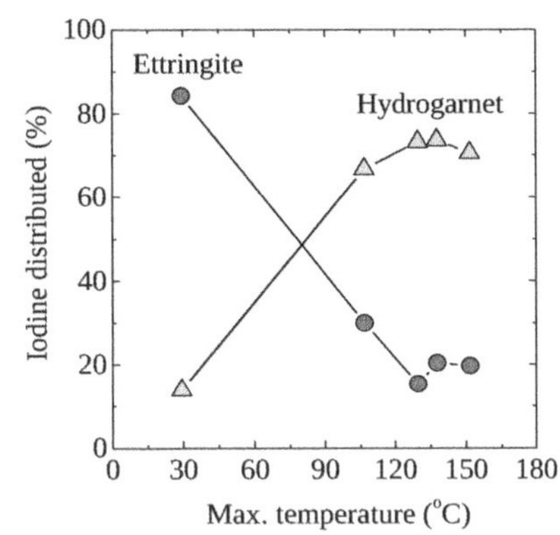

Figure 6. Change in iodine distribution to ettringite and hydrogarnet as a function of maximum temperature at curing.

Table II. Mineral composition and iodine distribution in IO_3-cement

Sample	Max. temp. (°C)	Mineral composition (wt%)						Iodine distributed (%)			
		AFt	AFm	HG	CA	AH	$CaCO_3$	AFt	AFm	HG	CA
200 mL-a[10]	-	12.0	23.1	18.3	2.2	44.4	0	78.6	5.6	15.8	0
200 mL-b	29.3	13.1	23.3	21.5	2.1	40.1	0	84.2	2.0	13.9	0
20 L-a, Center	151.8	9.4	17.1	25.8	1.1	46.5	0	19.7	9.8	70.5	0
20 L-a, Upper	-	11.2	17.1	24.3	1.2	46.1	0.2	25.4	16.6	58.0	0
20 L-a, Bottom	-	9.4	17.7	25.6	1	46.4	0	18.6	10.5	70.9	0
20 L-b, Center	137.9	6.2	21	25.7	1.2	45.8	0	20.4	6.0	73.6	0
20 L-b, Upper	106.8	7.5	20.4	24.1	1.2	45.5	1.3	29.9	3.4	66.7	0
20 L-b, Bottom	129.7	6.7	20.3	27.1	1.3	44.5	0	15.3	11.4	73.2	0

Iodine Fixation and Release Properties According to the Solubility Equilibrium Model

The performance of iodine immobilized solids with different thermal histories that correspond to the cement composition of the 200 mL and 20 L samples can be discussed with reference to a model. An iodine release model for the IO_3-cement has been developed based on the solubility equilibrium of cement minerals that come into contact with water.[10] PHREEQC 2[24] was used for the calculation. In this model the IO_3-bearing hydrates are assumed to be ideal solid solutions whose dissolution reactions are described in Table III together with their solubility product (log K). In this table, IS-AFt represents AFt in which half of the SO_4^{2-} ions have been exchanged for IO_3^- and OH^- in the oxyanion channel, IO_3-AFm is a form of AFm completely substituted for by IO_3^- with the entire amount of SO_4^{2-}, Cl-AFm is Friedel's salt[26], and I-HG represents a mineral in which the portion with OH^- is replaced by IO_3^-. The log K values for the IO_3-bearing hydrates are obtained from Haruguchi et al.[10] The log K data for the original hydrate minerals are also obtained from published studies.[25–27] A thermodynamic data-base of JNC-TDB (2005)[28] was used except for the specifically defined minerals mentioned above. The initial compositions of minerals are listed in Table IV in which case 1 and case 2 are the initial compositions of each of the experimental results for the 200 mL and 20 L samples as shown in Table II. The composition of the iodine-bearing hydrates was taken from their stoichiometry (using their composition formula).

Table III. Assumed solubility products of iodine-bearing minerals

Minerals	Chemical reaction	log K	References
IS-AFt	$Ca_6(Al(OH)_6)_2\cdot((IO_3)_{0.5}(OH)_{0.5}(SO_4)_{0.5})_3\cdot 24.5H_2O$ $= 6Ca^{2+} + 2Al(OH)_4^- + 1.5IO_3^- + 1.5SO_4^{2-} + 5.5OH^- + 24.5H_2O$	-51.29	10
AFt	$Ca_6Al_2O_6\cdot(SO_4)_3\cdot 32H_2O$ $= 6Ca^{2+} + 2Al(OH)_4^- + 3SO_4^{2-} + 4OH^- + 26H_2O$	-43.94	25
IO_3-AFm	$(CaO)_3\cdot Al_2O_3\cdot Ca(IO_3)_2\cdot 12H_2O$ $= 4Ca^{2+} + 2Al(OH)_4^- + 2IO_3^- + 4OH^- + 6H_2O$	-36.8	10
Cl-AFm	$(CaO)_3\cdot Al_2O_3\cdot CaCl_2\cdot 10H_2O$ $= 4Ca^{2+} + 2Al(OH)_4^- + 2Cl^- + 4OH^- + 4H_2O$	-27.1	26
AFm	$Ca_4Al_2O_6\cdot SO_4\cdot 12H_2O = 4Ca^{2+} + 2Al(OH)_4^- + SO_4^{2-} + 4OH^- + 6H_2O$	-29.25	25
I-HG	$Ca_3Al_2(OH)_8(IO_3)_4 = 3Ca^{2+} + 2Al(OH)_4^- + 4IO_3^-$	-33.20	10
HG	$Ca_3Al_2(OH)_{12} = 3Ca^{2+} + 2Al(OH)_4^- + 4OH^-$	-20.84	27

Table IV. Initial mineral composition of IO_3-cement assumed for model calculation (mol kg^{-1}). Case 1 and 2 are the initial compositions for the 200 mL and 20 L samples.

	IS-AFt	AFt	IO_3-AFm	AFm	I-HG	HG	CA	AH
Case 1 (200 mL)	0.0668	0.0219	0.0035	0.3672	0.0050	0.4708	0.0646	5.6933
Case 2 (20 L)	0.0192	0.0540	0.0072	0.2649	0.0258	0.6140	0.0326	5.9672

Figure 7 shows the model results concerning changes in iodine-bearing hydrates and released iodine in aqueous phase as a function of the liquid to solid ratio (L/S) resulting from contact with pure water. The previous calculation[10], which is identical with case 1, was stated only for AFt and aqueous data. For case 1 and case 2, iodine is immobilized by AFt and HG, respectively, up to an L/S of 800. Re-precipitation of HG was found at an L/S of around 1000. The iodine-bearing hydrates then disappeared at higher values of L/S. The iodine release behavior is therefore similar for both cases. However the conclusion that the scaling-up of IO_3-cement with different thermal histories and iodine-bearing minerals will have identical performance of iodine immobilization after disposal may not be robust. As stated above, the characteristics of iodine-bearing HG are not clear, so that the dissolution reaction and corresponding solubility product of I-HG were assumed by adjusting a model parameter to agree with the experimental results.[10] Therefore an issue for further study with respect to IO_3-cement is to obtain a detailed understanding of the physicochemical properties regarding iodine-bearing HG.

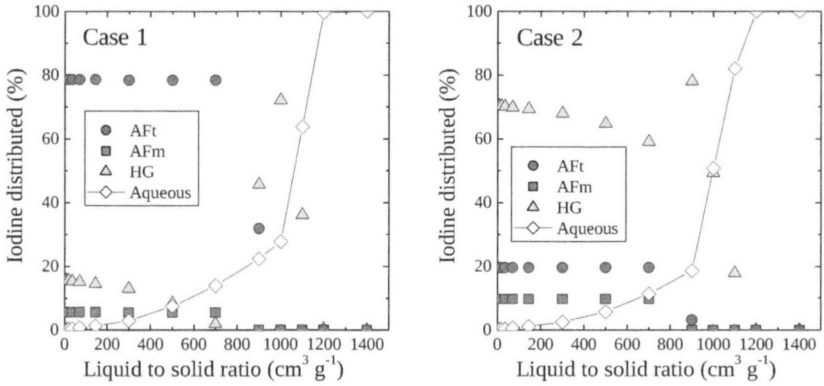

Figure 7. Calculation results of change in iodine leaching and iodine-bearing minerals as a function of the liquid to solid ratio by the solubility equilibrium model.[10] Case 1 and case 2 are the initial compositions of each of the experimental results for the 200 mL and 20 L samples (Table IV). The data of AFt and aqueous in the case1 are taken from the published studies.[10]

CONCLUSIONS

Using different sizes of IO_3-cement, the scale effect led to a rise in internal temperature up to a maximum of 150°C during the hardening process, which results in the change in the dominant iodine-bearing mineral from ettringite to hydrogarnet. This change was not significant for the iodine release model. However, further change through hydration heat in the dominant iodine-bearing mineral should be avoided. From the safety point of view in the operation of the processing plant, excess heat generation from the scaling-up of waste forms must not be permitted. Under the current scenario, a 200 L drum is considered to be too large. Optimization, either through applying an additional cooling system during the hardening process or reducing the waste form size, will be required for practical use. It can be concluded that in an engineered barrier system, temperature is an important factor affecting the interaction of iodine with cementitious materials in the form of immobilization and sorption/desorption phenomena through engineered barrier system.

ACKNOWLEDGMENTS

This research is a part of the “Research and development of processing and disposal technique for TRU waste (FY2014)” program funded by Agency for Natural Resources and Energy in Ministry of Economy, Trade and Industry of Japan.

REFERENCES

[1] Federation of Electric Power Companies (FEPC) and Japan Atomic Energy Agency (JAEA), Second Progress Report on Research and Development for TRU Waste Disposal in Japan (2007).

[2] IAEA Technical Reports Series No. 276, Treatment, Conditioning and Disposal of Iodine-129 (1987).

[3] F. Audubert, J. Carpena, J. L. Lacout, and F. Tetard, Elaboration of an Iodine-bearing Apatite Iodine Diffusion into a $Pb_3(VO_4)_2$ Matrix, *Solid State Ionics*, **95**, 113–119 (1997).

[4] M. C. Stennett, I. J. Pinnock, and N. C. Hyatt, Rapid Microwave Synthesis of $Pb_5(VO_4)_3X$ (X = F, Cl, Br and I) Vanadinite Apatite for the Immobilisation of Halide Radioisotopes, *Mater. Res. Soc. Symp. Proc.*, **1475**, 221-226 (2012).

[5] E. R. Vance and J. S. Hartman, Effects of Halides and Sulphates in Synroc-C, *Mater. Res. Soc. Symp. Proc.*, **556**, 41-46 (1999).

[6] N. C. Hyatt, J. A. Hriljac, A. Choudhry, L. Malpass, G. P. Sheppard, and E. R. Maddrell, Zeolite-Salt Occlusion: A Potential Route for the Immobilisation of Iodine-129?, *Mater. Res. Soc. Symp. Proc.*, **807**, 1-6 (2004).

[7] R. T. Jubin and S. H. Bruffey, High-Temperature Pressing of Silver-Exchanged Mordenite into a Potential Iodine Waste Form, *Proc. WM2014 Conference*, March 2–6, Phoenix, Arizona, USA (2014).

[8] T. Sakuragi, T. Nishimura, Y. Nasu, H. Asano, K. Hoshino, and K. Iino, Immobilization of Radioactive Iodine using AgI Vitrification Technique for the TRU Wastes Disposal: Evaluation of Leaching and Surface Properties, *Mater. Res. Soc. Symp. Proc.*, **1107**, 279-285 (2008).

[9] A. Mukunoki, T. Chiba, T. Kikuchi, T. Sakuragi, H. Owada, and T. Kogure, Corrosion and Alteration of Lead Borate Glass in Bentonite Equilibrated Water, *Mater. Res. Soc. Symp. Proc.*, **1518**, 15-20 (2013).

[10] Y. Haruguchi, S. Higuchi, M. Obata, T. Sakuragi, R. Takahashi, and H. Owada, A Study on Iodine Release Behavior from Iodine-Immobilizing Cement Solid, *Mater. Res. Soc. Symp. Proc.*, **1518**, 85-90 (2013).

[11] T. Sakuragi, S. Yoshida, O. Kato, and Kaoru Masuda, Effects of Hydrosulfide and pH on Iodine Release from an Alumina Matrix Solid Confining Silver Iodide, *Mater. Res. Soc. Symp. Proc.*, **1744**, (2015).

[12] T. R. Holland and D. J. Lee, Radionuclide Getters in Cement, *Cement and Concrete Research*, **22**, 247-258 (1992).

[13] M. L. Gougar, B. E. Scheetz, and D. M. Roy, Ettringite and C-S-H Portland Cement Phases for Waste Ion Immobilization: A Review, *Waste Management*, **16**, 295-303 (1996).

[14] F. Tomita, M. Wada, N. Mitsutsuka, M. Kaneko, M. Toyohara, T. Murase, and H. Fujihara, Development of Iodine Immobilization Process with Cementitious Materials, Proc. Radioactive Waste Management and Remediation ASME 1999 (1999).

[15] M. Toyohara , M. Kaneko , H. Ueda , N. Mitsutsuka, H. Fujiwara , T. Murase, and N. Saito, Iodine Sorption onto Mixed Solid Alumina Cement and Calcium Compounds, *J. Nucl. Sci. Technol.*, **37**, 970-978 (2000).

[16] T. Iwaida, K. Yano, S. Tanaka, and S. Nagasaki, Synthesis and Characterization of Sulfate-iodate Ettringite; and Their Potential Role in Immobilization of ^{129}I, *Journal of Nuclear Fuel Cycle and Environment*, 7, 57-63 (2001).

[17] M. Toyohara , M. Kaneko , N. Mitsutsuka, H. Fujiwara , N. Saito, and T. Murase, Contribution to Understanding Iodine Sorption Mechanism onto Mixed Solid Alumina Cement and Calcium Compounds, *J. Nucl. Sci. Technol.*, **39**, 950-956 (2002).

[18] I. Bonhoure, A. M. Scheidegger, E. Wieland, and R. Dahn, Iodine Species Uptake by Cement and CSH Studied by I K-edge X-ray Absorption Spectrometry, *Radiochim. Acta*, **90**, 647-651 (2002).

[19] Y. Niibori, T. Funabashi, and H. Miura, Some Experiments on Sorption Behavior of Iodide Ions into CSH Gel under the Condition Saturated with Saline Groundwater, *Mater. Res. Soc. Symp. Proc.*, **1518**, 225-230 (2013).

[20] K. Idemitsu, Y. Matsuki, M. Kishimoto, Y. Inagaki, T. Arima, Y. Haruguchi, Y. Yamashita, and M. Sasoh, Migration of Iodine Solidified in Ettringite into Compacted Bentonite, *Advances in Materials Science for Environmental and Energy Technologies II*, **241**, 23-34 (2013).

[21] D. Damidot and F. P. Glasser, Thermodynamic Investigation of the CaO-Al_2O_3-$CaSO_4$-H_2O System at 50°C and 85°C, *Cement and Concrete Research*, **22**, 1179-1191 (1992).

[22] H.F.W. Taylor, CEMENT CHEMISTRY 2nd edition, Thomas Telford Publishing, (1997).

[23] K. Suzuki, T. Nishikawa, Y. Yamada, I. Taniguchi, Analysis of Hydrated Phases for Evaluation the Durability of Concrete, *Concrete Research and Technology*, **1**, 39-49 (1990). (in Japanese)

[24] D. L. Parkhurst and C. A. J. Appele, User's Guide to PHREEQC (VERSION2), US. Geological Survey (1999).

[25] E. J. Reardon, Problems and Approaches to the Prediction of the Chemical Composition in Cement/water systems, *Waste Management*, **12**, 221-239 (1992).

[26] U. A. Birnin-Yauri and F. P. Glasser, Friedel's Salt, $Ca_2Al(OH)_6(Cl,OH)\cdot 2H_2O$: Its Solid Solutions and Their Role in Chloride Binding, *Cement and Concrete Research*, **28**, 1713–1723 (1998).

[27] T. Matschei, B. Lothenbach, and F. P. Glasser, Thermodynamic Properties of Portland Cement Hydrates in the System CaO–Al_2O_3–SiO_2–$CaSO_4$–$CaCO_3$–H_2O, *Cement and Concrete Research*, **37**, 1379-1410 (2007).

[28] JNC TN-8400, 2005-010, Development of Thermodynamic Database for Hyperalkaline, Argillaceous Systems, July 2005.

Materials and Processes for CO_2 Capture, Conversion, and Sequestration

PORPHYRIN-BASED CHEMISTRY FOR CARBON CAPTURE AND SEQUESTRATION

Lawrence P. Cook[1], Winnie Wong-Ng[2], and Greg Brewer[1]

[1] Department of Chemistry, The Catholic University of America, Washington, DC 20064
[2] Materials Measurement Science Division, National Institute of Standards and Technology, Gaithersburg, MD 20899

ABSTRACT

Two major types of porphyrinic framework structures offer promise for carbon capture and sequestration (CCS) applications: 1) metal-organic frameworks (MOFs) in which the porphyrins serve as struts linking vertices comprised of metal-based clusters; 2) covalently-bonded organic frameworks (COFs) in which the porphyrins are bonded together as part of a continuous structure forming the framework. Porphyrinic MOFs synthesized to date offer a somewhat greater range of internal porosities and structural moieties, while porphyrinic COFs may have greater stability under a variety of ambient conditions. For successful CCS applications, several factors related to the chemistry are important, including sorptive capacity, adsorption/desorption kinetics and thermodynamics, cycling lifetime, environmental compatibility, and cost-effective schemes for efficient synthesis. The possibility of using porphyrinic materials in multiple roles, e.g. for capture, storage, and electrolytic CO_2 reduction, is of special interest.

I. INTRODUCTION

It is well-known that there is a need to control the rising CO_2 emissions generated by the expanding world economy (Fig. 1). Over the past 50 years, the atmospheric CO_2 level has risen by about 20 %, to its current level of ~ 400 ppmv, a change attributed largely to anthropogenic sources. This increase may already be having adverse effects on global mean temperature, regional climatology, and ocean ecological systems; a longer-term effect on sea level will have far-reaching implications for coastal civilizations.

In the U.S. alone, it has been estimated that the per capita generation of CO_2 is approximately 18 000 kg (20 tons) annually [1]. While advances in the application of green technologies continue to expand the role of alternative energy sources in stemming the rise in CO_2 levels, it is likely that the burning of fossil fuels will be the major world energy source for several decades into the future.

Consequently, a great deal of attention has been focused on carbon capture and sequestration (CCS), and most recently, on CO_2 conversion. The development of advanced materials for CO_2 mitigation, especially solid adsorbents, is an essential part in these efforts. This

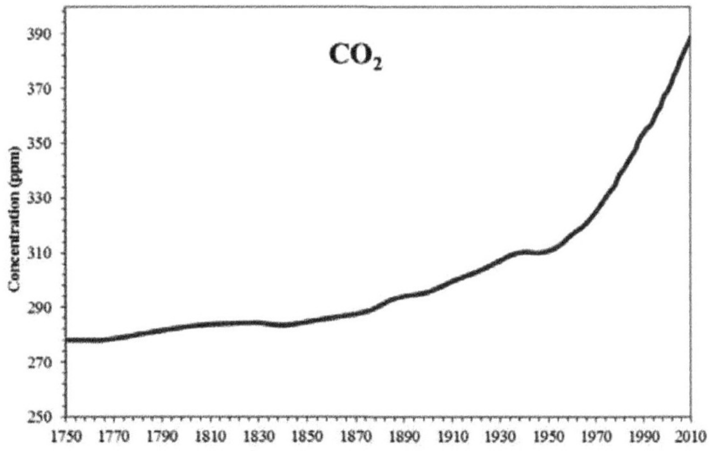

Figure 1. Atmospheric CO_2 concentration (ppmv) from1750 to 2010 [2].

paper outlines progress in the application of porphyrinic framework structures, a highly versatile group of compounds, to CCS and CO_2 conversion.

II. STRUCTURAL CHEMISTRY AND STABILITY OF PORPHYRINS

Porphyrins are a class of materials based on the porphyrin macrocycle, which is comprised of four linked pyrrole rings (Fig. 2). Pyrrole (C_4H_5N) is a stable aromatic compound, liquid at room temperature. When four pyrroles are linked to form a macrocycle, the electron delocalization is enhanced, further increasing the stability of the structure. The long-term thermodynamic

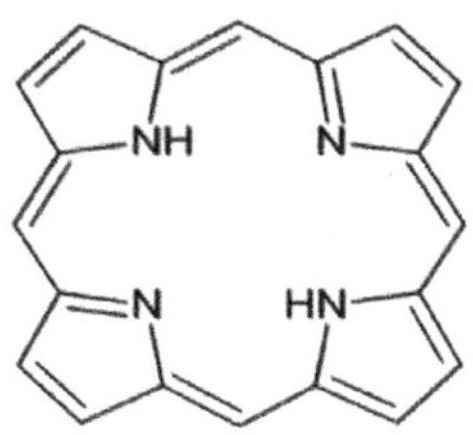

Figure 2. The basic porphyrin macrocycle (porphin, $C_{20}H_{14}N_4$) comprised of four pyrrole (C_4H_5N) rings linked together by four methine (=CH–) groups [3].

stability of porphyrins is evidenced by their natural occurrence in petroleum and in the naturally occurring mineral, abelsonite ($C_{31}H_{32}N_4Ni$) [4]. Porphyrins also have good thermal stability relative to many other organic compounds, with several being stable up to 300 °C to 400 °C [5]. Numerous possibilities exist for functionalization and metalation of porphyrins. Starting with the simplest porphyrin, known as porphin ($C_{20}H_{14}N_4$), a solid at room temperature, the macrocycle can be modified by adding various functional groups. For example, as shown in Fig. 3, ethyl (-CH_2CH_3) groups can be added to produce octaethyl porphyrin ($C_{36}H_{46}N_4$), phenyl (-C_6H_5) groups can be added to produce tetraphenyl porphyrin ($C_{44}H_{30}N_4$), and so on. Porphyrins can be metalated by deprotonation and coordination of a metal in the center of the macrocycle. Many of the resulting metalloporphyrins have useful chromophoric and catalytic properties.

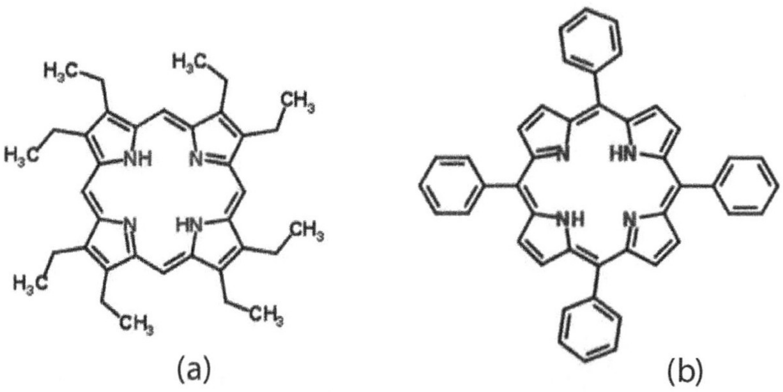

Figure 3. Porphyrin structures with simple sidearm additions;
(a) octaethyl porphyrin [6]; (b) tetraphenyl porphryin [7].

Metalloporphyrins are also important in living organisms and form the basis for the biochemical activity of hemoglobin and chlorophyll, which both have porphyrin derivatives at their core. Recently we have re-determined the structure of hemin ($C_{34}H_{32}ClFeN_4O_4$) [8], also known as α-chlorohemin or Fe(III) protoporphyrin IX, which is closely related to heme, the porphyrin in hemoglobin. Since the basic molecular structure of hemin is related to the porphyrins of relevance to this paper, it is illustrated in Fig. 4, and will be discussed briefly here. In hemin, the macrocycle is modified by addition of 4 methyl groups ($-CH_3$), 2 vinyl groups ($-CH=CH_2$), and 2 propionic acid groups ($-CH_2CH_2COOH$). The macrocycle is metalated by removing hydrogens from the interior nitrogens located on two of the pyrrole rings and bonding those nitrogens to Fe^{+3} introduced into the interior of the macrocycle; in actuality, the Fe bonds are delocalized among all four nitrogens of the macrocycle. A chlorine apically bonded to the Fe^{+3} provides charge balance. The crystallographic parameters of hemin, as determined by powder X-ray diffraction using synchrotron radiation, followed by Rietveld refinement of the results and Density Functional Theory (DFT) calculations to establish hydrogen positions, are given in Table 1. Noteworthy features of the structure are the internal planarity of the pyrrole groups, the "puckering" of the macrocycle due to the tilting of the pyrroles relative to one another, and the slight inclination of Fe-Cl bond axis relative to the average plane of the macrocycle. The position of the Fe is elevated by 0.48 Å relative to the plane of the macrocycle due to the electron spin. The functional groups are attached at varying angles to the macrocycle plane. The overall effect of

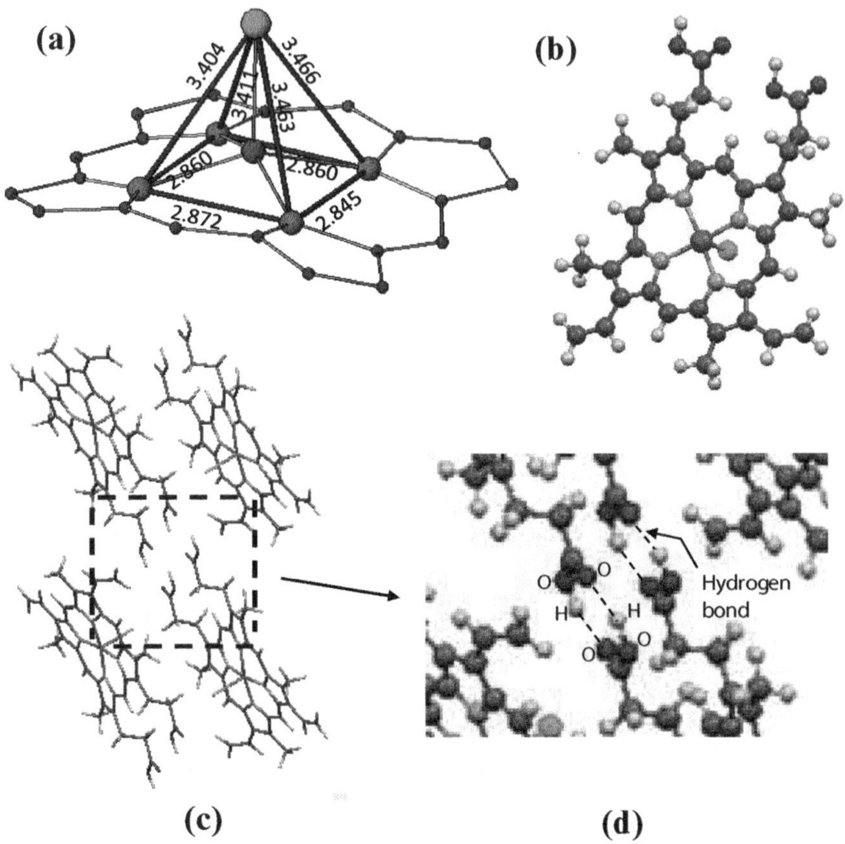

Figure 4. Structure of Fe(III) protoporphryin IX [8]; O-red, Fe-orange, N-blue, C-grey, Cl-green, H-white; (a) coordination of Fe^{+3}, with pyramidal (non-bonding) interatomic distances indicated; (Å); (b) ball-and-stick model; (c) 3-D perspective of structure; (d) location of hydrogen bonds.

these geometrical features is to produce a triclinic structure, with space group $P\bar{1}$. The porphyrin molecular units are held together in the structure by hydrogen bonding as indicated in Fig. 4d. Many porphyrins feature axial bonds extending in both directions from the central metal atom, creating further options for intermolecular bonding. In porphyrinic MOF and COF framework structures, the overall planarity and essentially square moiety provided by the macrocycle give many options for linking to other units and can produce structures with high symmetry.

Table 1. Refined Crystallographic Data for Fe(III) Protoporphyrin IX (Hemin) (triclinic, $P\bar{1}$, $Z = 2$) [8]

Parameter	100 K	Room Temp.
a (Å)	11.22468(6)	11.43217(7)
b (Å)	13.93930(8)	14.06412(10)
c (Å)	10.79818(9)	10.85390(9)
α (°)	99.6672(6)	98.6655(7)
β (°)	108.4124(8)	108.6294(8)
γ (°)	106.7175(6)	107.5025(7)
V (Å)^3	1471.71(10)	1517.21(2)
D_x (g cm^{-3})	1.471	1.427

III. PORPHYRINIC METAL-ORGANIC FRAMEWORKS

A. Structure and synthesis of MOFs

The interest in metal-organic frameworks (MOFs), also known as coordination polymers, has expanded greatly over the past two decades, as applications for gas capture and storage have become more important. The concept of a structural framework comprised of metal ions linked by ligands was suggested very early by Hofmann and Kuspert [9]. Their idea was confirmed experimentally by X-ray crystallography with the publication of the clathrate structure of $Ni(CN)_2{\cdot}NH_3{\cdot}C_6H_6$ in 1949 [10], which contains inorganic ligands. One of the first X-ray structures of a framework with organic ligands [11], that of bis(adiponitrilo)copper(I) nitrate ($C_{12}H_{16}N_5O_3Cu$), was published in 1959 [12], and a large number of these structures has been published since. Such compounds were termed metal-organic frameworks by Yaghi et al. in 1995 [13], who later applied a systematic approach known as reticular chemistry to the design and synthesis of these materials [14].

MOFs are generally defined as compounds consisting of metal ions or clusters (vertices or nodes) coordinated to organic molecules (linkers or struts) to form one-, two-, or three-dimensional structures that can be porous [15]. In porphyrinic MOFs, the porphyrin moieties are the linkers that connect to the metals. A review by Gao et al. [16] has summarized much of the work on porphyrinic MOFs, from 1991, when the first such compound was reported [17], up until 2013. They have tabulated 27 porphyrin ligands which have been used, or could be used, to build porphyrinic MOFs; of these, 14 contain carboxyl (COOH) groups. The growing interest in this field is illustrated by the rising number of publications, as shown in Fig. 5.

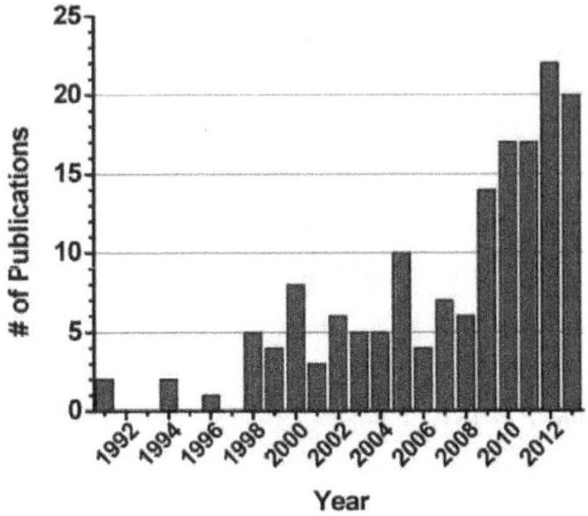

Figure 5. Number of publications on porphyrinic MOFs from 1991 to 2013 [16]. (Reprinted with permission from Chem. Soc. Rev. **43**, 5841-5866 (2014). Copyright (2014) The Royal Society of Chemistry.)

Gao et al. [16] have summarized four strategies for designing and synthesizing porphyrinic MOFs. In *strategy 1*, a crystal engineering approach is used in which preconceived structures with desirable properties are formulated on the basis of porphyrin building blocks and coordination bonding. In *strategy 2*, a pillared layer design is used, in which ditopic pillars such as dipyridine derivatives are used to link 2-D sheets, comprised of metal-containing secondary building units (SBUs) at the nodes linked by metalloporphyrin ligands. This approach has been used to construct the novel "flexible MOFs" of interest for tunable gas storage [18-20]. In *strategy 3*, nanoscopic metal-organic polyhedra are constructed, which provide porphyrin-constrained cages with π-electron density useful for interaction with guest molecules. These polyhedra are then bonded together along shared faces and edges to build up a continuous structure. An example of this approach follows in the next section. In *strategy 4*, post-synthetic modification is employed, by exchanging metal ions within the porphyrin macrocycle, or in some cases, by exchanging ligands.

B. CO_2 capture by porphyrinic MOFs

Gao et al. [16] have noted the existence of 33 porphyrin-based MOFs. Of these, the CO_2-sorptive properties of five are described, as indicated in Table 2. The synthesis and properties of one of these, MMPF-4, will be discussed in more detail, as follows. The design and synthesis of MMPF-4 was achieved by Wang et al. [21] with the aid of *strategy 3*, mentioned above. As the initial step, Wang et al. [21] were able to synthesize tetrakis(3,5-dicarboxyphenyl)porphyrin (tdcpp), which has a square topology, and provides a ligand useful for connecting the triangular moieties which would form the vertices of a high-symmetry framework. By combining the square tdcpp (Fig. 6a) with triangular $Zn_2(CO_2)_3$ "paddlewheels (Fig. 6b)," small cubicuboctahedron frameworks are formed, with six faces composed of tdcpp moieties linked by eight $Zn_2(CO_2)_3$ moieties at the corners (Fig. 6c). Eight of these frameworks can in turn be joined together by face-sharing to build up a continuous framework (Fig. 6d). This structure contains guest cages within the cubicuboctahedrons and an additional octahemioctahedron cage between the joined cubicuboctahedrons (Fig. 6e).

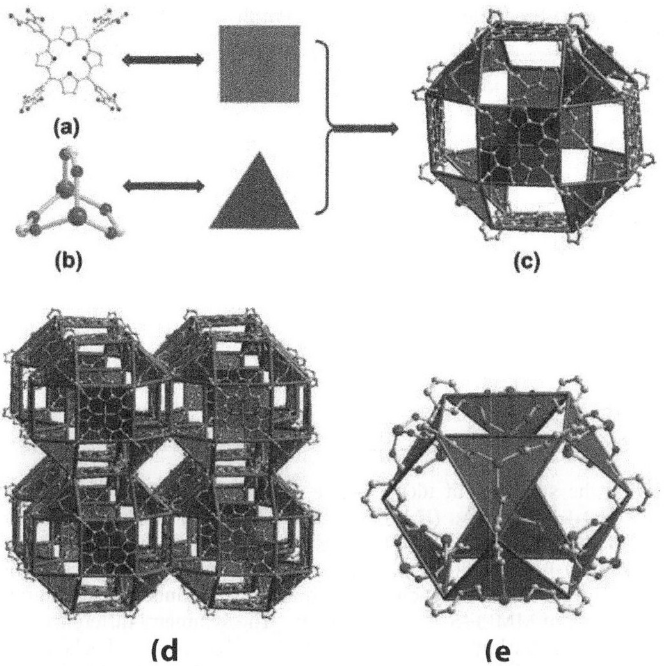

Figure 6. Construction of the MMPF-4 porphyrinic MOF [21]; combination of tdcpp square moieties (a) and $Zn_2(CO_2)_3$ paddlewheels (b) to form a small cubicuboctahedron (c); eight cubicuboctahedrons are linked together to form a framework (d) with an octahemioctahedron cage (e) between the

Table 2. Properties of CO_2-Sorptive Porphyrinic MOFs [16]

Name	Formula (w/o guest)	Brunauer-Emmett-Teller (BET) Surface Area (m^2g^{-1})	CO_2 Uptake (cm^3g^{-1}/ percentage by mass)	(T, P)
MMPF-1	$Cu_2(Cu\text{-}bdcpp)(H_2O)_2$	-----	80 / ---	195 K, 0.1 MPa
MMPF-2	$[Co(II)_3(OH)(H_2O)]_4(Co(II)tdcpp)_3$	1410	170 / 33.4	273 K, 0.1 MPa
MMPF-4	$[Zn_{19}(tdcpp)_3][(NO_3)_8]$	958	124 / 24.4 67 / 13.4	273 K, 0.1 MPa 298 K, 0.1 MPa
MMPF-7	$[In_{1.29}O_{0.57}(tcpp)]$	910	55 / 10.7 34 / 6.6	273 K, 0.1 MPa 298 K, 0.1 MPa
MMPF-8	[In[In-tcbpp)]	440	82 / 16.2 47 / 9.3	273 K, 0.1 MPa 298 K, 0.1 MPa

Following the synthesis of tdcpp, single crystals of MMPF-4, with formula stated as ($[Zn_{19}(tdcpp)_3][(NO_3)_8]\cdot(DMSO)_{61}\cdot(H_2O)_{25}$), and MMPF-5, with formula stated as ($[Cd_{11}(tdcpp)_3][(H_3O)_8]\cdot(DMSO)_{36}\cdot(H_2O)_{11}$), were synthesized [21] by solvothermal reaction of H_{10}tdcpp with $Zn(NO_3)_2$, and $Cd(NO_3)_2$, respectively, in dimethylsulfoxide (DMSO) at 135 °C. Single crystal X-ray studies using synchrotron radiation determined the space group of MMPF-4 as cubic $Ia\bar{3}$ and that of MMPF-5 as cubic $Pm\bar{3}m$. The symmetry differences between the two compounds are related to the higher symmetry of the $Zn_2(CO_2)_3$ paddlewheels, which have a mirror plane. The cubicuboctahedron in MMPF-4 has window openings of 7.831 Å x 8.048 Å with an internal diameter of 21.528 Å; in MMPF-5, the analogous dimensions are 8.004 Å x 8.195 Å with an internal diameter of 22.521 Å. The central octahemioctahedron cage in MMPF-4 has a window opening of 8.048 Å x 8.048 Å with an internal diameter of 11.189 Å; in MMPF-5, the analogous dimensions are 8.195 Å x 8.195 Å and 11.589 Å. The solvent-accessible volumes of MMPF-4 and MMPF-5 were calculated as 70.4 % and 75.5 %, respectively.

As noted in the above formulas, MMPF-4 and MMPF-5 have some DMSO guest content from the synthesis process, however this disappears upon heating to 200 °C. At small increments above this temperature, tdcpp ligands are lost, resulting in a collapse of the structures. From Table 2, MMPF-4 has a BET surface area of 958 m^2g^{-1}. The CO_2 sorption behavior of MMPF-4 is illustrated in Fig. 7. At 273 K, it has an uptake capacity of 124 cm^3g^{-1} (5.54 mmol g^{-1}; 24.4 percentage by mass), and at 298 K, the uptake capacity is 67 cm^3g^{-1} (3.0 mmol g^{-1}; 13.2 percentage by mass). These values are among the highest reported for MOFs. For applications to post-combustion CO_2 capture, CO_2/N_2 selectivity is important. This was calculated for a 15 %/85 % CO_2/N_2 mixture, as shown in the inset to Fig. 7. For MMPF-4, the calculated selectivity for CO_2 is 123 at 1 bar, also among the highest values reported for porous MOFs.

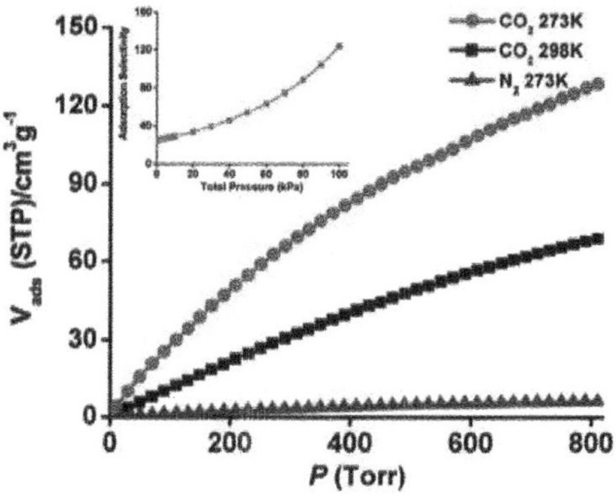

Figure 7. Sorption characteristics of MMPF-4 [21]; CO_2 adsorption isotherms at 273 K (red) and 298 K (black) and N_2 adsorption isotherm (blue) at 273 K; inset shows calculated CO_2/N_2 selectivity. (Reprinted with permission from Chem. Sci. **3**, 2823-2827 (2012). Copyright (2012) The Royal Society of Chemistry.)

IV. PORPHYRINIC COVALENT ORGANIC FRAMEWORKS (COFs)

A. Structure and synthesis of COFs

Côté et al. [22] were the first to use the "COF" acronym as shorthand for the covalently-bonded organic framework materials which they began investigating for applications in gas storage, photonics, and catalysis. COFs constitute micro- and mesoporous crystalline frameworks constructed from light elements such as H, B, C, N, and O, which form strong covalent bonds. Côté et al. [22] reported a general design strategy for COF synthesis. Their approach consisted of using simple reactions involving condensation of cyclic molecules to form extended 2-D sheet structures. Products of their first reported syntheses were designated as COF-1 $[(C_3H_2BO)_6{\cdot}(C_9H_{12})]$ and COF-5 ($C_9H_4BO_2$). The synthesis of COF-1, from the condensation of diboronic acid, is shown diagrammatically in Fig. 8; the synthesis of COF-5 follows a similar procedure. The presence, in the synthesis products, of structures comprised of layers of planar sheets with linked hexagonal rings, in accordance with results from structural modeling software, was confirmed by powder XRD, and FTIR and ^{11}B NMR spectroscopies. The structures are related to graphite, and the difference between COF-1 and COF-5 has to do with the stacking of the 2-D sheets; in COF-1 ($P6_3mmc$), the layers are stacked in a staggered AB arrangement, while in COF-5 ($P6_3/mmm$) the layers are superimposed directly over one another in an eclipsed mode. The C_9H_{12} in the formula of COF-1 indicates a mesitylene guest incorporated from the mesitylene/dioxane solution used to facilitate the condensation reaction; the mesitylene guest can be removed by heating to 200 °C. These 2-D COFs are thermally stable to 500 °C to 600 °C. As expected from the structure, COF-5 has the highest surface area of the two (see Table 3).

El-Kaderi et al. [23] used a reticular chemistry approach [14] to design COFs with a continuous 3-D framework, designated as COFs-102, -103, -105, and -108. These were synthesized by condensation reactions similar to, but more complicated than the reactions outlined above for 2-D COFs. The 3-D COFs were formed by linking together blocks of six-membered B_3O_3 (boroxine) and C_2O_2B (boronate ester) rings. These 3-D COFs are thermally stable to 400 °C to 500 °C, with very high surface areas (Table 3), and densities as low as 0.17 g cm^{-3}.

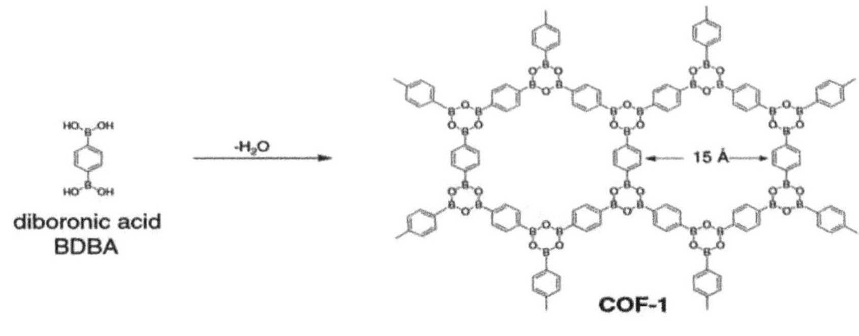

Figure 8. Synthesis of COF-1 by condensation of diboronic acid (BDBA) [22]. (Reprinted with permission from Science **310**, 1166-1170 (2005). Copyright (2005) American Association for the Advancement of Science.)

Furukawa and Yaghi [24] categorized COFs synthesized by boronic acid condensation into three groups: 1) COFs-1, and -6, 2-D structures with 1-D small pores; 2) COFs-5, -8, and -10, 2-D structures with 1-D large pores; 3) COFs-102, and -103, 3-D structures with 3-D medium pores. Properties of these boronic acid-derived COFs are given in Table 3.

Table 3. Properties of CO_2-Sorptive Boronic COFs [24]

Name	Formula	Structure (framework/ pores)	Pore Size (Å)	Total Pore Volume ($cm^3\ g^{-1}$)	BET Surface Area (m^2g^{-1})	Bulk Density ($g\ cm^{-3}$)	CO_2 Uptake (298 K, 55 bar) ($mg\ g^{-1}$)
COF-1	C_3H_2BO	2D/1D small	9	0.30	750	0.98	230
COF-5	$C_9H_4BO_2$	2D/1D large	27	1.07	1670	0.58	870
COF-6	$C_8H_3BO_2$	2D/1D small	9	0.32	750	1.1	310
COF-8	$C_{14}H_7BO_2$	2D/1D large	16	0.69	1350	0.71	630
COF-10	$C_{12}H_6BO_2$	2D/1D large	32	1.44	1760	0.48	1010
COF-102	$C_{25}H_{24}B_4O_8$	3D/3D med.	12	1.55	3620	0.43	1200
COF-103	$C_{24}H_{24}B_4O_8Si$	3D/3D med.	12	1.54	3530	0.43	1190

Wan et al. [25] designed and synthesized 2-D COF structures based on porphyrins in order to utilize the electrical conductivity associated with the close π-π distances of porphyrin macrocycles in stacked sheets. Using the condensation schemes of Fig. 9, they synthesized COF-366 ($C_{60}H_{38}N_8$) and COF-66 ($C_{72}H_{42}B_4N_4O_8$). To determine the nature of the stacking of the 2-D layers in these structures, they compared powder X-ray data with calculated patterns from simulated structures and determined that both COF-366 and COF-66 are best fitted by eclipsed

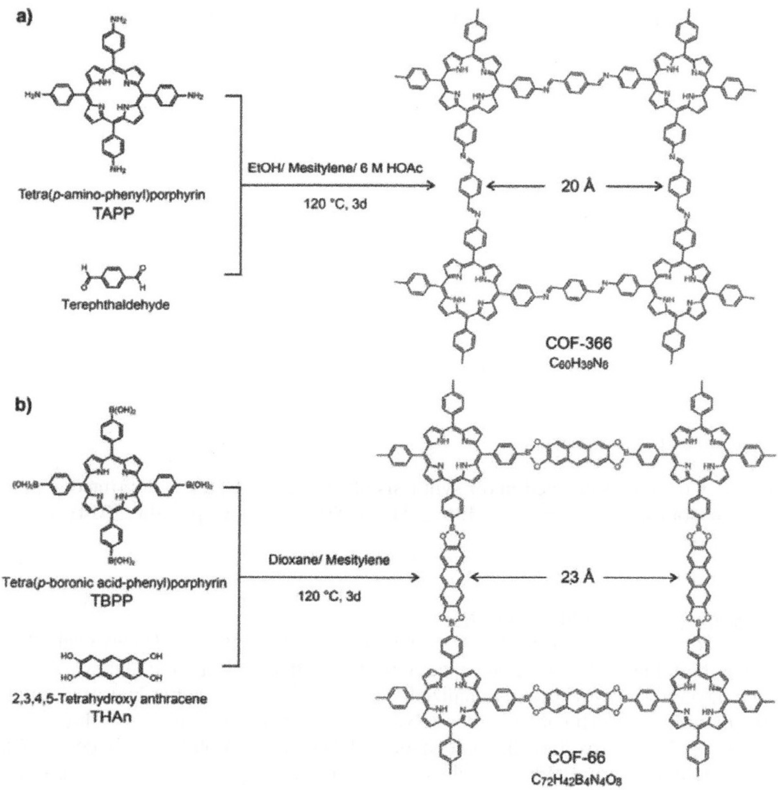

Figure 9. Condensation reactions for synthesis of COF-366 (a) and COF-66 (b) [25]. (Reprinted with permission from Chem. Mater. **23**, 4094-4097 (2011). Copyright (2011) American Chemical Society.)

models in which the stacked layers lie directly over one another. This results in stacking along [001], with square channels having diameters of 20.2 Å and 23.2 Å, respectively, for COF-366 (P_4/mmm) and COF-66 (P_4/m). Measured BET surface areas were 735 m^2 g^{-1} and 360 m^2 g^{-1}, respectively. Electrical properties were measured, and results indicate that COFs-366 and -66 have very high carrier mobilities, among the highest known among crystalline organic semiconductors.

Calik et al. [26] further explored the electrical properties of porphyrinic COFs by synthesizing a new layered 2-D COF, termed TP-Por COF using co-condensation of 5,15-bis(4-boronophenyl)porphyrin and 2,3,6,7,10,11-hexahydroxytriphenylene (HHTP) in a 7:3 mixture of acetonitrile and mesitylene. The TP-Por COF is a porous hexagonal framework with the trigonal triphenylenes at the vertices and the linear porphyrin molecules along the edges (Fig. 10). X-ray data indicated an eclipsed stacking topology, with an interlayer distance of 3.8 Å. The calculated surface area is 890 m^2 g^{-1}. Calik et al. [26] constructed a thin-film photovoltaic device based on this material.

3

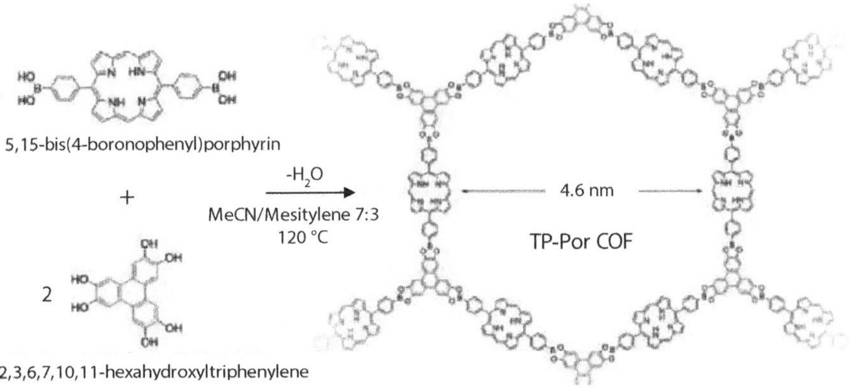

2

Figure 10. Condensation reaction for synthesis of TP-Por COF [26]. (Reprinted with permission from J. Am. Chem. Soc. **136**, 17802-17807 (2014). Copyright (2014) American Chemical Society.)

B. CO_2-capture by porphyrinic 2-D COFs

Relatively little data exists on CO_2 sorption by porphyrinic COFs. Huang et al. [27] sought to overcome the limited CO_2-sorptive capacity of 2-D COFs by designing and synthesizing new porphyrinic layered COFs with functionalized channel walls. The synthesis procedure is shown schematically in Fig. 11. The procedure involves: *Step 1*) condensation from a three-component mixture of 2,5-dihydroxyterephthalaldehyde (DHTA), 1,4-phthalaldehyde (PA), and tetraminophenylporphyrin (H_2P) to produce an imine-linked 2-D COF with porphyrin at the vertices and phenol linkages forming the pore walls; the completely reacted product is designated as ($[HO]_{100\%}$-H_2P-COF), where $[HO]_{x\%}$ refers to the fraction of phenol groups in the linkages; *Step 2*) addition of succinic acid ($C_4H_6O_4$), which opens up the phenol rings, and produces open carboxylic acid groups on the channel walls; the completely reacted product is designated as ($[HO_2C]_{100\%}$-H_2P-COF), where $[HO_2C]_{x\%}$ refers to the fraction of phenol groups which have undergone the ring-opening reaction; each ring-opening reaction results in functionalization with two carboxylic acid groups. The percentage of [HO] and $[HO_2C]$ substitution was controlled by adjusting the solution chemistry. Structure and composition of products was characterized by powder X-ray diffraction, elemental and thermogravimetric analysis, and IR, ^{11}H NMR, and energy-dispersive X-ray spectroscopies. These 2-D eclipsed layered structures were determined to have an interlayer spacing of 3.8 Å. The properties of the substituted H_2P-COFs are given in Table 4. The functionalization of the channel walls in the $[HO]_{x\%}$-H_2P-COFs is accompanied by a reduction in pore size and pore volume, due to the steric effects of the carboxylic acid groups.

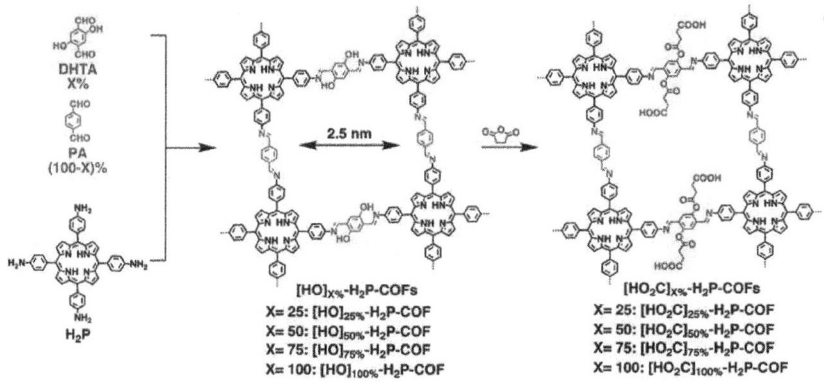

Figure 11. Synthesis of $[HO]_{X\%}$-H_2P-COFs and $[HO_2C]_{X\%}$-H_2P-COFs [27].

Table 4. Properties of Substituted CO_2-Sorptive Porphyrinic COFs [27]

Composition of H_2P-COF	BET Surface Area ($m^2\ g^{-1}$)	Pore Size (nm)	Pore Volume ($cm^3\ g^{-1}$)	CO_2 Uptake (273 K, 1 bar) ($mg\ g^{-1}$)	CO_2 Uptake (298 K, 1 bar) ($mg\ g^{-1}$)	Q_{ST} (isosteric heat of ads.) ($kJ\ mol^{-1}$)
$[HO]_{25\%}$	1054	2.5	0.89	54	31	32.2
$[HO]_{50\%}$	1089	2.5	0.91	46	34	29.4
$[HO]_{75\%}$	1153	2.5	0.96	52	32	31.5
$[HO]_{100\%}$	1284	2.5	1.02	63	35	36.4
$[HO_2C]_{25\%}$	786	2.2	0.78	96	58	38.2
$[HO_2C]_{50\%}$	673	1.9	0.66	134	67	39.6
$[HO_2C]_{75\%}$	482	1.7	0.54	157	72	41.2
$[HO_2C]_{100\%}$	364	1.4	0.43	174	76	43.5

In general, the substituted H_2P-COFs are stable in tetrahydrofuran, water, and in 1M acidic and basic solutions. Most importantly, according to Table 4, there is a large increase in CO_2 uptake associated with the $[HO_2C]$ functionalization of walls by carboxylic acid, whereas there is very little effect due to the presence of [HO] alone. The $[HO_2C]_{100\%}$-H_2P-COF exhibited sorptive capacity of 76 mg g^{-1} (1.7 mmole g^{-1}) at 298 K, which represents among the highest values for 2-D and 3-D COFs. Another important issue is that of selectivity. Huang et al. [27] measured relative adsorption in a 15 %/85 % CO_2/N_2 simulated flue gas mixture. Results are shown in Fig. 12, where the much greater selectivity of the fully functionalized $[HO_2C]_{100\%}$-H_2P-COF relative to the $[HO]_{100\%}$-H_2P-COF is evident. At 0.1 kPa, the adsorption selectivity for CO_2 is 323 for the former, but only about 18 for the latter; at 100 kPa, the selectivities are 77 and 8, respectively. Fig. 12 also compares selectivities with other adsorbants. $[HO_2C]_{100\%}$-H_2P-COF selectivity substantially exceeds that of NaX Zeolite over the entire pressure range, and surpasses that of

CuBTC (copper benzene-tricarboxylate) at low pressures, but overall is about a factor of 2X below that of MgMOF-74. Tests in a simulated packed-bed adsorber (Fig. 13) indicated negligible degradation in uptake capacity after 10 regeneration cycles under high vacuum at 353 K for 60 min.

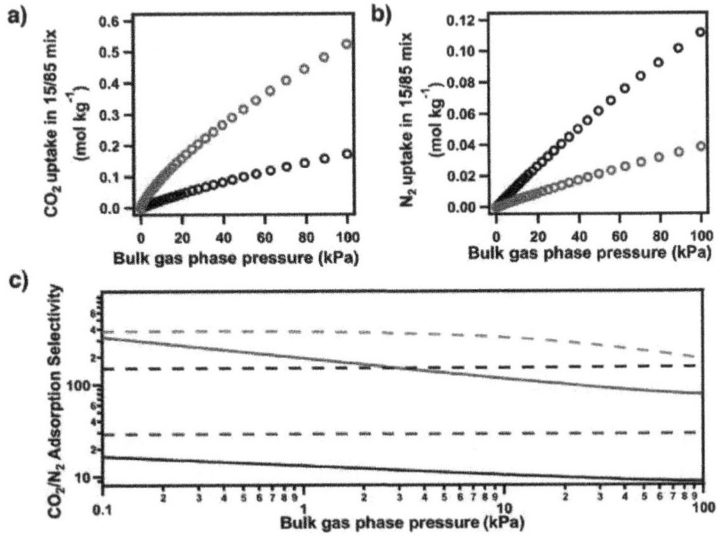

Figure 12. Sorption characteristics of H_2P-COFs [27]; (a) uptake of CO_2 and (b) N_2 by $[HO]_{100\%}$-H_2P-COF (black) and $[HO_2C]_{100\%}$-H_2P-COF (red) from a 15%/85% CO_2/N_2 gas mixture at 298 K; (c) CO_2/N_2 adsorption selectivity at 298 K relative to NaX zeolite (dashed blue curve), Cu-BTC (dashed black curve) and MgMOF-74 (dashed green curve).

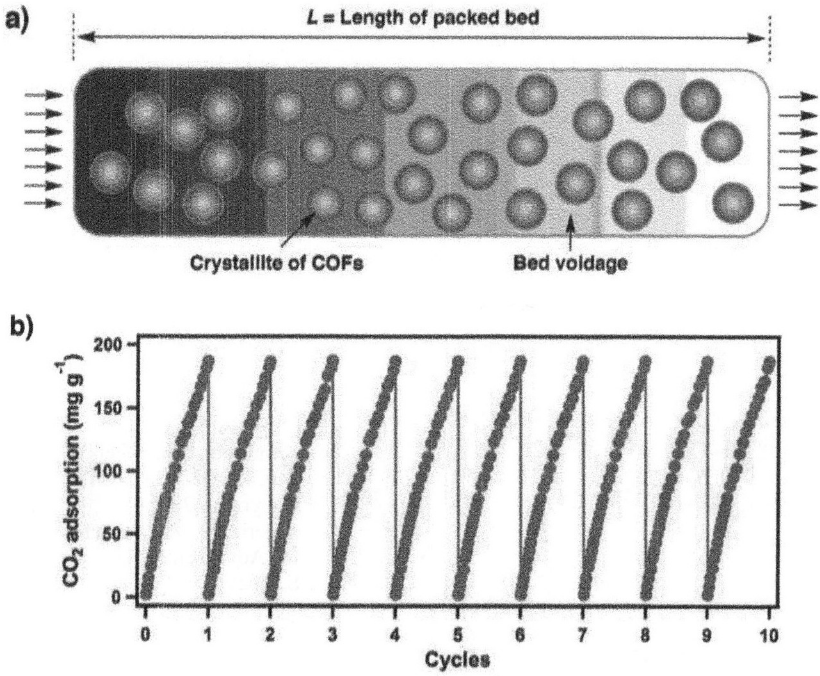

Figure 13. Simulated pressure-swing operating behavior of H_2P-COF adsorbents [27]; (a) schematic packed-bed adsorber; (b) cyclic behavior test of $[HO_2C]_{100\%}$-H_2P-COF at 273 K, with each saturation at 1 bar, followed by regeneration under high vacuum at 353 K for 60 min.

C. Electrochemical reduction of CO_2 by 2-D porphyrinic COFs

While much attention has been devoted to the development of porous materials for CO_2 capture and short-term storage, ultimately the issue of CO_2 disposal must be addressed. There are two distinct ways of accomplishing this: 1) by long-term underground storage; and 2) by conversion to useful products. Alternative (2) could promote a closed carbon cycle.

One possibility which is being actively explored is the electrochemical reduction of CO_2 to CO, an economically viable feedstock for the chemical industry. Porphyrinic COFs offer special promise for CO_2 conversion in two roles: first, as CO_2 capture and storage materials; and second, because of the electrical properties discussed above, as conductive media for suitably designed electrochemical processes. Lin et al. [28] have explored this possibility by synthesizing electrocatalytically active porphyrinic COFs. Starting with a metalated porphyrin, [5,10,15,20-tetrakis(4-aminophenyl)porphyrinato]cobalt [Co(TAP)], they completed an imine condensation with 1,4-benzenedicarboxaldehyde (BDA) or biphenyl-4,4'-dicarboxaldehyde (BPDA) to produce COF-366-Co or COF-367-Co, respectively (Fig. 14). Products of COF-367 with Co substituted by Cu were also produced in ratios Co/Cu = 10/90 (COF-367-Co(10 %)) and 1/99 (COF-367-Co(1

%)). FTIR spectroscopy and structural refinements of X-ray powder patterns of COF-366-Co produced reasonable fits to an orthorhombic *Cmmm* structure with 1-D, 21 Å channels running along the c-axis, a distance of 4.4 Å between the stacked layers, and a BET surface area of 1360 m^2g^{-1}. An analogous structure was obtained for COF-367-Co, with an increased channel width of 24 Å, an interlayer distance of 4.8 Å, and a BET surface area of 1470 m^2g^{-1}.

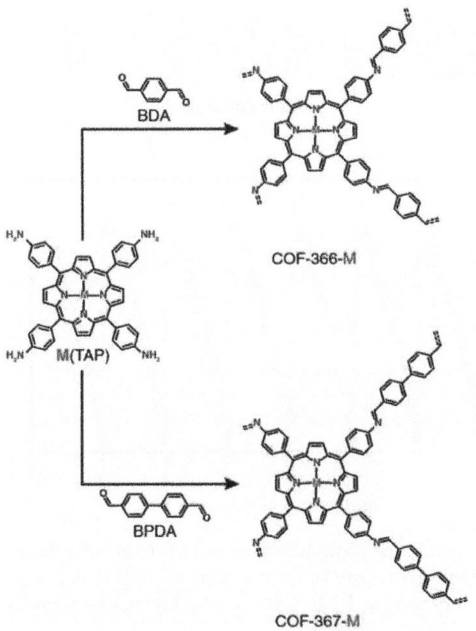

Figure 14. Synthesis of COF-366-M and COF-66-M (M=Co, Co/Cu (10/90), or Co/Cu (1/99)) [28]. (Reprinted with permission from Science **349**, 1208-1213 (2015). Copyright (2015) American Association for the Advancement of Science.)

Electrochemical measurements were made by Lin et al. [28] on COF powders deposited on a conductive carbon substate, in an aqueous CO_2-saturated, buffered solution. Results of long-term electrolysis experiments at -0.67 V, coinciding approximately with the Co(II)/Co(I) redox couple, are shown in Fig. 15, where CO and H production as a function of time are given for COF-366-Co, COF-367-Co, and Co(TAP). The voltage of -0.67 V, where the catalytic effect was optimized, corresponds to an overpotential of -0.55 V. As indicated, COF-367-Co showed greatly improved catalytic efficiency relative to COF-366-Co, and it produced more than 100 ml (4.5 mmol) of CO per mg of COF during a 24 h period, with 91 % Faradic efficiency. Apparently the lattice expansion of COF-367-Co relative to COF-366-Co enhanced the electrochemical process. The turnover frequency per electroactive Co (TOF_{EA}, defined as (mols product)(s^{-1})((mols catalyst)$^{-1}$)) for COF-366-Co was 34,000, and that for COF-367-Co was 48,000. These values were improved by partially substituting Cu for Co, demonstrating the tunability of the electrocatalysis. In comparison with the molecular Co(TAP) catalyst, the COFs showed marked improvement in TOF (26X), demonstrating the importance of the COF environment on the catalysis, perhaps due to enhanced electron delocalization. Also noteworthy is the reduced amount of the unwanted water electrolysis side reaction during the electrocatalytic process.

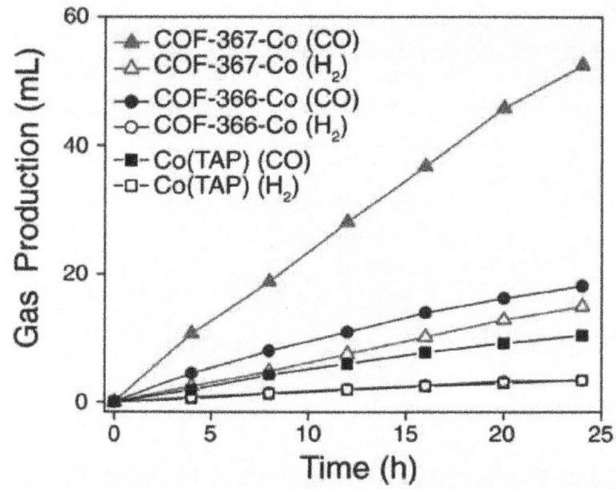

Figure 15. Electrochemical generation of CO by reduction of CO_2 using COF-366-Co and COF-367-Co, and comparison with the electrocatalytic performance of CoTAP [28]. (Reprinted with permission from Science **349**, 1208-1213 (2015). Copyright (2015) American Association for the Advancement of Science.)

V. APPLICATIONS OF PORPHYRINIC FRAMEWORKS TO CCS

For CCS, two of the principal methods for carbon capture by adsorbents are: 1) pressure swing adsorption (PSA), and 2) temperature swing adsorption (TSA). Regeneration of carbon capture materials is of critical importance to both methods. Although the development of porphyrinic materials for CCS is still in its infancy, on the basis of present knowledge it appears that there are potential candidates among both porphyrinic MOFs and porphyrinic COFs for PSA applications. For example MMPF-4 and $[HO_2C]_{100\%}$-H_2P-COF both show large low- to high-pressure differentials in CO_2 sorption capacity. Similarly, for TSA, both compounds also show a high sensitivity of sorption capacity to temperature over a relatively small range. However, a full economic analysis for any proposed CCS system will require a complete set of data on materials performance. For example, adsorbents with a high heat of adsorption may require higher temperatures or lower pressures for regeneration, resulting in higher regeneration costs, which will also be affected by the kinetics of the regeneration process. If longevity is an issue, then adsorbent cost, which includes the cost of synthesis as well as the cost of disposal, becomes more important.

Drage et al. [29] have given an outline of operating conditions for a proposed TSA process, which is reproduced in Table 5. According to this model a cyclic capacity of > 3 mmol g^{-1} is the target value. Several porphyrinic MOFs meet this goal, while there is a need for further improvement in this area for the porphyrinic COFs.

Table 5. Proposed Operating Conditions for a Temperature Swing Adsorption Process [29]

Parameter	Target
Operating T during adsorption	40 °C to 80 °C
Operating T during desorption	85°C to 140 °C
Cyclic capacity	> 3 mmol g^{-1}
Rate of adsorption	seconds
Operating pressure	1015 mbar
Pressure drop due to capture process	< 0.014 MPa (2 psi)
CO_2 product purity	> 95 %
CO_2 capture (of total emissions)	> 80 %

An equation given by Drage et al. [29] relates the key materials parameters for determining the regeneration energy for a TSA process. This equation is reproduced below. From the equation, as would be expected, adsorption capacity (L) and heat of adsorption (Q_r) have opposite effects on the requirements for regeneration energy. More energy-efficient CO_2 capture is favored by using materials with higher L, lower specific heat capacity (C_s), and lower Q_r. These parameters and others will need to be carefully considered in selecting solid adsorbents for CCS.

$$\frac{Q}{m_c} = \frac{m_e}{m_c} C_e \Delta T + \frac{B}{L} C_s \Delta T + C_{p.c} T_2 - C_s T_1 + \frac{Q_r}{m_c}$$

Q = regeneration energy (kJ)
m_c = mass of cyclic adsorbents (kg)
m_e = mass of equipment (kg)
C_e = specific heat of equipment (kJ $kg^{-1}K^{-1}$)
ΔT = temperature difference between capture and regeneration (K)
B = a dimensionless conversion factor
L = adsorption capacity (kg CO_2/kg adsorbent)
C_s = specific heat capacity of adsorbent (kJ $kg^{-1}K^{-1}$)
$C_{p.c}$ = constant pressure heat capacity for CO_2 (kJ $kg^{-1}K^{-1}$)
T_1 and T_2 = temperature of capture and regeneration, respectively (K)
Q_r = heat of adsorption (kJ)

A cost-mitigating factor, discussed briefly above, is the production of useful products from CO_2. An electrolytic process for CO_2 reduction could potentially be beneficial in this regard; a hypothetical electrolytic reduction system using porphyrinic 2-D COFs is shown schematically in Fig. 16.

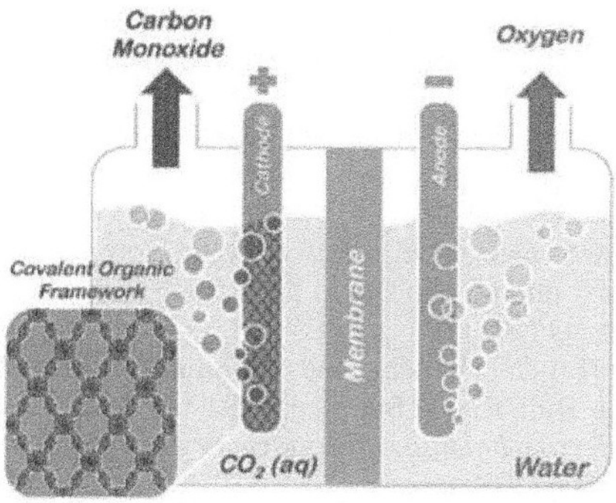

Figure 16. Schematic diagram of a proposed combined CO_2 capture/electrolysis process using sorptive, catalytic porphyrinic COFs [30]. (Reprinted with permission from Lynne Yarris and Omar Yaghi, Lawrence Berkeley National Laboratory).

VI. CONCLUSIONS

We have summarized some of the major developments in the chemistry of porphyrinic framework structures as applied to CCS. Both porphyrinic MOFs and porphyrinic COFs offer potential as effective CO_2 adsorbents. Given the large number of possible porphyrinic MOF topologies which have not yet been investigated, it is expected that the performance of these materials will continue to improve. For porphyrinic COFs, sorptive performance has been somewhat less, but no doubt will also continue to improve with further research; the synthesis of suitable 3-D porphyrinic COF frameworks may be helpful in this respect. Some porphyrinic COFs have the advantage of an intrinsic electrical conductivity, in addition to sorptive capacity and catalytic activity, opening up the possibility for combined capture/conversion processes using these materials.

VII. REFERENCES

1. Dennis Leaf, Hans J.H. Verolne, and William Hunt, Jr., *Overview of regulatory/policy/ economic issues related to carbon dioxide*, Environ. Internat., **29**, 303-310 (2003).

2. European Environmental Agency, http://www.eea.europa.eu/data-and-maps/figures/atmospheric-concentration-of-co2-ppm-1/csi013_fig04_co2_concentration.png/image_large (accessed Nov 15, 2015).

3. Wikipedia, https://en.wikipedia.org/w/index.php?title=Porphin&oldid=670545033 (accessed Nov 15, 2015)

4. Bo Zhang and Timothy D. Lash, *Total synthesis of the porphyrin mineral abelsonite and related petroporphyrins with five-membered exocyclic rings*, Tetrahedral Lett., **44**, 7253-7256 (2003).

5. German L. Perlovich, *Thermodynamics of porphyrin sublimation*, J. Porphyrins and Phthalo- cyanines, **4**, 699-706 (2000).

6. ChemSpider, http://www.chemspider.com/Chemical-Structure.10445887.html (accessed Nov 15, 2015)

7. ChemSpider, http://www.chemspider.com/Chemical-Structure.10291672. html (accessed Nov 15, 2015)

8. J. A. Kaduk, W. Wong-Ng, L.P. Cook, B. Chakraborty, and G. Brewer, *Synchrotron X-ray investigation of α-chlorohemin, $C_{34}H_{32}ClFeN_4O_4$, an Fe-porphyrin,* to be submitted to Powder Diffraction.

9. K.A. Hofmann and F.A. Kuspert, Z. Anorg. Allg. Chem., **15**, 204-224 (1897).

10. H.M. Powell and J.D. Rayner, *Clathrate compound formed by benzene with an ammonia-nickel cyanide complex, Nature*, **163**, 566-5679 (1949).

11. Satoshi Horike and Susumu Kitigawa, *Design of porous coordination polymers/metal-organic frameworks: past present and future*, pp. 3-21 *in* David Farrusseng, ed., Metal Organic Frameworks, Wiley-VCH, Weinheim, Germany (2011).

12. Yukio Kinoshita, Ikuo Matsubara, Taiichi Higuchi, and Yoshihiko Saito, *The crystal structure of bis(adiponitrilo)copper(I) nitrate*, Bull. Chem. Soc. Japan, **32**[11], 1221-1226 (1959).

13. O.M. Yaghi, G.M. Li, and H.L. Li, *Selective binding and removal of guests in a microporous metal-organic framework*, Nature, **378**, 703-706 (1995).

14. Omar M. Yaghi, Michael O'Keeffe, Nathan W. Ockwig, Hee K. Chae, Mohamed Eddaoudi, and Jaheon Kim, *Reticular synthesis and the design of new materials*, Nature, **423**,705-714 (2003).

15. Wikipedia, https://en.wikipedia.org/w/index.php?title=Metal-organic_framework&oldid=690283367 (accessed Nov 14, 2015)

16. Wen-Yang Gao, Matthew Chrzanowski, and Shengqian Ma, *Metal-metalloporphyrin frameworks: a resurging class of functional materials*, Chem. Soc. Rev., **43**, 5841-5866 (2014).

17. Brendan F. Abrahams, Bernard F. Hoskins, and Richard Robson, *A new type of infinite 3D polymeric network containing 4-connected, peripherally-linked metalloporphyrin building blocks*, J. Am. Chem. Soc., **113**, 3606-3607 (1991).

18. L. Espinal, D.L. Poster, W. Wong-Ng, A.J. Allen, and M. Green, *Standards, Data, and MetrologyNeeds for CO_2 Capture Materials – A Critical Review*, Environ. Sci. and Technol., **47**, 11960-11975 (2013).

19. W. Wong-Ng, J.T. Culp, Y. S. Chen, P. Zavalij, L. Espinal, D.W. Siderius, A.J. Allen, S. Scheins, and C. Matranga, *Improved Synthesis and Crystal Structure of the Flexible Pillared Layer Porous Coordination Polymer: Ni(1,2-bis(4-pyridyl)ethylene)[Ni(CN)$_4$]*, Crystal Eng. Comm., **15**, 4684-4693 (2013).

20. Kristi L. Kauffman, Jeffrey T. Culp, Andrew J. Allen, Laura Espinal-Thielen, Winnie Wong-Ng, Thomas D. Brown, Angela Goodman, Mark P. Bernardo, Russel J. Pancoast, Danielle Chirdon, and Christopher Matranga, *Selective adsorption of CO_2 from light gas mixtures using a structurally dynamic porous coordination polymer*, Angew Chem. Int. Ed., **50**, 10888-10892 (2011).

21. Xi-Sen Wang, Matthew Chrzanowski, Wen-Yang Gao, Lukasz Wojtas, Yu-Sheng Chen, Michael J. Zaworotko, and Shengqian Ma, *Vertex-directed self-assembly of a high-symmetry supermolecular building block using a custom-designed porphyrin*, Chem. Sci, **3**, 2823-2827 (2012).

22. Adrien P. Côté, Annabelle I. Benin, Nathan W. Ockwig, Michael O'Keeffe, Adam J. Matzger, and Omar M. Yaghi, *Porous, Crystalline Covalent Organic Frameworks*, Science, **310**, 1166-1170 (2005).

23. Hani M. El-Kaderi, Joseph R. Hunt, José L. Mendoza-Cortés, Adrien P. Côté, Robert Taylor, Michael O'Keeffe, and Omar M. Yaghi, *Designed Synthesis of 3D Covalent Organic Frameworks*, Science, **316**, 268-272 (2007).

24. Hiroyasu Furukawa and Omar M. Yaghi, *Storage of hydrogen, methane, and carbon dioxide in highly porous covalent organic frameworks for clean energy applications*, J. Am. Chem. Soc., **131**, 8875-8883 (2009).

25. Shun Wan, Felipe Gándara, Atushi Asano, Hiroyasu Furukawa, Akinori Saeki, Sanjeev K. Dey, Lei Liao, Michael W. Ambrogio, Youssry Y. Botros, Xiangfeng Duan, Shu Seki, J. Fraser

Stoddart, and Omar M. Yaghi, *Covalent Organic Frameworks with high charge carrier mobility*, Chem. Mater., **23**, 4094-4097 (2011).

26. Mona Calik, Florian Auras, Laura M. Salonen, Kathrin Bader, Irene Grill, Matthias Handloser, Dana D. Medina, Mirjam Dogru, Florian Löbermann, Dirk Trauner, Achim Hartschuh, and Thomas Bein, *Extraction of photogenerated electrons and holes from a covalent organic framework integrated heterojunction*, J. Am. Chem. Soc., **136**, 17802-17807 (2014).

27. Ning Huang, Xiong Chen, Rajamani Krishna, and Donglin Jiang, *Two-dimensional covalent organic frameworks for carbon dioxide capture through channel-wall functionalization*, Angew. Chem. Int. Ed., **54**, 2986-2990 (2015).

28. Song Lin, Christian S. Diercks, Yue-Biao Zhang, Nikolay Kornienko, Eva M. Nichols, Yingho Zhao, Aubrey R. Paris, Dohyung Kim, Peidong Yang, Omar M. Yaghi, and Christopher J. Chang, *Covalent organic frameworks comprising cobalt porphyrins for catalytic CO_2 reduction in water*, Science, **349**, 1208-1213 (2015).

29. Trevor C. Drage, Colin E. Snape, Lee A. Stevens, Joseph Wood, Jiawei Wang, Andrew I. Cooper, Robert Dawson, Xiao Guo, Christopher Satterley, and Robin Irons, *Materials challenges for the development of solid sorbents for post-combustion carbon capture*, J. Mater. Chem., **22**, 2815-2823 (2015).

30. Lynne Yarris, *Soaking up carbon dioxide and turning it into valuable products*, Lawrence Berkeley National Laboratory News Release, August 27, 2015 (diagram courtesy of Omar Yaghi).

THERMAL STABILITY OF NOVEL MULTILAYER LANTHANUM ZIRCONATE BASED THERMAL BARRIER COATINGS

Xingye Guo[a], Zhe Lu[b], Yeon-Gil Jung[b], Li Li[c], James Knapp[c], Jing Zhang[a]*,

[a] Department of Mechanical Engineering, Indiana University-Purdue University Indianapolis, IN 46202, USA
[b] School of Nano and Advanced Materials Engineering, Changwon National University, Changwon, Gyeongnam 641-773, Republic of Korea
[c] Praxair Surface Technologies Inc., Indianapolis, IN 46222, USA

*Corresponding author: jz29@iupui.edu

ABSTRACT

$La_2Zr_2O_7$ is one of the proposed next generation thermal barrier coating materials due to its low thermal conductivity and superior high temperature stability. In this work, $La_2Zr_2O_7$ and conventional 8 wt% yttria stabilized zirconia (8YSZ) coatings were deposited using plasma spray technique. Both single-layer $La_2Zr_2O_7$ and double-layer 8YSZ/$La_2Zr_2O_7$ coating architectures were designed in the matrix for evaluation. Heat treatment, thermal cycling, and thermal shock tests were conducted on the samples to evaluate their thermal stability. All the $La_2Zr_2O_7$ coats were delaminated in the furnace cycle test in less than 20 cycles. This is because residual thermal stresses accumulated during thermal cycling. In the heat treatment and the jet engine thermal shock (JETS) tests, the SCL $La_2Zr_2O_7$ and DCL dense 8YSZ/$La_2Zr_2O_7$ coatings were fully delaminated. In contrast, the DCL porous 8YSZ/$La_2Zr_2O_7$ coating sample only partially delaminated on the edge. This is because of the porous 8YSZ layer provides strain compliance. The findings will help to propose efficient thermal barrier coating systems.

INTRODUCTION

Thermal barrier coatings (TBCs) provide thermal insulation to metallic substrate of gas turbine engine parts. The TBCs allow the gas turbine designer to increase the temperature and consequently the engine efficiency[1, 2]. TBC systems are typically multi-layer type coatings, which consist of a metallic bond coat and at least one ceramic topcoat layer. The metallic bond coat protects the superalloy substrate from oxidation and corrosion, and improves the adhesive bond between the ceramic topcoat and the superalloy substrate [3]. The bond coat has stress compliance to minimize the thermal-expansion-mismatch stresses between the superalloy substrate and the ceramic topcoat in heating and cooling process [1]. The ceramic topcoat has an extremely lower thermal conductivity than superalloy substrate, which establishes a large temperature drop (more than 100 K) in the ceramic topcoat layer [3]. Currently, the most preferred technique to fabricate thermal barrier coatings is air plasma spray (APS) process, because it is capable of producing durable coatings economically. TBC powders are fed into the high temperature arc plasma, where the powders melt into "splat" as they are propelled to the target substrate. Alternatively, the electron beam–physical vapor deposition (EB-PVD) process produces columnar structure, which has outstanding test lifetime [4, 5].

The criteria for TBCs material selection include high melting point, low thermal conductivity, high coefficient of thermal expansion, good thermal and chemical stability, no phase change, and low sintering activity. Nowadays, state-of-the-art TBCs are 8YSZ ceramic topcoat with a NiCrAlY bond coat. The 8YSZ material has a significantly lower thermal conductivity than the metallic substrate [6, 7]. However, the maximum surface temperatures of 8YSZ coating a limited to 1473 K for long time operation, because of the phase transformation and the sintering of the coating [7]. So $La_2Zr_2O_7$ TBCs coats have been proposed as an alternative TBCs material operated at higher temperatures. $La_2Zr_2O_7$ has an even lower thermal conductivity than that of 8YSZ, and it has no phase change from room temperature to its melting temperature (2573 K) [8-10]. The coefficient of thermal expansion of $La_2Zr_2O_7$ is $9.0{\sim}10.0 \times 10^{-6}\ K^{-1}$, which is a little bit lower than that of 8YSZ ($10{\sim}11 \times 10^{-6}\ K^{-1}$) [8, 10, 11].

EB-PVD produced $La_2Zr_2O_7$ coating and APS produced dense $La_2Zr_2O_7$ and 8YSZ multi-layer coatings were investigated by other researchers. Bobzin *et al.* studied single-ceramic (SCL) and double-ceramic layer (DCL) EB-PVD TBC top coats consisting of $La_2Zr_2O_7$, $Gd_2Zr_2O_7$ and 7 wt% YSZ. The temperature-dependent phase behavior and change of thermal conductivity of SCL and DCL top coats were investigated [12]. Vassen *et al.* studied thermal and mechanical properties and thermal stability of hot pressed fully dense $La_2Zr_2O_7$ materials [8]. Cycling performances between double and SCL $La_2Zr_2O_7$ TBC samples were compared. It concluded that highly dense $La_2Zr_2O_7$ material has a lower thermal conductivity than that of YSZ. The DCL TBC samples, which were composed of $La_2Zr_2O_7$ and YSZ, have similar results with SCL YSZ coating at temperatures below 1300 °C. At higher temperatures, the DCL TBCs samples revealed excellent thermal cycling behavior [8, 10, 13]. Zhu et al. investigated the durability and failure mechanism of plasma-sprayed $La_2Zr_2O_7$/mullite/Si coatings on mullite/SiC or SiC/SiC ceramic matrix composite substrates at 1650 °C under thermal gradient cyclic conditions [14]. The $La_2Zr_2O_7$ coating showed significant thermal conductivity increase in the laser flux condition at 1575 °C, indicating the coating was undergoing substantial sintering at elevated temperatures. The plasma-sprayed $La_2Zr_2O_7$/mullite/Si systems on SiC substrate showed severe spallation on heating under high heat flux (95-100 W/cm^2) at 1650 °C [14].

The goal of this work is to study the thermal stability of SCL and DCL porous $La_2Zr_2O_7$ coats. In this work, three types of SCL and DCL $La_2Zr_2O_7$ and 8YSZ structures were deposited. Heat treatment, cyclic furnace cycle test and jet engine thermal shock test were conducted for the TBC samples. The failure mechanisms of these TBC systems were discussed.

METHODS

SCL $La_2Zr_2O_7$ and DCL 8YSZ/$La_2Zr_2O_7$ ceramic coatings were deposited onto a NiCrAlY bond coat (thickness of 229 μm) using APS technique. Haynes 188 superalloy round buttons with 1 inch diameter and 1/8 inch thickness were used as the substrate. The porosity of porous 8YSZ and porous $La_2Zr_2O_7$ coating layers is about 12%. The dense 8YSZ has a porosity less than 8%. Three types of samples are listed in Table 1.

Table 1: List of $La_2Zr_2O_7$ and YSZ samples

Sample	Top coat	Top coat thickness
1	SCL porous $La_2Zr_2O_7$	432 μm
2	DCL porous 8YSZ/porous $La_2Zr_2O_7$	127/305 μm
3	DCL dense 8YSZ/porous $La_2Zr_2O_7$	127/305 μm

Furnace heat treatment experiments were conducted for all of the as sprayed coating samples in argon atmosphere at 1353 K (1080 °C) for 4 hours. Microstructure images of heat treated samples were taken using a scanning electron microscope (SEM, JEOL Model JSM-5610, Japan).

The FCT experiments were conducted between high temperature 1366 K (1093 °C) and room temperature. In each cycle, samples were heated at 1366 K for 50 minutes, and then cooled by compressed air for 10 minutes.

The JETS experiments create a large temperature gradient (several hundred degrees Celsius) across the TBC layer [15, 16]. This experiment impacts mainly the ceramic layer and bond coat layer thermomechanically [16]. In this work, the TBC samples were heated at 1505 K (1232 °C) for 20 s. Then, the TBC samples were cooled by compressed nitrogen jet for 20 s to provide a maximum thermal gradient. Thereafter the samples were cooled by ambient cooling for 40 s. The heating and cooling cycles were repeated until cracks or spallation were spotted on the TBC surface. Both the front and back surface temperatures were measured using pyrometer.

RESULTS AND DISCUSSION

Heat Treatment Tests

After 4 hours of heat treatment, the SCL porous $La_2Zr_2O_7$ and the DCL dense 8YSZ/$La_2Zr_2O_7$ coatings were fully delaminated. However, DCL porous 8YSZ/$La_2Zr_2O_7$ coating only delaminated on the edge, due to the CTE difference between porous 8YSZ and dense 8YSZ. Porous 8YSZ has a lower CTE than that of dense 8YSZ. As a result, the thermal stress between $La_2Zr_2O_7$ and porous 8YSZ is lower than that between $La_2Zr_2O_7$ and dense 8YSZ. Figure 1 shows the SEM images of the heat treated TBC samples. From the microstructure images, all the delamination occurred in the area of $La_2Zr_2O_7$ coating layer that was close to the interface between $La_2Zr_2O_7$ and bond coat (Figure 1a), between $La_2Zr_2O_7$ and porous 8YSZ (Figure 1b) and between $La_2Zr_2O_7$ and dense 8YSZ (Figure 1c). However, the porous and dense 8YSZ coating layers were all intact. The fracture toughness of $La_2Zr_2O_7$ coating is about 1.1 $MPa \cdot m^{1/2}$, which is lower than that of 8YSZ (2.0 $MPa \cdot m^{1/2}$) [8, 18]. So $La_2Zr_2O_7$ layer is weaker than 8YSZ layer.

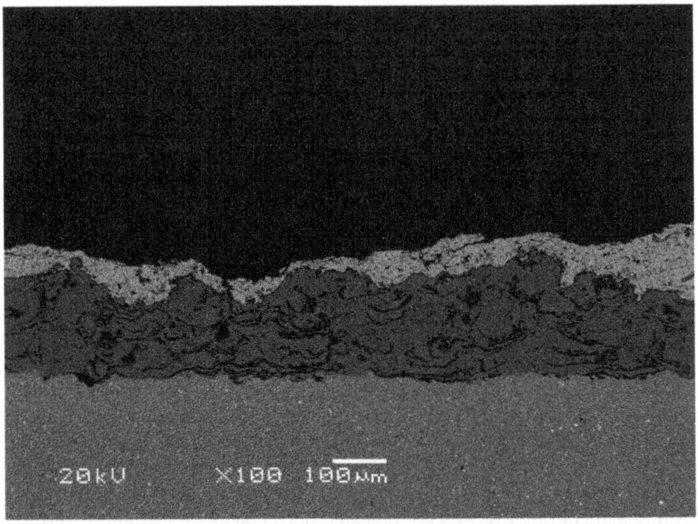

(a)

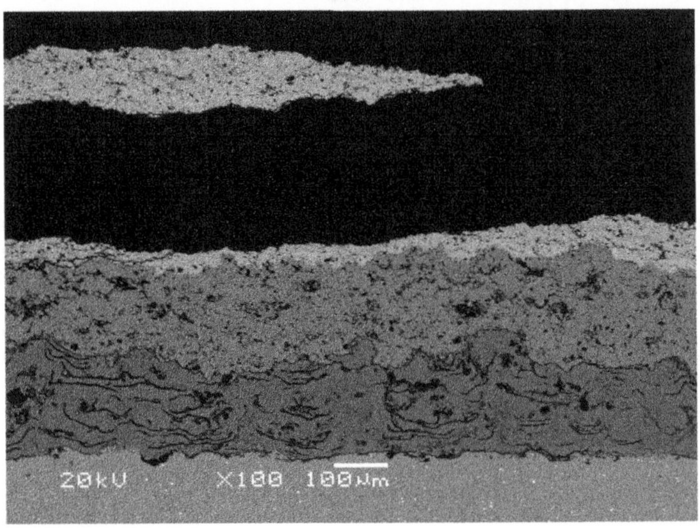

(b)

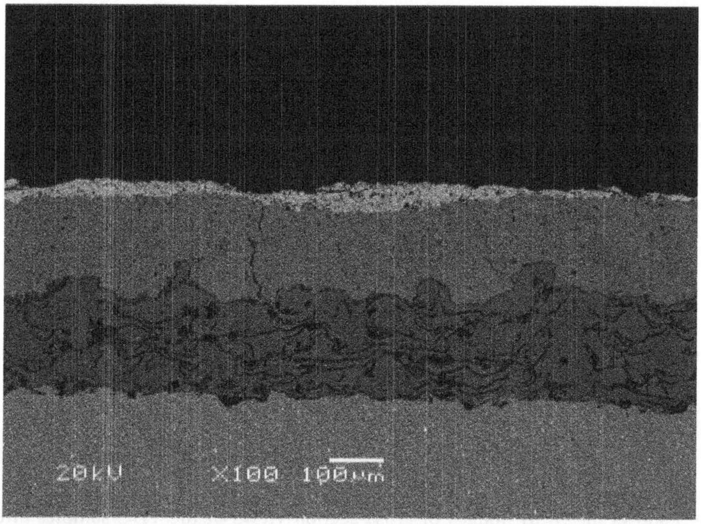

(c)

Figure 1: SEM images of cross-section views of heat treated samples (a) SCL porous $La_2Zr_2O_7$ coating; (b) DCL porous 8YSZ/$La_2Zr_2O_7$ coating, and (c) DCL dense 8YSZ/$La_2Zr_2O_7$ coating.

Furnace Cycle Tests

All $La_2Zr_2O_7$ coats layers were delaminated in the first 20 cycles in the FCT experiments. However, no significant delamination occurred in porous and dense 8YSZ sublayers in DCL samples after 2000 cycles. Because the coefficient of thermal expansion (CTE) of $La_2Zr_2O_7$ (9-10 $\times 10^{-6}$ /K) is lower than that of 8YSZ (10-11 $\times 10^{-6}$ /K), and bond coats have even higher CTE (about 15×10^{-6} /K). The CTE difference between $La_2Zr_2O_7$ and bond coat is larger than that between 8YSZ and bond coat, which leads to higher thermal stresses in $La_2Zr_2O_7$ layers [10, 17]. The failure is ultimately occurred due to the large residual stress at the interface between $La_2Zr_2O_7$ and sublayer.

Jet Engine Thermal Shock Tests

In the JETS experiments, the SCL $La_2Zr_2O_7$ coating sample was fully delaminated in 40 cycles. Figure 2 (a) shows the optical image of the SCL $La_2Zr_2O_7$ coating after JEST experiments. The delamination occurred at the interface between the bond coat and topcoat. Figure 2 (b) shows that the DCL dense 8YSZ/$La_2Zr_2O_7$ coating sample was partially delaminated on the edge after 2000 cycles. Therefore, the porous 8YSZ/$La_2Zr_2O_7$ coating sample has the best thermal stability in all samples. Figure 2 (c) shows the DCL dense 8YSZ/$La_2Zr_2O_7$ coating. The $La_2Zr_2O_7$ layer was completely delaminated from the dense 8YSZ layer.

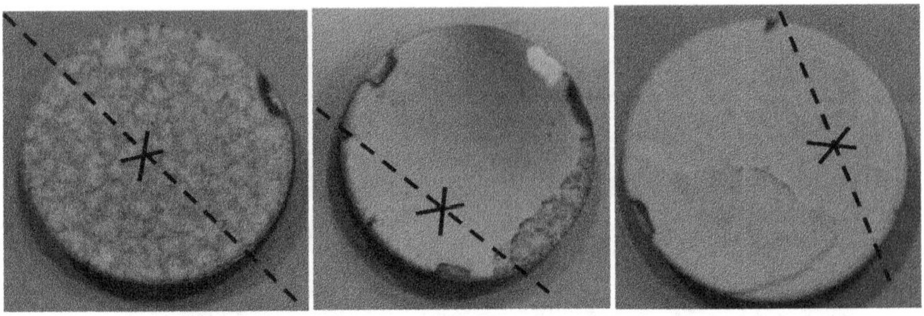

(a) (b) (c)

Figure 2: Optical images of TBCs sample after JETS (a) SCL $La_2Zr_2O_7$; (b) Porous 8YSZ+ $La_2Zr_2O_7$; (c) Dense 8YSZ+ $La_2Zr_2O_7$. The dash line shows the cutting plane of the cross-section view in Figure 3. The cross shows the position where the SEM images were taken in Figure 3.

Figure 3 shows the SEM images of cross-section views of the TBC samples after JETS experiments. The images were taken from the positions shown in Figure 2. For the SCL $La_2Zr_2O_7$ TBC coating (Figure 3a), since the CTE difference between $La_2Zr_2O_7$ layer (9-10 $\times 10^{-6}$ /K) and bond coating layer (about 15×10^{-6} /K) is large, there is a relatively large volume mismatch between $La_2Zr_2O_7$ layer and bond coat [10]. In addition, the fracture toughness of $La_2Zr_2O_7$ is low. So SCL $La_2Zr_2O_7$ sample tends to delaminate near the interface between $La_2Zr_2O_7$ and bond coat. Figure 3b shows that the DCL porous 8YSZ/$La_2Zr_2O_7$ sample was intact after 2000 JETS cycles. The porous 8YSZ layer provides strain compliance due to its high porosity. Figure 3c shows that the DCL dense 8YSZ/ $La_2Zr_2O_7$ sample was fully delaminated at the interface between $La_2Zr_2O_7$ layer and dense 8YSZ layer. The delamination is caused by the thermal residual stress induced by CTE difference between $La_2Zr_2O_7$ and 8YSZ. In summary, the DCL porous 8YSZ/$La_2Zr_2O_7$ sample had a better performance in JETS experiments than the DCL dense 8YSZ/$La_2Zr_2O_7$ and SCL $La_2Zr_2O_7$ samples.

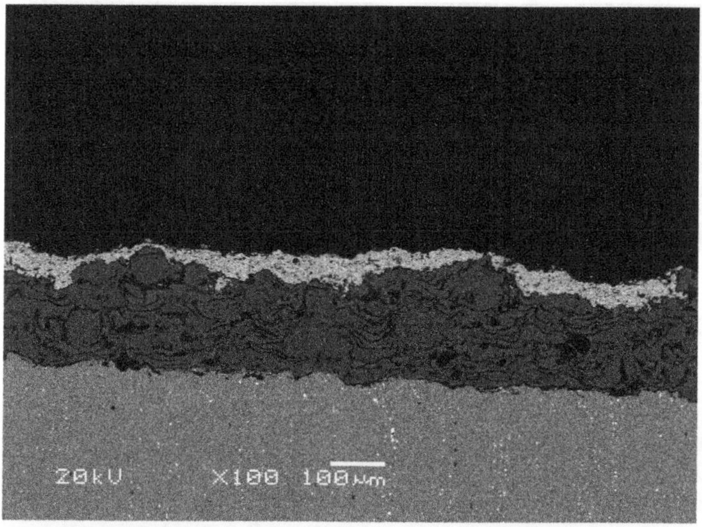

(a)

(b)

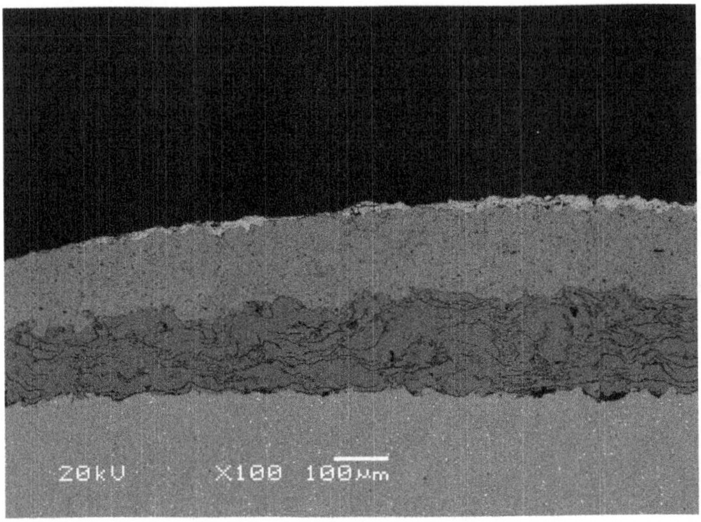

(c)

Figure 3: SEM images of (a) Interface $La_2Zr_2O_7$ and bond coat, (b) Interface $La_2Zr_2O_7$ and porous 8YSZ, (c) Interface $La_2Zr_2O_7$ and dense 8YSZ

CONCLUSIONS

The thermal stabilities of SCL $La_2Zr_2O_7$, DCL porous 8YSZ/$La_2Zr_2O_7$ and dense 8YSZ/$La_2Zr_2O_7$ samples were investigated using heat treatment, FCT and JETS tests. The conclusions are summarized as follows:

1. After 4 hours of heat treatment, the SCL $La_2Zr_2O_7$ and DCL dense 8YSZ/$La_2Zr_2O_7$ coatings were fully delaminated. In contrast, the DCL porous 8YSZ/$La_2Zr_2O_7$ coating sample only partially delaminated on the edge. Porous 8YSZ has a similar CTE as $La_2Zr_2O_7$. As a result, the thermal stress between $La_2Zr_2O_7$ and porous 8YSZ is low.

2. All the $La_2Zr_2O_7$ coats were delaminated in the furnace cycle test in less than 20 cycles. This is because residual thermal stresses accumulated during thermal cycling.

3. In the JETS experiments, both the SCL $La_2Zr_2O_7$ and DCL dense 8YSZ/$La_2Zr_2O_7$ coatings were fully delaminated. In contrast, the DCL porous 8YSZ/$La_2Zr_2O_7$ coating sample only partially delaminated on the edge. This is because of the porous 8YSZ layer provides strain compliance. The results are consistent with the heat treatment data, which shows that the DCL porous 8YSZ/$La_2Zr_2O_7$ coating has the best thermal stability.

REFERENCES

[1] D. R. Clarke, M. Oechsner, and N. P. Padture, "Thermal-barrier coatings for more efficient gas-turbine engines," *MRS Bulletin,* vol. 37, pp. 891-898, 2012.

[2] D. Clarke and C. Levi, "Materials design for the next generation thermal barrier coatings," *Annual Review of Materials Research,* vol. 33, pp. 383-417, 2003.

[3] R. Vaßen, M. O. Jarligo, T. Steinke, D. E. Mack, and D. Stöver, "Overview on advanced thermal barrier coatings," *Surface and Coatings Technology,* vol. 205, pp. 938-942, 11/15/ 2010.

[4] R. A. Miller, "Current status of thermal barrier coatings — An overview," *Surface and Coatings Technology,* vol. 30, pp. 1-11, 1987.

[5] J. R. Davis, *Handbook of thermal spray technology*: ASM international, 2004.

[6] R. Vassen, X. Cao, and F. Tietz, "Zirconates as New Materials for Thermal Barrier Coatings," *J. Am. Ceram. Soc.,,* vol. 83, pp. 2023–2028, 2000.

[7] X. Q. Cao, R. Vassen, and D. Stoever, "Ceramic materials for thermal barrier coatings," *Journal of the European Ceramic Society,* vol. 24, pp. 1-10, 2004.

[8] R. Vassen, X. Cao, F. Tietz, D. Basu, and D. Stöver, "Zirconates as New Materials for Thermal Barrier Coatings," *Journal of the American Ceramic Society,* vol. 83, pp. 2023-2028, 2000.

[9] F. H. Brown and P. O. L. Duwez, "The Systems Zirconia-Lanthana and Zirconia-Neodymia," *Journal of the American Ceramic Society,* vol. 38, pp. 95-101, 1955.

[10] R. Vaßen, F. Traeger, and D. Stöver, "New Thermal Barrier Coatings Based on Pyrochlore/YSZ DCL Systems," *International Journal of Applied Ceramic Technology,* vol. 1, pp. 351-361, 2004.

[11] J. Feng, B. Xiao, R. Zhou, W. Pan, and D. R. Clarke, "Anisotropic elastic and thermal properties of the double perovskite slab–rock salt layer Ln2SrAl2O7 (Ln = La, Nd, Sm, Eu, Gd or Dy) natural superlattice structure," *Acta Materialia,* vol. 60, pp. 3380-3392, 5// 2012.

[12] K. Bobzin, N. Bagcivan, T. Brögelmann, and B. Yildirim, "Influence of temperature on phase stability and thermal conductivity of single- and double-ceramic-layer EB–PVD TBC top coats consisting of 7YSZ, Gd2Zr2O7 and La2Zr2O7," *Surface and Coatings Technology,* vol. 237, pp. 56-64, 12/25/ 2013.

[13] X. Q. Cao, R. Vassen, W. Jungen, S. Schwartz, F. Tietz, and D. Stöver, "Thermal Stability of Lanthanum Zirconate Plasma-Sprayed Coating," *Journal of the American Ceramic Society,* vol. 84, pp. 2086-2090, 2001.

[14] D. Zhu, N. P. Bansal, and R. A. Miller, "Thermal Conductivity and Stability of HfO2-Y2O3 and La2Zr2O7 Evaluated for 1650 C Thermal/Environmental Barrier Coating Applications," in *Proceedings of the 105th Annual Meeting and Exposition of the American Ceramic Society, The American Ceramic Society, Nashville, TN,* 2003.

[15] T. Taylor, "Testing of stability and thermal properties of thermal barrier coatings," *ASM International, Member/Customer Service Center, Materials Park, OH 44073-0002, USA, 1994.,* pp. 654-659, 1994.

[16] A. Bolcavage, A. Feuerstein, J. Foster, and P. Moore, "Thermal shock testing of thermal barrier coating/bondcoat systems," *Journal of Materials Engineering and Performance,* vol. 13, pp. 389-397, 2004/08/01 2004.

[17] H. Lehmann, D. Pitzer, G. Pracht, R. Vassen, and D. Stöver, "Thermal Conductivity and Thermal Expansion Coefficients of the Lanthanum Rare-Earth-Element Zirconate System," *Journal of the American Ceramic Society,* vol. 86, pp. 1338-1344, 2003.

[18] G. K. Beshish, C. W. Florey, F. J. Worzala, and W. J. Lenling, "Fracture toughness of thermal spray ceramic coatings determined by the indentation technique," *Journal of Thermal Spray Technology,* vol. 2, pp. 35-38, 1993/03/01 1993.
[19] A. G. Evans, D. R. Mumm, J. W. Hutchinson, G. H. Meier, and F. S. Pettit, "Mechanisms controlling the durability of thermal barrier coatings," *Progress in Materials Science,* vol. 46, pp. 505-553, 2001.

Hybrid Organic-Inorganic Materials for Alternative Energy

ELECTROCHEMICAL PROPERTIES OF MELTING GEL COATINGS

L. C. Klein[1], A. Degnah[1], K. Al-Marzoki[1], G. Rodriguez[2], A. Jitianu[2,3], J. Mosa[4], and M. Aparicio[4],
[1]Department of Materials Science and Engineering, Rutgers University, 607 Taylor Road, Piscataway, New Jersey 08854 USA
[2]Department of Chemistry, Lehman College, CUNY, Davis Hall, 250 Bedford Park Boulevard West, Bronx, New York 10468 USA
[3]PhD Program in Chemistry, The Graduate Center of the City University of New York, New York, NY 10016 USA
[4]Instituto de Cerámica y Vidrio, Consejo Superior de Investigaciones Científicas (CSIC) Kelsen 5 (Campus de Cantoblanco) 28049 MADRID, SPAIN

ABSTRACT

Organic–inorganic melting gel coatings were applied to titanium alloy substrates to decrease their corrosion and extend their lifetime in an electrochemical application. Melting gels containing phenyl substitutions were coated onto the substrates. The ratio of diphenyl dimethoxysilane (DPhDMS) to methanol was varied to change the thickness of the coatings. The coatings were inspected visually to see that there is good adhesion between the coating and the substrate. Nanoindenter tests were performed to determine hardness.

INTRODUCTION

Organic-inorganic hybrid sol-gel coatings have been formulated to protect a variety of metals[1]. In particular, stainless steel has been a common subject of these coating studies because it is used in a variety of electrochemical systems, often as the current collector in fuel cells[2,3]. The emphasis of many studies of metal corrosion is to reduce the need for chromate-based chemical conversion coatings[4,5,6]. The protection mechanism of the coatings are one or more of the following: barrier properties, cathodic protection, and inhibition and passivation. The primary function of the melting gel coatings in this investigation is a barrier layer[7,8].

In a previous study using melting gels as barrier materials[7], lithium metal was mounted on mica and sealed with melting gel. After consolidation, a melting gel with molar ratio 1.0 mole PhTMS (phenyltrimethoxysilane)-0.15 mole DPhDMS (diphenyl dimethoxysilane) was monitored for degradation of the lithium metal. This was determined visually. Initially, the lithium metal was shiny, but gas transport through the gel caused the surface of the metal to change due to the formation of Li_3N. Using an areal calculation, a degradation constant was estimated (K_{diff} = 0.00192 (%) $sec^{-1/2}$), which allowed the estimate of a total time to decay of greater than 68 years. The benefits of melting gel coatings depend on their hydrophobic character, their absence of porosity, and their adhesion[9,10,11]. Also, the melting gels are applied at low temperature, which precludes oxidation of the metal substrate during the application[12].

Previous efforts to treat metal surfaces with sol-gel materials include mineralization treatments with sodium silicates[4]. These coatings are effective on galvanized steel. Other treatments have been applied in a spray operation with mixtures of tetraethylorthosilicate (TEOS) and methyltriethoxysilane (MTES) on 304 stainless steel[5] or dipped[6]. The sprayed layers were followed by dip coatings to decrease the surface roughness. In addition to the organically modified sol-gel coatings, some coatings have been prepared with TEOS, 3-methacryloxy propyl trimethoxysilane (MPTS) followed by thermal curing with methyl methacrylate on mild steel[13].

Besides the studies on steels, there are studies of organic-inorganic coatings for aluminum alloys. For example, TEOS and triethoxy octylsilane (TEOCS) has been tested on aluminum alloys such as AA2024[14]. Another study reported coatings based on TEOS and 3-glycidoxypropyl trimethoxysilane (GPTS), combined with cerium phosphate inhibitors[15]. Organic-inorganic coatings have been prepared with TEOS and 3-methoxy silylpropyl methacrylate (TSPM) with TiO_2-CeO_2 nanoparticles as inhibitors[16] and with SiO_2 nanoparticles for reinforcement[17]. These coatings were applied to AA2024 and tested in a 3.5% NaCl solution. In addition, there are organic-inorganic hybrids developed especially for copper that incorporate clay particles in phenyl trimethoxysilane (PhTMS)[18]. A final example used GPTS and incorporated microcontainers with poly(ethylene imine) and poly(styrene sulfonate), which could be released to enable self-healing[19].

Recently, a study was carried out with melting gels in the methyl-substituted family, using methyltriethoxysilane (MTES) and dimethyldiethoxysilane (DMDES)[2]. Thick (> 0.5 mm) hybrid glass coatings were obtained by consolidation of methyl substituted melting gels on AISI 304 Stainless Steel. The presence of methyl moieties and a Si-O-Si backbone were confirmed by FT-IR and Raman spectra. Both SEM analysis of the cross-section and micro-scratch tests showed homogeneous, thick, crack-free and well-bonded coatings with largely plastic deformation behavior. Electrochemical characterization (potentiodynamic polarization and electrochemical impedance spectroscopy) in NaCl 3.5% wt. solution indicated very stable surface films that showed no breakdown after four months of immersion. Current densities in polarization curves were as low as 10^{-3} nA/cm^2, six orders of magnitude lower in comparison to the bare substrate. Impedance measurements also presented an excellent behavior with values around 10^{11} ohms, six orders of magnitude higher than the bare substrate.

In this study, a formulation similar to the one tested with lithium metal has been used. Because the melting gel in this case is applied directly to metal substrates, a systematic variation of the solvent to water ratio was investigated. That is the water to PhTMS ratio was held constant, and the amount of methanol used to dilute the DPhDMS component was varied. A previous composition was duplicated[11], and then the amount of methanol was doubled and doubled again. The reason for increasing the methanol was to control the viscosity of the solution. Previous studies changed the viscosity by changing the precursor, using phenyl and methyl substitutions, and changing the reactive groups from methoxy to ethoxy. Based on these previous studies, it was expected that the methanol acts to dilute, which can delay cross-linking or facilitate condensation[20]. Moreover, the higher amount of solvent allows longer time of reaction and a better formation of the inorganic network. In this way, it is possible to adjust the glass transition temperature. In addition, the methanol content may help to control the thickness of coating in dipping or pouring, and its careful removal can help prevent cracking during drying and cooling from the consolidation temperature.

EXPERIMENTAL PROCEDURE

Synthesis

The preparation of the melting gel was carried out using a pair of mono- and disubstituted alkoxysilanes, PhTMS (Gelest Inc., Morrisville, PA) with DPhDMS (Gelest Inc.). All of the substituted alkoxysilanes were used without further purification. Hydrochloric acid (Fisher Scientific, Atlanta, GA) was used as a catalyst. As a solvent, anhydrous methanol (Sigma-Aldrich, Milwaukee, WI) was used with PhMTMS–DPhDMS. These gels were prepared using 1 mole of mono- to 0.25 moles of di-substituted alkoxysilane. A two-step synthesis was used. First, the PhTMS and water were mixed together. The molar ratio of PhTMS:H_2O was 1:1.5. The pH of the water was adjusted to 2.5 using a 0.1 M HCl solution.

After the water was added to the PhTMS, the mixture was stirred continuously at 40°C in a covered beaker. In the absence of a cosolvent, the mixture initially showed immiscibility between water and PhTMS. A clear solution was obtained after 30 min of stirring, and stirring was continued for 6 hours.

In the second step, DPhDMS was diluted with methanol. The molar ratio was 4, 8 or 16 moles alcohol to 1 mole DPhDMS. The DPhDMS-methanol mixture was added dropwise to the clear solution of PhTMS and water. This mixture was stirred for another two hours in a closed beaker at 40°C. Then, the clear solution was cooled to room temperature and stirred overnight in an open system until it gelled. The gels were then dried at 70°C for 24 h, followed by a heat treatment at 110°C. After this heat treatment, the gels were rigid at room temperature, but they resoftened at ~100°C.

Sample Coating

The titanium alloy substrate was prepared by gently rubbing with SiC abrasive paper (P2400/6.5 μm), followed by cleaning in isopropanol in an ultrasonic bath. The previously prepared melting gel was softened at 100°C. Approximately the same amount of gel was poured each time onto the substrate. The coated samples were placed in an oven and heated to 150°C for 24, 48 or 96 hours before cooling back to room temperature, which took about 4 hours. A typical sample is shown in Figure 1.

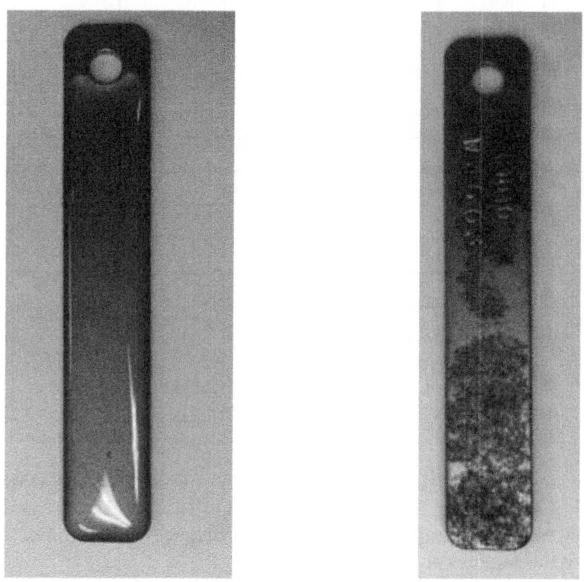

Figure 1. Comparison of the coated side (a) and the uncoated side (b) and stored in ambient conditions for 6 months

Nanoindentation

A Hysitron Triboindenter (Minneapolis, MN) was used with a Berkovich tip with a roughly 100 nm tip radius. As a nanoindenter, it is instrumented allowing for continuous measurement of both depth and load. In operation, the pyramidal tip is gently pushed into the sample, and load and depth are recorded. During unloading, the modulus and hardness are calculated. The instrument is calibrated using a fused quartz standard. The loading conditions were 1 mN and 3 mN, and a loading cycle of 10 s load and 10 s unload. For each indentation load, three indents were made, averaged and the standard deviation was calculated.

RESULTS

The composition with 4 moles of methanol was duplicated from an earlier study (T_g = 21°C)[11]. The consolidation temperature, after which the gels no longer soften, was determined experimentally. The consolidation temperature for this melting gel is 180°C, as previously determined for gels with 4 moles methanol. That value was determined in a bulk sample. In the case of the coatings on titanium alloy, the consolidation occurred by heating for 1 day at 150 or 200°C, followed by a slow cool of about 4 hrs. The results of the indentation hardness measurements are shown in Table 1.

Table 1. Hardness measurements on melting gels after consolidation

	Mole Ratio						
PhTMS	DPhDMS	Methanol	Heat Treatment	Time	Hardness 1mN	Hardness 3mN	Cracked
			°C	hours	MPa	MPa	
1	0.25	4	150	48	169±1.2	170±0.9	No
1	0.25	8	150	48	211±5.8	220±3.0	Yes
1	0.25	16	150	48	113±0.8	121±0.8	No
1	0.25	4	150	96	-	192±5.3	No
1	0.25	8	150	96	207±3.6	211±1.8	Yes
1	0.25	16	150	96	164±0.2	169±4.6	No
1	0.25	4	200	24	165±4.3	150±17.7	Yes
1	0.25	8	200	24	214±8.8	221±5.4	Yes
1	0.25	16	200	24	144±2.8	-	Yes

A total of 9 samples are listed in Table 1. The molar ratio of PhTMS to DPhDMS is the same in all cases. The level of methanol is increased from 4 to 8 to 16. The heat treatment temperature is either 150 or 200°C. For the samples at 150 °C, the difference is the time of treatment, either 48 hr or 96 hr. All of the samples at 200°C cracked.

Indentation curves for loading and unloading are shown in Figure 2. The samples in this case were heated for 96 hr at 150 C, and the curves were recorded for 1 mN load. The 3 curves for the samples with 8 and 16 moles of methanol are almost indistinguishable. There is considerable variation in the sample with 4 moles, reflecting a nonuniform surface. Each sample was tested at 1 mN and 3 mN (not shown). The hardness appears to be independent of load.

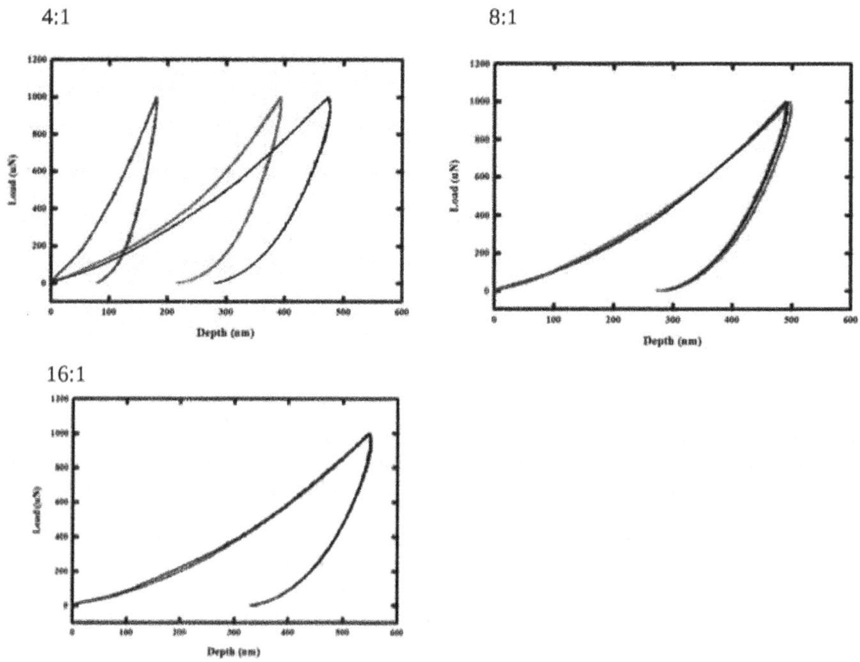

Figure 2. Indentation hardness for 96 hours at 150°C and 1 mN load for 4:1, 8:1 and 16:1 moles methanol to DPhDMS

Doubling the time from 48 hr to 96 hr at 150°C increased the hardness somewhat. Doubling the methanol made the gel more fluid for pouring, but harder when consolidated. The samples with 8 moles of methanol cracked. Increasing methanol to 16 moles introduced surface "stickiness", but these samples were less likely to crack than at 8 moles.

The best crack-free coatings were formed using a treatment of 1 day at 150°, followed by 6 days at 110°C and a 4 hour furnace cool to room temperature. To see the benefits of the coating, compare the samples side by side in Figure 1. The coated side of the sample (a) is still shiny, while the uncoated side of the sample (b) is showing serious corrosion damage after 6 months.

SUMMARY

As seen previously, melting gels provide a barrier mechanism for corrosion protection. The surface is hydrophobic. The layer is nonporous. Diffusion through the layer is limiting. Further electrochemical testing is being carried out at CSIC. To further improve the protection provided by the melting gels addition of inhibitors (e.g., Ce) is being planned.

ACKNOWLEDGEMENTS

We acknowledge the financial support of the National Science Foundation (NSF) through the Award #1313544, Materials World Network, SusChEM: Hybrid Sol-Gel Route to Chromate-free Anticorrosive Coatings and by Ministerio de Economia y Competividad PCIN-2013-030 (Spain); support from King Abdulaziz City for Science and Technology (KACST, Saudi Arabia) Fellowship (Degnah); support from the HCED (Iraq Education Initiative Consortium of American Universities) (Al-Marzoki); and financial support for the symposium from Aldrich.

REFERENCES

[1]R. B. Figueira, C. J. R. Silva, E. V. Periera, Organic-inorganic hybrid sol-gel coatings for metal corrosion protection: a review of recent progress, *J. Coat. Technol. Res.*, **12**, 1-35 (2015).

[2]M. Aparicio, A. Jitianu, G. Rodriguez, A. Degnah, K. Al-Marzoki, J. Mosa, L. C. Klein, Corrosion Protection of 304 Stainless Steel with Melting Gel Coatings, to appear in *Electrochimica Acta*, (2015).

[3]W. Chen, L. C. Klein and C. Huang, "Solution preparation of $Li(Co,Fe)O_2$ coatings for molten carbonate fuel cell components", *J. Sol-Gel Science & Tech.*, **27**, 203-211 (2001).

[4]J. J. Hahn, N. G. McGowan, R. L. Heimann, T. L. Barr, Modification and characterization of mineralization surface for corrosion protection, *Surface and Coatings Technology* **108/109**, 403-407 (1998).

[5]T. Hwang, H. Y. Lee, H. Kim, G. T. Kim, Two layered silica protective film made by a spray-an-dip coating method on 304 stainless steel, *J. Sol-Gel Sci. Technol.* **55** 207-212 (2010).

[6]B. Nikrooz, M. Zandrahimi, H. Ebrahimifar, High temperature oxidation resistance and corrosion properties of dip coated silica coating by sol gel method on stainless steel *J. Sol-Gel Sci. Technol.* **63** 286-293 (2012).

[7]A. Jitianu and L. C. Klein, "Encapsulating Battery Components with Melting Gels" Ceramic Transactions 250: Advances in Materials Science for Environmental and Energy Technologies III, eds. T. Ohji,, J. Matyas, N. J. Manjooran, G. Pickrell, A. Jitianu,, American Ceramic Soc., Westerville, OH, 2014, pp. 279-286.

[8]A. Jitianu, G. Amatucci, L. C. Klein, Organic-inorganic Sol-Gel Thick Films for Humidity Barriers, *J. Materials Research*, **23**, 2084-2090 (2008).

[9]A. Jitianu, J. Doyle, G. Amatucci, L. C. Klein, Methyl modified siloxane melting gels for hydrophobic films, *J. Sol-Gel Science & Tech.*, **53**, 272-279 (2010).

[10]H. Kakiuchida, M. Takahashi, Y. Tokuda, H. Masai, M. Kuniyoshi, T. Yoko, Viscoelastic and Structural properties of the Phenyl-Modified Polysiloxane System with a three-dimensional Structure, *J. Phys. Chem. B*, **110**, 7321-7327 (2006).

[11]A. Jitianu, G. Amatucci, L. C. Klein, Phenyl-substituted siloxane hybrid gels that soften below 140°C, *J. Am. Ceram. Soc.*, **92**, 36-40 (2009).

[12]M. Kuniyoshi, M. Talahashi, Y. Tokuda, T. Yoko, Preparation of phenyl-modified siloxane glasses with softening temperature controlled by rapid heat treatment (RHT), *Chemistry Letters* **35** 1384-1385 (2006).

[13]P. Hammer, F. C. dos Santos, B. M. Cerrutti, S. H. Pulcinelli, C. V. Santilli, Highly corrosion resistant siloxane-polymethyl methacrylate hybrid coatings, *J. Sol-Gel Sci. Technol.* **63** 266-274 (2012).

[14]X. Pan, J. Wu, Y. Ge, K. Xiao, H. Luo, S. Gao, X. Li, Preparation and characterization of anticorrosion Ormosil sol-gel coatings for aluminum alloy, *J. Sol-Gel Sci. Technol.* **72** 8-20 (2014).

[15]A. Wittmar, M. Wittmar, A. Ulrich, H. Caparrotti, M. Veith, Hybrid sol-gel coatings doped with transition metal ions for the protection of AA 2024-T3, *J. Sol-Gel Sci. Technol.* **61**, 600-612 (2012).

[16]M. Zaharescu, L. Predoana, A. Barau, D. Raps, F. Gammel, N. C. Rosero-Navarro, Y. Castro, A. Duran, M. Aparicio, SiO_2 based hybrid inorganic-organic films doped with TiO_2-CeO_2 nanoparticles for corrosion protection of AA2024 and Mg-AZ31B alloys, *Corrosion Science* **51**, 1998-2005 (2009).

[17]N. C. Rosero-Navarro, S. A. Pellice, A. Duran, M. Aparicio, Effects of Ce-containing sol-gel coatings reinforced with SiO_2 nanoparticles on the protection of AA2024, *Corrosion Science* **50**, 1283-1291 (2008).

[18]D. Sefzadeh, E. Golmoghani-Ebrahimi, Formation of Novel and crack free nanocomposites based on sol gel process for corrosion protection of copper, *Surface and Coatings Technology*, **210**, 103-112 (2012).

[19]M. Zheludkevich, D. G. Shchukin, K. A. Yasakau, H. Möhwald, M. G. S. Ferreira, Anticorrosion coatings with self-healing effect based on nanocontainers impregnated with corrosion inhibitor, *Chem. Mater.* **19**, 402-411 (2007).

[20]A. Jitianu, G. Gonzalez and L. C. Klein, Hybrid Sol-Gel Glasses with Glass Transition Temperatures below Room Temperature, *J. Am. Ceram. Soc.* DOI: 10.1111/jace.13798 (2015).

Author Index

Printed and bound by CPI Group (UK) Ltd, Croydon, CR0 4YY
17/09/2026
14964240-0002